OPERATION OF MUNICIPAL WASTEWATER TREATMENT PLANTS

Volumes in Manual of Practice No. 11

OPERATION OF MUNICIPAL WASTEWATER TREATMENT PLANTS

Manual of Practice No. 11
Sixth Edition

Volume III   Solids Processes

Prepared by
Operation of Municipal Wastewater Treatment Plants Task Force
of the Water Environment Federation

WEF Press

New York   Chicago   San Francisco   Lisbon   London   Madrid
Mexico City   Milan   New Delhi   San Juan   Seoul
Singapore   Sydney   Toronto

Cataloging-in-Publication Data is on file with the Library of Congress.

McGraw-Hill books are available at special quantity discounts to use as premiums and sales promotions, or for use in corporate training programs. For more information, please write to the Director of Special Sales, Professional Publishing, McGraw-Hill, Two Penn Plaza, New York, NY 10121-2298. Or contact your local bookstore.

Operation of Municipal Wastewater Treatment Plants, Volume III

1 2 3 4 5 6 7 8 9 0   DOC/DOC 0 1 2 1 0 9 8 7

ISBN: P/N 978-0-07-154370-5 of set
        978-0-07-154367-5

MHID: P/N 0-07-154370-8 of set
        0-07-154367-8

This book is printed on acid-free paper.

IMPORTANT NOTICE

The material presented in this publication has been prepared in accordance with generally recognized engineering principles and practices and is for general information only. This information should not be used without first securing competent advice with respect to its suitability for any general or specific application.

The contents of this publication are not intended to be a standard of the Water Environment Federation (WEF) and are not intended for use as a reference in purchase specifications, contracts, regulations, statutes, or any other legal document.

No reference made in this publication to any specific method, product, process, or service constitutes or implies an endorsement, recommendation, or warranty thereof by WEF.

WEF makes no representation or warranty of any kind, whether expressed or implied, concerning the accuracy, product, or process discussed in this publication and assumes no liability.

Anyone using this information assumes all liability arising from such use, including but not limited to infringement of any patent or patents.

Task Force

Prepared by the Operation of Municipal Wastewater Treatment Plants Task Force of the Water Environment Federation

Michael D. Nelson, *Chair*

Douglas R. Abbott

George Abbott

Mohammad Abu-Orf

Howard Analla

Thomas E. Arn

Richard G. Atoulikian, PMP, P.E.

John F. Austin, P.E.

Elena Bailey, M.S., P.E.

Frank D. Barosky

Zafar I. Bhatti, Ph.D., P. Eng.

John Boyle

William C. Boyle

John Bratby, Ph.D., P.E.

Lawrence H. Breimhurst, P.E.

C. Michael Bullard, P.E.

Roger J. Byrne

Joseph P. Cacciatore

William L. Cairns

Alan J. Callier

Lynne E. Chicoine

James H. Clifton

Paul W. Clinebell

G. Michael Coley, P.E.

Kathleen M. Cook

James L. Daugherty

Viraj de Silva, P.E., DEE, Ph.D.

Lewis Debevec

Richard A. DiMenna

John Donnellon

Gene Emanuel

Zeynep K. Erdal, Ph.D., P.E.

Charles A. Fagan, II, P.E.

Joanne Fagan

Dean D. Falkner

Charles G. Farley

Richard E. Finger

Alvin C. Firmin

Paul E. Fitzgibbons, Ph.D.

David A. Flowers

John J. Fortin, P.E.

Donald M. Gabb

Mark Gehring

Louis R. Germanotta, P.E.

Alicia D. Gilley, P.E.

Charlene K. Givens

Fred G. Haffty, Jr.

Dorian Harrison

John R. Harrison

Carl R. Hendrickson

Webster Hoener

Brian Hystad

Norman Jadczak

Jain S. Jain, Ph.D., P.E.

Samuel S. Jeyanayagam, Ph.D., P.E., BCEE

Bruce M. Johnston

John C. Kabouris

Sandeep Karkal

Gregory M. Kemp, P.E.

Justyna Kempa-Teper

Salil M. Kharkar, P.E.

Farzin Kiani, P.E., DEE

Thomas P. Krueger, P.E.

Peter L. LaMontagne, P.E.

Wayne Laraway

Jong Soull Lee

Kurt V. Leininger, P.E.

Anmin Liu

Chung-Lyu Liu

Jorj A. Long

Thomas Mangione

James J. Marx

Volker Masemann, P. Eng.

David M. Mason

Russell E. Mau, Ph.D., P.E.

Debra McCarty

William R. McKeon, P.E.

John L. Meader, P.E.

Amanda Meitz

Roger A. Migchelbrink

Darrell Milligan

Robert Moser, P.E.

Alie Muneer

B. Narayanan

Vincent L. Nazareth, P. Eng.

Keavin L. Nelson, P.E.

Gary Neun

Daniel A. Nolasco, M. Eng., M. Sc., P. Eng.

Charles Norkis

Robert L. Oerther

Jesse Pagliaro

Philip Pantaleo

Barbara Paxton

William J. Perley

Beth Petrillo

Jim Poff

John R. Porter, P.E.

Keith A. Radick

John C. Rafter, Jr., P.E., DEE

Greg Ramon

Ed Ratledge

Melanie Rettie

Kim R. Riddell

Joel C. Rife

Jim Rowan

Hari Santha

Fernando Sarmiento, P.E.

Patricia Scanlan

George R. Schillinger, P.E., DEE

Kenneth Schnaars

Ralph B. (Rusty) Schroedel, Jr., P.E., BCEE

Pam Schweitzer

Reza Shamskhorzani, Ph.D.

Carole A. Shanahan

Andrew Shaw

Timothy H. Sullivan, P.E.

Michael W. Sweeney, Ph.D., P.E.

Chi-Chung Tang, P.E., DEE, Ph.D.

Prakasam Tata, Ph.D., QEP

Gordon Thompson, P. Eng.

Holly Tryon

Steve Walker

Cindy L. Wallis-Lage

Martin Weiss

Gregory B. White, P.E.

George Wilson

Willis J. Wilson

Usama E. Zaher, Ph.D.

Peter D. Zanoni

Under the direction of the MOP 11 Subcommittee of the Technical Practice Committee

Water Environment Federation
Improving Water Quality for 75 Years

Founded in 1928, the Water Environment Federation (WEF) is a not-for-profit technical and educational organization with members from varied disciplines who work toward the WEF vision of preservation and enhancement of the global water environment. The WEF network includes water quality professionals from 79 Member Associations in more than 30 countries.

For information on membership, publications, and conferences, contact

Water Environment Federation
601 Wythe Street
Alexandria, VA 22314-1994 USA
(703) 684-2400
http://www.wef.org

Contents

Volume I: Management and Support Systems

Chapters 11, 23, and 32 were not updated for this edition of MOP 11. Chapters 11, 23, and 32 included herein are taken from the 5th edition of the manual, which was published in 1996.

Volume II: Liquid Processes

Volume III: Solids Processes

Manuals of Practice of the Water Environment Federation

The WEF Technical Practice Committee (formerly the Committee on Sewage and Industrial Wastes Practice of the Federation of Sewage and Industrial Wastes Associations) was created by the Federation Board of Control on October 11, 1941. The primary function of the Committee is to originate and produce, through appropriate subcommittees, special publications dealing with technical aspects of the broad interests of the Federation. These publications are intended to provide background information through a review of technical practices and detailed procedures that research and experience have shown to be functional and practical.

Water Environment Federation Technical Practice
Committee Control Group
B. G. Jones, *Chair*
J. A. Brown, *Vice-Chair*
S. Biesterfeld-Innerebner
R. Fernandez
S. S. Jeyanayagam
Z. Li
M. D. Nelson
S. Rangarajan
E. P. Rothstein
A. T. Sandy
A. K. Umble
T. O. Williams
J. Witherspoon

Acknowledgments

This manual was produced under the direction of Michael D. Nelson, *Chair*.

The principal authors and the chapters for which they were responsible are

George Abbott (2)

Thomas P. Krueger, P.E. (2)

Thomas E. Arn (3)

Greg Ramon (3)

Gregory M. Kemp (4)

Roger A. Migchelbrink (5)

Melanie Rettie (6)

Michael W. Sweeney, Ph.D., P.E. (6)

David Olson (7)

Russell E. Mau, Ph.D., P.E. (8)

Peter Zanoni (9)

Louis R. Germanotta, P.E. (10)

Norman Jadczak (10)

James Kelly (11)

Richard Sutton, P.E. (12)

Alicia D. Gilley, P.E. (13)

Jorj Long (14)

Keavin L. Nelson, P.E. (15)

Ed Ratledge (16)

Kathleen M. Cook (17)

Frank D. Barosky (17)

Jerry C. Bish, P.E., DEE (18)

Ken Schnaars (19)

Donald J. Thiel (20, Section One)

William C. Boyle (20, Section Two)

Donald M. Gabb (20, Section Three)

Samuel S. Jeyanayagam (20, Section Four)

Kenneth Schnaars (20, Section Five)

Alan J. Callier (20, Section Six)

Jorj Long (20, Appendices)

Howard Analla (21)

John R. Harrison (22)

Anthony Bouchard (23)

Sandeep Karkal, P.E. (24)

Daniel A. Nolasco, M. Eng., M. Sc., P. Eng. (25)

Donald F. Cuthbert, P.E., MSEE (26)

Carole A. Shanahan (27)

Kim R. Riddell (28)

Peter L. LaMontagne, P.E. (29)

Patricia Scanlan (30)

Webster Hoener (30)

Gary Neun (30)

Jim Rowan (30)

Hari Santha (30)

Elena Bailey, M.S., P.E. (31)

Rhonda E. Harris (32)

Peter L. LaMontagne, P.E. (33)

Wayne Laraway (33)

Contributing authors and the chapters to which they contributed are

David M. Mason (3)   Jerry Miles (23)
David W. Garrett (9)   Michael Richard (23)
Earnest F. Gloyna (23)   John Bratby, Ph.D., P.E. (29)
Roy A. Lembcke (23)   Curtis L. Smalley (32)

Michael D. Nelson served as the Technical Editor of Chapters 10 and 25.

Tony Ho; Dr. Mano Manoharan, Standards Development Branch, Ontario Ministry of the Environment, Canada; and Dr. Glen Daigger, CH2M HILL collaborated in the successful development of the methodologies presented in Chapter 25.

Efforts of the contributors to this manual were supported by the following organizations:

Abbott & Associates, Chester, Maryland
Alberta Capital Region Wastewater Commission, Fort Saskatchewan, Alberta,
    Canada
Allegheny County Sanitary Authority, Pittsburgh, Pennsylvania
Alliant Techsystems, Lake City Army Ammunition Plant, Independence, Missouri
American Bottoms Regional Wastewater Treatment Facility, Sauget, Illinois
Archer Engineers, Lee's Summit, Missouri
Baxter Water Treatment Plant, Philadelphia, Pennsylvania
Bio-Microbics, Inc., Shawnee, Kansas
Biosolutions, Chagrin Falls, Ohio
Black & Veatch, Kansas City, Missouri
Brown and Caldwell, Seattle, Washington; Walnut Creek, California
Burns and McDonnell, Kansas City, Missouri
Capitol Environmental Engineering, North Reading, Massachusetts
Carollo Engineers, Walnut Creek, California
CDM, Albuquerque, New Mexico; Manchester, New Hampshire
CH2M Hill, Cincinnati, Ohio; Santa Ana, California
City of Delphos, Delphos, Ohio
City of Edmonton, Alberta, Canada
City of Fairborn Wastewater Reclamation District, Fairborn, Ohio
City of Frankfort, Frankfort Sewer Department, Frankfort, Kentucky
City of Phoenix Water Service Department, Phoenix, Arizona
City Of Tulsa Water Pollution Control Section, Tulsa, Oklahoma
City of Tulsa, Tulsa, Oklahoma
Clayton County Water Authority, Jonesboro, Georgia
Consoer Townsend Envirodyne Engineers, Nashville, Tennessee
Consolidated Consulting Services, Delta, Colorado
CTE Engineers, Spokane, Washington
David A. Flowers, Cedarburg, Wisconsin
DCWASA, Washington, D.C.
Department of Biological Systems Engineering, Washington State University,
    Pullman, Washington
Earth Tech Canada, Inc., Markham, Ontario, Canada
Earth Tech, Sheboygan, Wisconsin
East Norriton-Plymouth-Whitpain Joint Sewer Authority, Plymouth Meeting,
    Pennsylvania

EDA, Inc., Marshfield, Massachusetts

Eimco Water Technologies Division of GLV, Austin, Texas

El Dorado Irrigation District, Placerville, California

Electrical Engineer, Inc., Brookfield, Wisconsin

EMA, Inc., Louisville, Kentucky; St. Paul, Minnesota

Environmental Assessment and Approval Branch, Ministry of the Environment, Toronto, Ontario, Canada

Eutek Systems, Inc., Hillsboro, Oregon

Fishbeck, Thompson, Carr & Huber, Inc., Farmington Hills, Michigan; Grand Rapids, Michigan

Floyd Browne Group, Delaware, Ohio

Gannett Fleming, Inc., Newton, Massachusetts

Givens & Associates Wastewater Co., Inc., Cumberland, Indiana

Grafton Wastewater Treatment Plant, South Grafton, Massachusetts

Greater Vancouver Regional District, Burnaby, British, Columbia, Canada

Greeley and Hansen, L.L.C., Chicago, Illinois; Philadelphia, Pennsylvania; Phoenix, Arizona

Hazen and Sawyer, PC, New York, New York; Raleigh, North Carolina

HDR Engineering, Charlotte, North Carolina

Hillsborough County Water Department, Tampa, Florida

Hubbell, Roth, & Clark, Inc., Detroit, Michigan

Infrastructure Management Group, Inc., Bethesda, Maryland

ITT/Sanitaire WPCC, Brown Deer, Wisconsin

Kennedy/Jenks Consultants, San Francisco, California

Los Angeles County Sanitation Districts, Whittier, California

MAGK Environmental Consultants, Manchester, New Hampshire

Malcolm Pirnie, Inc., Columbus, Ohio

Massachusetts Institute of Technology, Cambridge, Massachusetts

Metcalf & Eddy, Inc., Philadelphia, Pennsylvania

Metro Wastewater Reclamation District, Denver, Colorado

Mike Nelson Consulting Services LLC, Churchville, Pennsylvania

Ministry of Environment, Toronto, Ontario, Canada

MKEC Engineering Consultants, Wichita, Kansas

MWH, Cleveland, Ohio

New York State Department of Environmental Conservation, Albany, New York

NOLASCO & Assoc. Inc., Ontario, Canada

Nolte Associates, San Diego, California

Northeast Ohio Regional Sewer District, Cleveland, Ohio

Novato Sanitary District, Novato, California

NYSDEC, Albany, New York

Orange County Sanitation District, Fountain Valley, California

Parsons Engineering Science, Inc., Tampa, Florida

Pennoni Associates, Inc., Philadelphia, Pennsylvania

Peter LaMontagne, P.E., New Britain, Pennsylvania

Philadelphia City Government, Philadelphia, Pennsylvania

Philadelphia Water Department, Philadelphia, Pennsylvania

PSG, Inc., Twin Falls, Idaho

R.V. Anderson Associates, Limited, Toronto, Ontario, Canada

Red Oak Consulting—A Division of Malcolm Pirnie, Inc., Phoenix, Arizona

Research and Development Department, Metropolitan Water Reclamation District of Greater Chicago, Cicero, Illinois

Rock River Water Reclamation District, Rockford, Illinois

Rohm and Haas Co., Croydon, Pennsylvania

Sanitary District of Hammond, Hammond, Indiana

Seacoast Environmental, L.L.C., Hampton, New Hampshire

Siemens Water Technologies, Waukesha, Wisconsin

Simsbury Water Pollution Control, Simsbury, Connecticut

Tata Associates International, Naperville, Illinois

Thorn Creek Basin Sanitary District, Chicago Heights, Illinois

Trojan Technologies Inc., London, Ontario, Canada

U. S. Environmental Protection Agency, Philadelphia, Pennsylvania

U.S. Filter Operating Services, Indian Harbour Beach, Florida

United Water, Milwaukee, Wisconsin

University of Wisconsin, Madison, Wisconsin

USFilter, Envirex Products, Waukesha, Wisconsin

USFilter, Inc., Norwell, Massachusetts

Veolia Water North America, L.L.C., Houston, Texas; Norwell, Massachusetts

West Point Treatment Plant, Seattle, Washington

West Yost & Associates, West Linn, Oregon

Weston & Sampson Engineers, Inc., Peabody, Massachusetts

Chapter 27

Solids Management

Solids management refers to the use or disposal of solids that have been removed from wastewater, processed, and are ready to leave the treatment plant. The solids in this context are typically biosolids but could also be scum, grease, screenings, grit, or ash. [The Water Environment Federation defines *biosolids* as "primarily organic solids produced by wastewater treatment processes that can be beneficially recycled." Before treatment, this material is called *wastewater solids* or *sludge*. To be classified as biosolids, the material must meet state and U.S. Environmental Protection Agency (U.S. EPA) criteria for beneficial use.]

TYPES OF SOLIDS

Wastewater treatment plants receive and generate many types of solids, including ash, scum, grease, screenings, grit, and sludge (e.g., raw, treated, primary, secondary, tertiary, combined, and chemical). Some also receive septage from onsite treatment systems and liquid sludge from other treatment plants. (For more information on solids, see Chapter 28.)

Treatment plants also generate solid wastes, including paper, garbage, automotive residuals, equipment oils and greases, chemical-impregnated rags, lab wastes, office wastes (e.g., printer cartridges and copy- and fax-machine chemicals), light bulbs, batteries, and barrels. These wastes must be disposed or recycled according to local regulations. [Given that wastewater treatment plants are supposed to protect the health of both the public and the environment, it behooves them to recycle as much of their wastes as possible (Fisichelli, 1992).]

CHARACTERIZATION

The physical, chemical, and biological characteristics of biosolids affect both its use and the public's perception of it. For example, federal and state regulators base land-application requirements on biosolids' pollutant and pathogen concentrations. A land-application site's neighbors, on the other hand, base their acceptance on the material's consistency and odor.

Biosolids should be sampled for metals and organics after they have been processed so the analytical results will describe the actual material to be used or disposed. Once the chemical and biological standards have been met, the biosolids characteristics that will affect its management options are

- nutrient concentrations, which can affect surface water quality;
- micronutrient concentrations, which can affect soils and plant growth (a certified agronomist may be needed to pinpoint the source of abnormal plant growth); and

- concentrations of various constituents (e.g., mercury and sulfur), which can affect an incinerator's air-pollution controls.

A wastewater treatment plant should keep a large database containing multiple years' worth of chemical analysis results, so staff can prove that the plant's biosolids have not contained harmful levels of constituents such as metals, organics, pathogens, dioxin, radioactive components, polychlorinated biphenyls (PCBs), and so is unlikely to contain them in the future. To be convincing, the data should show that metals and organics concentrations are stable or decreasing—no wild variations—and every result should have been promptly checked by experts and quality-control personnel.

Also, the treatment plant should have an active pretreatment program to minimize unwanted constituents in its influent. Such constituents tend to end up in the solids, and the presumption may be that the resulting biosolids should not be land-applied if these constituents might cause the soil to need remediation for certain uses. However, biosolids typically are land-applied before many test results are available, so plant staff and regulators need to be confident that the material will not harm humans or the environment. They should be able to base this confidence on historical data and a successful pretreatment program.

To demonstrate that a wastewater treatment plant's biosolids management program is based on best management practices (BMP), staff should:

- Do the sampling and analysis required to demonstrate that regulatory standards are being met;
- Do additional sampling and analysis (e.g., priority pollutant analysis, radioactivity, dioxin, and possibly others) on a regular basis (because telling the public that the plant's biosolids meet all federal and state standards does not inspire confidence; specifying how the biosolids surpass those standards does);
- Maintain an effective pretreatment program (even if only to show that there are no industries in the plant's service area) and an active pollution prevention program to help industries and the public reduce their use of pollutants;
- Develop and use a written sampling plan and chain-of-custody form;
- Review every analysis, including those done by a contract lab, to verify that the appropriate test was done, the results are accurate, the appropriate quality-control measures were taken (if the tests were done onsite, the quality-control procedures should be written based on appropriate industry protocols); and
- Ensure that the biosolids look innocuous (e.g., no pieces of fast-food wrappers, condiment pouches, or personal sanitary products should be seen in the material).

HANDLING

GENERAL. While all wastewater solids and septage must be treated to meet the relevant federal, state, and local regulations, differences in processing methods can result in biosolids with widely varying consistencies, which will affect how they can be transported and beneficially used. Handling methods, for example, can greatly affect the consistency of biosolids: Screw conveyors can turn biosolids into a paste that is difficult to apply via manure spreaders. Belt conveyors alter the material less.

For example, two modifications to the solids treatment process at the Allegheny County Sanitary Authority in Pittsburgh, Pa., had unexpected results. Originally, the biosolids were dewatered via belt presses and then mixed with lime. This produced a crumbly mixture that was easily applied via standard manure spreaders. When plant staff replaced the lime-addition equipment with mixers and screw conveyors, the biosolids became a "pudding" that was difficult to spread. However, when staff replaced the belt presses with centrifuges, the material reverted back to the crumbly, spreadable mixture. (Each treatment plant's sludge is unique, so the best process train will be site-specific.)

CHEMICAL ADDITIVES. Any compound added during wastewater treatment typically ends up in the biosolids and may affect its physical, chemical, or handling characteristics. It also may affect the biosolids' permitted use and disposal options.

Lime and sodium bicarbonate, which are added during anaerobic digestion to control pH, should not be problematic; they only increase the concentration of compounds that occur naturally in biosolids and are typically harmless. Likewise, the acids and caustics used to control pH in aerobic digesters should not affect the biosolids' ability to meet regulatory standards. Calcium, on the other hand, can change the soil chemistry and nutrient uptake at land-application sites, so agronomists may recommend occasional soil tests (e.g., once every 3 years) to check the macro- and micronutrient levels at the sites.

Some treatment plants use organic polymers or inorganic chemicals to thicken or dewater sludge. Some regulators are concerned that nitrogen-based polymers may decompose to produce amines and other intensely odorous compounds. However, research has shown that the polymer may adsorb to particulate, making it less biodegradable and significantly less reactive—effectively, an inert material (Dental et al., 2000). The portions of the polymer that do break off are typically innocuous and biodegradable. So, polymer additions should not prohibit the biosolids from being land applied safely.

Solids-conditioning chemicals [e.g., lime, ferric chloride, ferrous sulfate, ferrous chloride, and aluminum sulfate (alum)], which may be added before or after thickening or dewatering processes, should not affect the biosolids' ability to meet regulatory standards. However, if the ferric chloride solution is a byproduct of another process (e.g., spent pickle liquor from an industrial source), then the solution should be analyzed for metals and possibly organics.

Some treatment plants use potassium permanganate to control odors; this also should not affect the biosolids' ability to meet regulatory standards.

TREATMENT PROCESSES. Most sludge-stabilization methods (e.g., anaerobic digestion, aerobic digestion, composting, lime stabilization, thermal treatment, and heat drying) produce land-applicable biosolids, but their products' characteristics and required handling methods will vary. Anaerobic digestion, for example, produces biosolids with low levels of organics and bacteria. However, the material may be somewhat odorous and difficult to dewater.

Aerobic digestion, which is mostly used at small treatment plants, produces a land-applicable biosolids that may be harder to thicken than other types of biosolids.

Composting, when done properly, produces a relatively dry, biologically stable, odor-free biosolids that may be stored outside without developing odors or attracting insects. Some treatment plants sell this material, thereby reducing solids-handling costs. Composted biosolids typically look better and emit less odor than other types of biosolids.

Lime stabilization produces biosolids whose quality some state regulators question, because they are not convinced that this method produces "stabilized" solids. Also, the biosolids are likely to be odorous (primarily ammonia). Research has shown, however, that if enough alkaline (or lime and heat) is added, then the resulting biosolids meet the pathogen-reduction and vector-attraction-reduction standards for Exceptional Quality (EQ) biosolids and are less likely to be odorous because the odorous gases are driven off during processing and can be collected and treated then.

Thermal treatment processes destroy pathogens, improve the resulting biosolids' thickening and dewatering characteristics, and greatly reduce its volume. Although odors are produced during treatment, the biosolids have little odor (as long as the material remains dry). However, the biosolids may have higher metals concentrations and could spontaneously combust.

Mechanical drying methods and drying beds have potential odor problems and may be costly if covers are needed to abate odors or because of temperature extremes.

Incineration destroys organic matter, reduces the sludge volume, and produces an ash that may be beneficially used if its metals levels do not exceed regulatory standards. At temperatures well above 816 °C (1500 °F), the ash may become a hard frit that can be used as a fill, roadbed material, or ceramics ingredient. This option may be attractive to larger wastewater treatment plants that have difficulty finding enough land for composting or land application.

Emerging sludge-stabilization technologies include a microwave technology and methods involving quicklime and heat. For more information on sludge-stabilization methods, see Chapter 32.

MANAGEMENT

USE OPTIONS. Biosolids that meet the requirements of U.S. EPA's Standards for the Use or Disposal of Sewage Sludge (40 *CFR* 503), as well as state and local standards, can be beneficially used. Metals levels in U.S. biosolids have dropped since U.S. EPA's pretreatment regulations were promulgated, and most treatment plants now have no trouble meeting 40 *CFR* 503's metals requirement, according to various studies. The pathogen and vector-attraction-reduction levels, however, depend on the processing method used.

Land Application. The most common method for using biosolids is land applying it on farmland. In 1988 (before ocean disposal stopped), 33% of sewage sludge was land applied, according to the preamble of 40 *CFR* 503. A more recent U.S. EPA study gives a rough estimate that 60% of biosolids are now land applied. Before a treatment plant can begin spreading (or injecting) biosolids on farmland, however, all federal, state, and local paperwork must be completed and approved. Some states require every site to be permitted, while others require the biosolids generator to obtain a general permit and maintain a register of individual sites.

To obtain a general permit, the biosolids generator may need the following:

- An open and ongoing relationship with regulators (they should be familiar with the treatment plant and its operations because they have toured the site at the plant staff's invitation);
- A database of biosolids test results (at least 3 years' worth, but preferably since plant startup because regulators will be more comfortable about letting a plant land apply biosolids if extensive physical, chemical, and pathogenic data demonstrate its safety);
- A history of the treatment plant's operations and a detailed description of its solids processes;
- A quality-control and quality-assurance plan, standard operating procedures for each analysis, and chain-of-custody forms or a logbook of all samples;

- A public relations plan and ongoing outreach efforts to inform the public about the biosolids management program;
- A written description of all land-application activities and personnel, along with the contact information for the treatment plant staff and regulators who should be notified if a question arises or emergency occurs;
- A spill prevention and control plan, a contingency plan, and a long-range plan for future improvements to the biosolids management program; and
- A written procedure (and form or logbook) for handling complaints from each land-application site's neighbors.

Some states have training programs for biosolids generators and land appliers. Biosolids personnel should take such programs (whether mandatory or optional) because they will learn how to interact with regulators and what their expectations are. Biosolids personnel also should become active members of biosolids-related professional organizations [e.g, the Water Environment Federation (WEF), its member associations (MAs), and regional associations], which are sources of useful information, help, and support. In addition to committees devoted to biosolids issues, WEF and its MAs also publish technical materials.

Once a treatment plant has met the general permit requirements, its personnel can establish the land-application program by doing the following:

- Find suitable farmland. The site must have enough acres to justify the expenditure of time and resources that go into obtaining a permit. The vicinity must be checked for wetlands, wind direction, remoteness, location of neighbors, local ordinances on biosolids use, and the general feeling about biosolids.
- Determine which farmers want the material, explain the approval requirements, verify ownership, obtain the necessary data and signatures, and notify the neighbors and any local officials (e.g., township or borough supervisors, county officials, conservation districts, and regional state regulators);
- Contact all of the relevant biosolids regulators; prepare the necessary paperwork; do the required field work [e.g., soil sampling every 10.1 ha (25 ac) or so] and lab tests; procure aerial photographs of the area and mark the fields, boundaries, and all water-related landmarks (e.g., waterways, wetlands, water supplies, and wells); determine the delivery truck's travel route; and complete any other work required by local agencies;
- Inform local agencies and the public about the benefits of beneficially using biosolids (especially if land application or the treatment plant is new to the area), ensure that all program-related personnel (e.g., generators, transporters,

and land appliers) understand the benefits of biosolids and can explain them properly to the public, and maintain an ongoing local outreach program for the benefit of anyone who inquires about land application; and

- Inspect the land-application sites regularly to verify that the work is done correctly (e.g., the trucks are not causing problems, mud is not tracked onto roads, odors are controlled, the site is well-maintained, and any unspread biosolids are neatly stored) and that a treatment plant representative will be able to answer all questions during regulatory inspections.

Land Reclamation. Land reclamation is similar to land application except that the biosolids-application rates are higher because the application is not just to provide nutrients for the current growing season but to establish long-term growth on disturbed and nutrient-poor soils. The land-reclamation sites are usually acidic, strip-mined areas (Sopper, 1993; Pennsylvania Organization for Watersheds and Rivers, 2003). However, biosolids also have successfully reclaimed rangeland in New Mexico by providing organic content that helps retain moisture in the soil.

Lime-stabilized biosolids work well on strip-mined ground where the pH may be low across the site—the lime content of the biosolids provides the pH adjustment needed to allow grasses to grow—but other types of biosolids are also effective as long as appropriate pH adjustments are made. The material's organic matter and slow-release nitrogen will provide a good base for most seed mixtures. However, if the seed mixture includes warm-season grasses, which develop slowly (over 2 to 3 years), they can be choked by faster-growing fescue and other grasses, so multiple applications of biosolids at lower rates may be necessary to avoid this problem.

Biosolids have also successfully revegetated capped cells at landfills. Once a cell has been filled, it is typically covered with a geomembrane, or filter fabric covered with soil to keep water from infiltrating into the cell and to minimize leachate from the cell. The biosolids are then applied in the same manner as for reclaiming strip-mined land.

Forestry. The King County (Washington) Wastewater Treatment Division has been applying biosolids to tree farms since 1987 and to state forests since 1995. The biosolids make an excellent soil amendment and source of nutrients for trees, as illustrated by tree rings from biosolids-fertilized trees, which are wider after the applications. The forestry projects help protect and enhance forests and wildlife habitat along the scenic highway that leads from Seattle to the mountains.

Commercial Products. Some treatment plants (on their own or with a private company) compost biosolids for use in landscaping and gardening. Biosolids that can be bagged and sold may be used by the public for their gardens and potted plants.

Incineration. In 1993, 381 wastewater treatment plants (2.8%) incinerated their sludge (16% of the total volume produced nationwide), and seven plants co-incinerated sludge with municipal solid waste in municipal waste combustors, according to the preamble of 40 *CFR* 503. Some treatment plants choose to incinerate biosolids because they are in colder areas and need an effective management method in winter, when land application is infeasible. Some plants in large metropolitan areas incinerate solids to avoid the odor complaints related to hauling Class B biosolids long distances through city neighborhoods. The incineration process is equipment- and energy-intensive and requires air-pollution-control devices; the leftover ash must also be used or disposed. (For more information on incineration processes, see Chapter 32.)

Some wastewater treatment plants recover the heat generated during incineration and use it to heat other treatment processes, to generate electricity, or to make steam for heating or use in other treatment processes. Incinerator ash is typically land-filled, but it could be beneficially used if its metal concentrations are within regulatory limits (all pathogens were burned off during incineration). For example, the ash could be used as

- An amendment to soil that will be dug up and sold by topsoil producers (the ash-and-soil mixture produced must meet regulators' criteria);
- A component of cement, concrete, or asphalt;
- An alkaline addition to sludge (with other alkaline additives, perhaps);
- A component of house shingles;
- A component of plastic;
- A ceramic-like material (if further treated at higher temperatures); and
- Landscaping bricks (if further processed at higher temperatures).

To qualify for most of these uses, the ash must meet certain criteria and be available on demand at specified volumes. Not all treatment plants can meet such requirements, and they must compete with other industries that produce large volumes of suitable ash.

Heat Drying and Other Thermal Processes. Together, heat drying and pelletizing produce a marketable fertilizer that meets the 40 *CFR* 503 requirements for EQ biosolids, and so has fewer regulatory recordkeeping and reporting requirements if used for land application. This biosolids management option is a proven technology in which odors can be contained and controlled. The resulting biosolids pellets have much less volume and weight than the influent solids and are easily handled, conveyed, and stored. They can be delivered to consumers in bulk or in bags or other containers.

The disadvantages of heat drying and pelletizing include the dust's explosive potential and the potential for overheating and fires. Also, the equipment is expensive, complex, and maintenance-intensive, requiring qualified operators. Air-emissions-control equipment also is required, especially because drying certain types of solids can result in more odorous pellets.

Solids also may be treated via the following thermal processes:

- phased thermophilic digestion;
- heat pasteurization, which produces a marketable fertilizer;
- residuals gasification, which produces liquid, gaseous, and hydrogen-derived fuels;
- treating biomass thermo-chemically to produce a low- to medium-grade heat content gas;
- Enersligle™ (a process that produces a fuel-grade oil); and
- Cambi™ (thermal hydrolysis using temperature and pressure to heat and cool sludge).

DISPOSAL OPTIONS. Treatment plants typically decide whether to use or dispose of biosolids based on the cost of each evaluated option. Sludge that will ultimately be landfilled may need as much treatment as if it were to be land applied, so the related labor and trucking costs may be the overriding factors.

The disposal methods—"dumping", landfilling, monofilling, surface application, and co-disposal with municipal solids waste—all typically involve putting the biosolids in a hole in the ground. These holes are subject to federal, state, and sometimes local regulations, which require liners; daily and final covers; collection of leachate and methane gas; groundwater monitoring; control of odors, vectors, and nuisances; and caps and other closure methods.

DIVERSIFICATION AND CONTINGENCY PLANS. The Allegheny County Sanitary Authority's wastewater treatment plant in Pittsburgh, Pennsylvania, produces approximately 154 000 wet tonne/a (170 000 wet ton/yr) for the following purposes:

- Almost one-half is incinerated, producing steam for use in various plant processes as an energy-recovery method;
- Almost one-half is land applied on farms and strip mines by contractors; and
- The rest is landfilled.

Treatment plant staff chose this combination because the plant is within city limits, surrounded by businesses and residences, and cannot store solids onsite because of potential odors. This diversity enables the treatment plant to operate efficiently and without odors, regardless of maintenance needs or weather conditions.

Treatment plants do not just process wastewater, and solids management is not a small part of the operation. It requires at least as much time, effort, and financial commitment as wastewater does. So, all treatment plants should have a contingency plan for unexpected events (e.g., severe weather). Planning for regulatory changes, however, is another matter. Plant staff should establish a good working relationship with regulators and keep abreast of the regulatory changes under development. Changes that will result in more solids treatment will involve more money and equipment, which can take months or years to secure, install, and start up.

FACTORS AFFECTING THE DECISION. The most important factor in choosing a biosolids management option is the applicable laws, regulations, and local ordinances. Check carefully: even if land application is technically allowed, obtaining permits and site approvals can be difficult and time-consuming. The regulators who permit generators and approve sites may have many duties and, therefore, little time for these tasks or for complaints from organizations that vehemently oppose land application. In some states, local governments have established local ordinances that impose fees and requirements doubling the cost and effort of land application. The current trend is that regulators want treatment plants to produce odorless Class A biosolids.

Another important factor is whether to treat the plant's solids to meet Class A or Class B biosolids requirements. Typically, Class A treatment processes cost more than Class B processes. Most land-application programs use Class B biosolids, which work well on farms and strip mines and are acceptable to farmers and miners. Class A biosolids have lower pollutant and pathogen levels, and can be sold or given away to the public. (The sale price typically does not cover the entire production cost, but does help defray it.)

Other factors that affect the choice of biosolids management option include:

- The treatment plant's location;
- Local weather conditions;
- Past solids-management practices (everyone is more comfortable with something they know);
- Distance from available beneficial use sites;
- The number and strength of local anti-biosolids groups;
- Support of local government officials;
- Estimated costs;

- Available funding; and
- Local landfill availability and pricing.

Each option has advantages and disadvantages, and plant staff should avoid making the decision based solely on cost. Choosing to land apply Class B biosolids, for example, may ultimately fail if the anti-biosolids groups complain daily to regulators, who then become overwhelmed with public relations efforts and cannot keep up with permit applications and site-approval requests. Also, switching from a long-used method to a new one will be stressful, especially if the old method was simple (e.g., landfilling) and the new one is complex (e.g., composting or heat drying). Treatment plant personnel also need to decide whether to manage solids in-house or to outsource the work.

OUTSOURCING. *Outsourcing* is the process of hiring another entity to perform some of an organization's work. It enables treatment plant personnel to decide how much of the solids management work they want to do and how much they prefer to let someone else do. For example, the treatment plant could produce the biosolids and then hire a contractor to land apply it (e.g., perform all the work related to permitting, transporting, land applying, reporting, and communicating with the public and regulators). Alternatively, the treatment plant could hire a contractor to handle all of the solids-management processes.

One advantage of outsourcing is that the treatment plant does not have to expand its workforce. Also, plant personnel can rely on the expertise of a biosolids management specialist rather than "starting from scratch" themselves. Nor does the plant have to invest in a fleet of specialized vehicles (e.g., tractors, spreaders, and front-end loaders) or—if the entire solids-management program is outsourced—any solids or biosolids handling, processing, or storage facilities.

One disadvantage of outsourcing is that the treatment plant is still responsible for its contractor's actions, because as far as regulators are concerned, the biosolids generator remains their owner. So, plant personnel should be familiar with all aspects of the contractor's work, monitor the operations, and confirm that the program is effective and meets all requirements. Plant personnel also should maintain a suitable public outreach program and keep in contact with contractor staff, beneficial-use site owners, and regulators.

Once treatment plant personnel have decided to hire a solids management contractor, they need to draft an agreement that covers all usual and exceptional matters affecting the services that they expect the contractor to perform. The bid specifications or request for proposal should be developed by a team that includes legal counsel, environmental scientists, engineers, and operations and maintenance staff, as well as input from regulators. (Most outsourcing treatment plants are willing to share their bid specifications.) Any contractor hired to manage a treatment plant's solids should be well-known in their field and have experienced staff, a good compliance record, and a good work history.

ODORS AND PUBLIC ACCEPTANCE. Another important factor is the problem of odor complaints. Some complaints may be justified. Others will have nothing to do with the treatment plant's biosolids; they may be the result of odors from neighbors down the road. The complaints may even be more general: "We don't want your sewage sludge here." Justified or not, these complaints will cost the treatment plant personnel much time and energy combating their effects on regulators and the news media.

Although odor complaints cannot be completely avoided, treatment plant personnel can address the problem by

- Producing as odorless a product as possible;
- Monitoring conditions at beneficial-use sites (e.g., remoteness, wind direction, possibility of inversions, humidity, topography, and occurrence of holiday weekends);
- Encouraging regulators to educate the public about biosolids;
- Supporting environmental programs in schools and teaching students about biosolids;
- Establishing and maintaining public outreach programs on biosolids;
- Inviting local government and media representatives to tour the sites so they become familiar with biosolids and their benefits for farms and strip mines; and
- Following the suggestions of biosolids organizations [e.g., WEF and the National Biosolids Partnership (NBP)] concerning the public and the media.

Making the public familiar with the beneficial use of biosolids will hopefully counteract their instinctive fear of and aversion to the material.

RECORDKEEPING. Treatment plant personnel should keep the required records for as long as the regulations dictate. All permits, lab analyses, and regulatory reports should be maintained in a well-organized file or bookshelf to help regulators or third-party verifiers when they inspect the solids facilities.

Plant personnel also should set up an archive system. For example, paper files could be copied onto new storage media (e.g., CDs or optical disks) and then stored off-site so the hardcopies are backed up, but the data remain at staff's fingertips. (For more information on management information systems, see Chapter 6.)

REGULATORS AND INSPECTIONS. Regulators and treatment plant personnel are "in this together": both answer to the public, their supervisors, and elected officials. The regulators who inspect facilities or approve permits want a treatment plant's solids management program to succeed because it makes their jobs easier. If regulators approve a permit and the treatment plant does a poor job, then both look bad. On the other hand, if the solids management program is a success, then both look good.

So, if regulators ask for information that is not "required" by the regulations, they probably need the data to help them approve the plant's paperwork. Treatment plants will get their work done faster if their personnel are cooperative.

EMS AND ISO 14001. The National Association of Clean Water Agencies (NACWA), U.S. EPA, and WEF formed the NBP in 1997 "to advance environmentally sound and accepted biosolids management practices". For example, the NBP encourages treatment plants and solids contractors to develop environmental management systems (EMSs) to improve their production and use of biosolids. The partnership's *EMS Guidance Manual*, which is based on ISO 14001, is designed to help treatment plants develop EMSs tailored for their solids management programs. The systems will help treatment plants do a good job, develop a good reputation, and improve working relations with regulators and environmental groups. However, they may or may not help public relations efforts, because some people will never accept any material that comes from a wastewater treatment plant.

OTHER SOLIDS

Other solids separated from wastewater may be beneficially used but are typically landfilled. The use of grit is being explored; it is washed to remove rubbish and slime, and then separated by size into sand and gravel for use as aggregate in asphalt mixtures and sub-base course materials.

COSTS

The costs of various solids management options are shown in Tables 27.1 through 27.4. These costs are from estimates, demonstration projects, and full-scale operations. Most are from 2002 or later, but costs from the 1990s are included for historical comparison. To make the figures comparable, some discretion was used (judgment as to number of days of operation per week, etc.).

TABLE 27.1   Costs of various biosolids process chains.

Process	Includes	Cost ($/wet ton)	Cost ($/dry ton)	Reference
Heat drying and pelletizing	Dewatering; privatized	92.47	462.39	Frankos, 2003
Composting in-vessel Paygro	Dewatering and trucking; privatized	105.36	526.80	Frankos, 2003
Land application	Privatized	32.69	163.45	Frankos, 2003
Class A lime stabilization	Thickening		1100	Leininger and Nester, 2003
Class A lime stabilization	Dewatering		900–920	Leininger and Nester, 2003
Class A ATAD	Dewatering		900–920	Leininger and Nester, 2003
Class A ATAD	No dewatering		770	Leininger and Nester, 2003
Class A indirect steam drying	Dewatering		980–1040 1600–1700 thru land application	Leininger and Nester, 2003
JVAP™ frame press and Class A heating/drying	Dewatering		820–1335 storage and land application	Leininger and Nester, 2003
Upgrading and keeping Class B process	Aerobic digestion		750	Leininger and Nester, 2003
Class A thermal drying process	Fully allocated costs; digestion, dewatering, land application		average: 257 range: 187–328	Bullard, 2002
Land application demonstration	Dewatering through land application; centrifuges		172–220	Sloan et al., 2002
	Dewatering thru land application; belt filter press		154–160	Sloan et al., 2002
Land application current system	Dewatering thru land application; gravity belt and belt filter press		202	Sloan et al., 2002
Land application and landfill daily cover	Aerated static pile composting		405	Van Der March et al., 2002
	In-vessel composting		399	Van Der March et al., 2002

TABLE 27.1   Costs of various biosolids process chains (*continued*).

Process	Includes	Cost ($/wet ton)	Cost ($/dry ton)	Reference
	Rotary heat drying		93	Van Der March et al., 2002
	Pre-pasteurization and RDP-Cambi		224	Van Der March et al., 2002
	Chemical addition—Bioset		217	Van Der March et al., 2002
	Dedicated landfill		300	Van Der March et al., 2002
Class A lime stabilization; pre-1996 study survey of 10 facilities	Thicken-digest-dewater; En-Vessel Pasteur. (RDP-EVP)		312 – 10 mgd 228 – 20 mgd 158 – 40 mgd 139 – 60 mgd	Rothberg, Tamburini & Winsor Inc., 1996
	Thicken-digest-dewater; Biofix with lime only		~345 – 10 mgd ~250 – 20 mgd ~180 – 40 mgd ~160 – 60 mgd	Rothberg, Winsor Inc., 1996 Tamburini &
Aerated static pile composting; pre-1996 study survey of 14 facilities	Thicken-digest-dewater		360 – 10 mgd 277 – 20 mgd 204 – 40 mgd 184 – 60 mgd	Rothberg, Tamburini & Winsor Inc., 1996
Thermal drying; pre-1996 study survey of 9 facilities	Thicken-digest-dewater		493 – 10 mgd 372 – 20 mgd 276 – 40 mgd 247 – 60 mgd	Rothberg, Tamburini & Winsor Inc., 1996

ATAD = autothermal thermophilic anaerobic digestion.

TABLE 27.2   Unit cost snapshot for various processes.

Process	Cost ($/wet ton)	Cost ($/dry ton)	Comments/Reference
Belt filter press dewatering	76		3% to 23.5% total solids (Hagaman, 1998)
Belt press dewatering	30	187	to 15.9% (Lee et al., 2003)
Centrifuge dewatering	33.25	187	Lee et al., 2003
Anaerobic digestion	2.25 per gallon (4.25–1.90/gal)		Cost depends on size of digester (Potts et al., 2003)
	3.85 per gallon		Concrete egg-shaped digester (Potts et al., 2003)
	1.88–4.24 per gallon		Costs include everything but storage (Marx, 2002)
Centrifuge dewatering		172.63	Drury et al., 2002
Belt press dewatering		185.77	Drury et al., 2002
Aerobic digestion without thickening	6.07 per gallon	146	Curley et al., 2002
Aerobic digestion with thickening	6.23 per gallon	150	Curley et al., 2002
Anaerobic digestion without thickening	6.87 per gallon	165	Curley et al., 2002
Anaerobic digestion with thickening	7.31 per gallon	176	Curley et al., 2002
Composting	50 (38 after revenues)		Hogan, 2003
Heat drying and pelletizing		250–350 412	Hogan, 2003 with dewatering and bond repayment pelletizing (Hogan, 2003)
Multiple hearth incinerator		192.30	at 20.80% total solids (Leger, 1998)
Lime stabilization— RDP EnVessel process		41.33	at 22–25% total solids (Hagaman, 1998)
Composting	22-own facility 29-outside facility		Lee et al., 2003
Multiple hearth incinerator		183.78	Sherodkar and Baturay, 2003
Fluidized bed tray dryer and pelletizer		190 for O&M	Janses et al., 2003
Incinerate scum and grease		248	6-year average (Dominak and Stone, 2002)

TABLE 27.2   Unit cost snapshot for various processes (*continued*).

Process	Cost ($/wet ton)	Cost ($/dry ton)	Comments/Reference
Land application of biosolids	20/ton – by facility 25/ton – contractor 55/ton – to landfill		agriculture application; Chambersburg, Pennsylvania (Hook, 2003)
	0.055 per gallon		Class B; North Carolina (Hagaman, 1998)
		88	includes lagoon drying and subsurface injection; Calgary, Canada (Tatem, 1998)
Landfill biosolids	58	233	Hampton, New Hampshire (Berkel, 2003)
	approx. 59		by contractor; Cleveland, Ohio (Dominak and Stone, 2002)
Landfill ash		approx. 50	Cleveland, Ohio (Dominak and Stone, 2002)
Landfill grit and screenings	approx. 53		Cleveland, Ohio (Dominak and Stone, 2002)

TABLE 27.3   Sales value of biosolids.

Material	Value	Comments
Biosolids	$6–7 per percentage point of nitrogen	Florida (Maestri, 1998)
Compost	$10 per dry metric ton or $6.50 per cubic meter	Palo Alto, California (Nichols, 1998)
Bulk dry pellets	$63 per dry metric ton ($28.22 after marketing and distribution)	Palo Alto, California (Nichols, 1998)
Compost	$5.45 per cubic yard	Baltimore, Maryland

TABLE 27.4   Incineration costs.

Time	Operations and maintenance ($/dry ton)	Amortized capital installed capacity ($/dry ton)	Total ($/dry ton)	Reference
1990 actual[1]	70–90	100–125	170–215	Walsh et al., 1990
1990 actual[2]	180–200	200–230	380–430	Walsh et al., 1990
2003 estimate[3]	105–187	37–55	114–187[5]	Welp and Lundberg, 2003
2003 estimate[4]	135–247	37–55	144–247[5]	Welp and Lundberg, 2003

[1] Based on a well-operated sludge incineration system operating at capacity; starts with dewatered sludge and includes furnaces, heat recovery, air-pollution control, and ash disposal systems; includes some reserve capacity; 20 years for equipment, 40 years for buildings, and 8%.
[2] Same as 1 and also including thickening and dewatering.
[3] Does not include thickening and dewatering; fluid bed incinerator; 20 years, 6%.
[4] Does not include thickening and dewatering; multiple hearth incinerator; 20 years, 6%.
[5] Total includes $27–55 of energy credit.

REFERENCES

Author Unknown (2003) Dedication Ceremony Held for Rattler Mountain Reclamation Project. *Pennsylvania Organization for Watersheds and Rivers*, **4**, 24, June 13.

Berkel, P. F. (2003) Rotary Press—Innovative New Technology for Dewatering Municipal Biosolids—A Case Study of Hampton, N.H. *Proceedings of Residuals and Biosolids Management: Partnering for a Safe, Sustainable Environment* [CD-ROM]; Baltimore, Maryland, Feb. 19–22; Water Environment Federation: Alexandria, Virginia.

Bullard, C. M. (2002) Accommodating Process Variation in Design and Sizing of Thermal Drying Facilities. *Proceedings of the 75th Annual Water Environment Federation Technical Exhibition and Conference* [CD-ROM]; Chicago, Illinois, Sept. 28–Oct 2; Water Environment Federation: Alexandria, Virginia.

Curley, S. N.; et al. (2002) Aerobic Digesters in an Oxidation Ditch Configuration: An Unequivocal Success. *Proceedings of the 75th Annual Water Environment Federation Technical Exhibition and Conference* [CD-ROM]; Chicago, Illinois, Sept. 28–Oct 2; Water Environment Federation: Alexandria, Virginia.

Dentel, S. K.; Chang, L.-L.; Raudenbush, D. L.; Junnier, R. W.; Abu-Orf, M. M. (2000) *Analysis and Fate of Polymers in Wastewater Treatment*, WERF Project 94-REM-2; Water Environment Research Federation: Alexandria, Virginia.

Dominak, R. R.; Stone, L. A. (2002) Residuals Disposal Costs—A Detailed Analysis. *Proceedings of the 75th Annual Water Environment Federation Technical Exhibition and Conference* [CD-ROM]; Chicago, Illinois, Sept. 28–Oct. 2; Water Environment Federation: Alexandria, Virginia.

Drury, D. D.; et al. (2002) Comparison of Belt Press vs. Centrifuge Dewatering Characteristics for Anaerobic Digestion. *Proceedings of the 75th Annual Water Environment Federation Technical Exposition and Conference* [CD-ROM]; Chicago, Illinois, Sept. 28–Oct. 2; Water Environment Federation: Alexandria, Virginia.

Fisichelli, A. P. (1992) Operators as Environmentalists. *Oper. Forum*, **9**(1), 14–15.

Frankos, N. (2003) City of Baltimore's Back River Wastewater Treatment Plant Biosolids Program: How You Can Learn from Our Mistakes. *Proceedings of Residuals and Biosolids Management: Partnering for a Safe, Sustainable Environment* [CD-ROM]; Baltimore, Maryland, Feb. 19–22; Water Environment Federation: Alexandria, Virginia.

Garland, S.; Kester, G.; Walker, J.; et al. (2000) *Summary of State Biosolids Programs*, Draft Report, EPA-832/D-00-002; U.S. Environmental Protection Agency: Washington, D.C., December.

Hagaman, L., Jr. (1998) Beneficial Co-Utilization—Lenoir, North Carolina. *Proceedings of the 12th Annual WEF Residuals and Biosolids Management Conference*; Water Environment Federation: Alexandria, Virginia.

Hogan, R. S. (2003) PowerPoint presentation by R. C. Delo. *Proceedings of the MABA Seminar on Design–Build–Operate Procurement Issues*; Linthicum Heights, Maryland, Aug. 19.

Hook, J. (2003) Farms Here Are Accepting More Sludge—Old Disposal Method Draws Some New Fears. *Public Opinion* [Online] http://www.publicopiniononline.com/news/stories/20030201/local news/893644.html, Feb. 1.

Janses, U.; et al. (2003) Drying–Pelletizing and Fluid Bed Combustion: A Flexible and Sustainable Solution for the City of Brugge, Belgium. *Proceedings of Residuals and Biosolids Management: Partnering for a Safe, Sustainable Environment* [CD-ROM]; Baltimore, Maryland, Feb. 19–22; Water Environment Federation: Alexandria, Virginia.

Lee, S.; et al. (2003) Comparison of Belt Press Vs. Centrifuge Dewatering Characteristics for Anaerobic Thermophilic Disgestion. *Proceedings of Residuals and Biosolids Management: Partnering for a Safe, Sustainable Environment* [CD-ROM]; Baltimore, Maryland, Feb. 19–22; Water Environment Federation: Alexandria, Virginia.

Leger, C. B. (1998) Oxygen Injection for Multiple Hearth Sludge Incinerator. *Proceedings of the 12th Annual WEF Residuals and Biosolids Management Conference*; Water Environment Federation: Alexandria, Virginia.

Leininger, K. V.; Nester, W. R. (2003) Evaluating Liquid to Dried "Class A" Biosolids. *Proceedings of Residuals and Biosolids Management: Partnering for a Safe, Sustainable Environment* [CD-ROM]; Baltimore, Maryland, Feb. 19–22; Water Environment Federation: Alexandria, Virginia.

Maestri, T. J. (1998) Seghers Sludge Drying and Pelletizing. *Proceedings of the 12th Annual WEF Residuals and Biosolids Management Conference*; Water Environment Federation: Alexandria, Virginia.

Marx, J. J. (2002) An International Comparison of Anaerobic Digester Designs and Costs. *Proceedings of the 75th Annual Water Environment Federation Technical Exhibition and Conference* [CD-ROM]; Chicago, Illinois, Sept. 28–Oct. 2; Water Environment Federation: Alexandria, Virginia.

Nichols, C. E. (1998) Palo Alto Analyzes the Tradeoffs—Incineration vs. Pellets vs. Composts: Markets and Economics. *Proceedings of the 12th Annual WEF Residuals and Biosolids Management Conference*; Water Environment Federation: Alexandria, Virginia.

Potts; L., et al. (2003) An International Comparison of Anerobic Digester Designs and Costs. *Proceedings of Residuals and Biosolids Management: Partnering for a Safe, Sustainable Environment* [CD-ROM]; Baltimore, Maryland, Feb. 19–22; Water Environment Federation: Alexandria, Virginia.

Rothberg, Tamburini & Winsor, Inc. (1996) *Biosolids Stabilization—Which 'Class A' Stabilization Method Is Most Economical?*, Bulletin 334; National Lime Association: Arlington, Virginia.

Sherodkar, N. M.; Baturay, A. (2003) Maximizing Incineration Rates in Multiple Hearth Furnaces. *Proceedings of Residuals and Biosolids Management: Partnering for a Safe, Sustainable Environment* [CD-ROM]; Baltimore, Maryland, Feb. 19–22; Water Environment Federation: Alexandria, Virginia.

Sloan, D. S.; et al. (2002) A Comparison of Sludge Dewatering Methods: A High Tech Demo Project. *Proceedings of the 75th Annual Water Environment Federation Technical Exhibition and Conference* [CD-ROM]; Chicago, Illinois, Sept. 28–Oct. 2; Water Environment Federation: Alexandria, Virginia.

Sopper, W. E. (1993) *Municipal Sludge Use in Land Reclamation*; Lewis Publishers: Boca Raton.

Tatem, T. (1998) The City of Calgary's Biosolids Disposal Program. *Proceedings of the 12th Annual WEF Residuals and Biosolids Management Conference*; Water Environment Federation: Alexandria, Virginia.

Van Der March, M. M.; et al. (2002) An Evaluation of Biosolids Management Strategies: Selecting an Effective Approach to an Uncertain Future. *Proceedings of the 75th Annual Water Environment Federation Technical Exhibition and Conference* [CD-ROM]; Chicago, Illinois, Sept. 28–Oct. 2; Water Environment Federation: Alexandria, Virginia.

Walsh, M. J.; et al. (1990) *Fuel-Efficient Sewage Sludge Incineration*, EPA-600/S2-90-038; U.S. Environmental Protection Agency Risk Reduction Engineering Laboratory: Cincinnati, Ohio, September.

Welp, J., Lundberg, L. (2003) Thermal Oxidation. *Proceedings of the First Bioenergy Technology Summit*, Aug. 14–15; Water Environment Federation: Alexandria, Virginia.

SUGGESTED READINGS

Clapp, C. E.; Larson, W. E. (1994) *Sewage Sludge: Land Utilization and the Environment*; Dowdy, R. H., Ed.; American Society of Agronomy: Madison, Wisconsin.

Crites, R. W.; Reed, S. C., Bastian, R. (2000) *Land Treatment Systems for Municipal and Industrial Wastes*; McGraw-Hill: New York.

Dalton, P. (2003) How People Sense, Perceive, and React to Odors. *BioCycle*, November.

Hill, D. M. (2003) Success Is a Process and a Goal. *Proceedings of Residuals and Biosolids Management: Partnering for a Safe, Sustainable Environment* [CD-ROM]; Baltimore, Maryland, Feb. 19–22; Water Environment Federation: Alexandria, Virginia.

IFC Inc. (1995) *An Introduction to Environmental Accounting as a Business Management Tool*, EPA-742/R-95-001; U.S. Environmental Protection Agency: Washington, D.C., June.

Kilian, R. E.; et al. (2003) How to Put One Egg in Multiple Baskets—EMWD's Regional Biosolids Management Approach Makes Sense. *Proceedings of Residuals and Biosolids Management: Partnering for a Safe, Sustainable Environment* [CD-ROM]; Baltimore, Maryland, Feb. 19–22; Water Environment Federation: Alexandria, Virginia.

King County Department of Natural Resources and Parks, Wastewater Treatment Division Web site. http://dnr.metrokc.gov (accessed Feb 2007).

Montag, G. M. (1984) Life-Cycle Cost Analysis Versus Payback for Evaluating Project Alternatives. *Heating/Piping/Air Conditioning*, September.

Patrick, R.; et al. (1997) *Benchmarking Wastewater Operations—Collection, Treatment, and Biosolids Management*, Project 96-CTS-5; Water Environment Research Foundation: Alexandria, Virginia.

Roxburgh, R.; et al. (2005) Better than a Crystal Ball. *Oper. Forum*, April.

Slattery, L.; et al. (2003) Biosolids Management Alternatives Study for Arlington Water Pollution Control Plant. *Proceedings of Residuals and Biosolids Management: Partnering for a Safe, Sustainable Environment* [CD-ROM]; Baltimore, Maryland, Feb. 19–22; Water Environment Federation: Alexandria, Virginia.

SNF Floerger. Sludge Dewatering Brochure; SNF Inc.: Riceboro, Georgia.

Stehouwer, R. C. (1999) *Land Application of Sewage Sludge in Pennsylvania: Biosolids Quality*; Penn State College of Agricultural Sciences, Cooperative Extension: University Park, Pennsylvania.

Suzuki, S.; et al. (2002) Treatment and Beneficial Reuse of Grit. *Proceedings of the 75th Annual Water Environment Federation Technical Exhibition and Conference* [CD-ROM]; Chicago, Illinois, Sept. 28–Oct. 2; Water Environment Federation: Alexandria, Virginia.

Chapter 28

Characterization and Sampling of Residuals

INTRODUCTION

Sludges, biosolids, and other residuals typically are sampled and analyzed to ensure that various solids handling and treatment processes are performing properly. Residuals management may entail meeting a number of pollutant parameters established to protect human health and prevent nutrient overloading of the environment. U.S. and other regulatory requirements have increased the frequency and scope of analysis for residuals. Additionally, information obtained from sampling and analysis can improve the treatment facility's bottom line.

This chapter describes the qualitative and quantitative aspects of various residuals found in wastewater treatment plants (WWTPs). Parameters, properties, and miscellaneous analyses used to characterize residuals are also described. In addition, it notes the sampling procedures necessary for process control and regulatory purposes.

TYPES OF RESIDUALS

Residuals are produced by the removal of soluble and insoluble constituents during wastewater treatment. Of the various residuals, sludges and biosolids are composed primarily of water, but contain dissolved and suspended solids that can range from 0.5 to 7% of total unthickened solids. Thickening, stabilization, and dewatering processes are used to make these residuals acceptable for disposal (Figure 28.1). Utilities should remove as much water as possible to minimize transportation and disposal costs. (For more information on thickening, stabilization, and dewatering equipment and systems, see Chapters 29 through 33.)

Sludges can be classified broadly as primary, secondary, mixed, and chemical.

PRIMARY. Raw sludges are those that have not been chemically or biologically treated to reduce pathogens. Because the volatile solids content has not been stabilized, they tend to putrefy readily. Primary sludges are produced by sedimentation processes and consist mainly of fecal material, food wastes, fibers, silt, and, to lesser degrees, heavy metals and other trace minerals and substances. Properly settled primary sludge typically will have a total solids concentration between 1 and 3%; if gravity thickened, it can contain up to 10% total solids. This thickening can effectively cut the sludge volume in half, thereby reducing the sizing requirements (hydraulically) and costs of downstream units. Primary sludge may be easily thickened and dewatered if not allowed to become septic. This sludge is highly putrescible, generates unpleasant odors, and contains enough pathogens to cause disease in humans on contact with unprotected skin and mucous membranes, especially if the skin is cut or abraded.

SECONDARY. Secondary sludge is composed of the solids that were not readily removed via primary clarification and those produced during the consumption of soluble and insoluble organics in the secondary treatment system. Secondary sludge is either returned to the treatment process (return activated sludge) or wasted for further

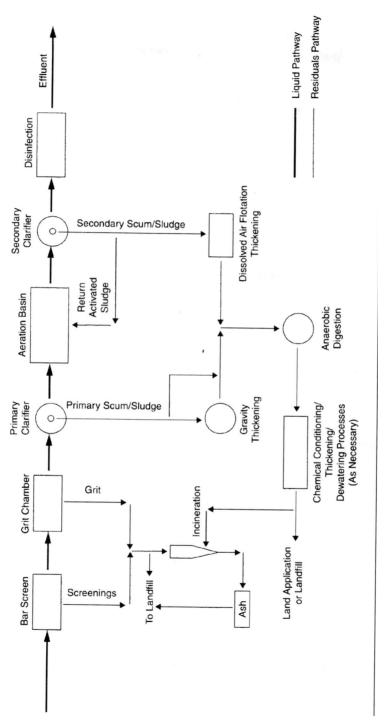

FIGURE 28.1.   Residuals pathways in primary and secondary treatment systems.

treatment. Whether produced by aerobic or anaerobic decomposition, these microbial masses can account for most of the sludge volume at a treatment facility. Properly managed activated sludges will contain 0.5 to 1.5% total solids; attached growth (biofilm) sloughings will contain 1 to 4% total solids.

MIXED. Primary and secondary sludges may be combined after their respective sedimentation processes. They may be individually thickened before combination or combined first and then further processed. Secondary sludges at smaller WWTPs are sometimes wasted to the primary clarifier, thus producing a mixed sludge. Plants with no primary treatment, such as extended aeration "package plants", produce only a secondary sludge.

CHEMICAL. Chemical sludges are produced when chemicals are added to precipitate otherwise hard-to-remove substances from the wastewater. A typical form of this sludge is when lime, alum, ferrous or ferric chloride, or ferrous or ferric sulfate is used to remove phosphorus from wastewater via precipitation. These chemical sludges contain a considerable amount of primary and secondary sludges. While some chemicals improve the thickening and dewatering characteristics in downstream processes, others reduce dewaterability and have unwanted side effects (e.g., low pH and alkalinity). These effects may require the addition of alkaline chemicals (typically lime or caustic) to adjust these parameters before nitrification or discharge to the receiving stream.

STABILIZED. Sludges that have been stabilized (have reduced pathogen concentrations and vector attraction reductions) are called *biosolids* and can be used beneficially as soil amendment. Biosolids are typically land-applied, although they could be landfilled or incinerated and then landfilled. Maximal reuse of these solids is achieved through agricultural application, forest application (silviculture), land reclamation, composting, and other such applications.

Approximately one-third of the solids produced in the United States were, in some way, land-applied (U.S. EPA, 2004). More solids are expected to be beneficially used and less landfilled, as landfill sites become an increasingly expensive disposal method. Restrictions on ocean dumping in the United States and the dramatic shift in the number of U.S. plants providing secondary or better treatment by 2012 can only exacerbate the problem of landfill disposal, particularly in the most densely populated areas of the United States, according to a U.S. Environmental Protection Agency (U.S. EPA) assessment (U.S. EPA, 1993a).

Sludges can be stabilized anaerobically (e.g., in an anaerobic digester where sludge is heated; organic compounds are hydrolyzed into methane, carbon dioxide, and water; and volatile solids are reduced). They also can be stabilized aerobically in an unheated digester, producing carbon dioxide and water, and reducing volatile solids.

Stabilization is defined by the pathogen and vector-attraction reductions set forth in 40 *CFR* 503 and the guidance document for issuance of U.S. EPA permits (U.S. EPA, 1993b, 2004). The regulations limit the amount of pathogens that sludge can contain. The agency allows pathogens to be reduced via composting, heat drying, heat treatment, thermophilic aerobic digestion, irradiation, pasteurization, reduction of fecal coliform density, anaerobic digestion, and lime stabilization (U.S. EPA, 1993b, 2004). Similarly, vector attraction reduction can be achieved by a 38% volatile solids reduction via aerobic or anaerobic digestion (U.S. EPA, 1993b, 2004). It also can be done via thermophilic aerobic digestion, reducing the specific oxygen uptake rate to less than 1.5 (mg oxygen/h)/g of total solids at 20°C, raising the percent solids concentration, or raising the pH to 12 or higher for 2 hours using alkali addition and remaining at a pH greater than 11.5 for 22 more hours without more alkali (U.S. EPA, 1993b, 2004). The U.S. EPA regulations also set land-application ceiling concentrations and cumulative pollutant loadings for nine metals that can cause adverse human or environmental effects in excessive amounts (Tables 28.1 and 28.2).

OTHER RESIDUALS. Wastewater treatment plants also produce several other types of residuals. While they have significantly less volume and weight than sludges, their disposal is by no means a small concern.

Screenings. Screenings typically are removed by bar screens or bar racks. They are relatively large debris consisting of rags, plastic, cans, leaves (especially in autumn in the

TABLE 28.1   Pollutant ceiling concentrations for selected heavy metals.

Pollutant	Ceiling concentration (mg/kg[a])
Arsenic	75
Cadmium	85
Copper	4300
Lead	840
Mercury	57
Molybdenum	75
Nickel	420
Selenium	100
Zinc	7500

[a]All concentrations are on a dry weight basis.

TABLE 28.2   Cumulative pollutant loading rates for selected heavy metals.

Pollutant	Cumulative pollutant loading rate (kg/ha)
Arsenic	41
Cadmium	39
Copper	1500
Lead	300
Mercury	17
Molybdenum	—[a]
Nickel	420
Selenium	100
Zinc	2800

[a]Limit deleted by a U.S. Federal Court action.

northern hemisphere), rocks, and similar items. Quantities vary, although they are typically small (4 to 40 mL/m^3). Screenings can be removed and disposed of either by hauling them to a landfill or by incineration. Macerating or comminuting the screenings and returning them to the liquid stream are not recommended because many downstream processes and units (e.g., aeration diffusers, mixer shafts, and electronic probes) are subject to fouling from reintroduced rags, strings, and similar debris.

Grit. Grit is composed of many heavy or coarse materials (e.g., sand, cinders, gravel, and similar inorganic matter). The organic fraction of grit may include coffee grounds, eggshells, corn, and seeds. Grit also includes any items that passed through the bar screens. Grit in downstream processes is undesirable, because it can wear out pump impellers, piping, and other equipment. It builds up in vessels with lower flowthrough velocities (e.g., channels, aeration basins, digesters, piping, and wet wells), thereby reducing their working volumes.

Grit can be removed via grit chambers (typically aerated) or centrifugal separation. It may be washed to remove organics that will readily putrefy. In aerated grit systems, cyclonic or vortex-type separators, the amount of organics in the grit can be controlled somewhat by varying the amount of air supplied to the air diffusers in a grit chamber or limiting the air to the air scour unit in a vortex. Volumes of grit removed vary (4 to 200 mL/m^3 is typical) (Metcalf and Eddy, Inc., 2003). Grit, like screenings, varies in quantity because of variations in influent flow. After a dry weather period followed by a wet weather event, both residuals typically will increase because of flushing action in the collection system. Grit is almost always landfilled. If grit is being stored for removal, some facilities cover it with a layer of lime to reduce odors and provide a modicum of vector control.

Scum. Scum is the product of clarifier or tank skimmings. Primary clarifier skimmings are composed of fats, oils, grease, and floating debris (e.g., plastic and rubber products). It can often plug the scum trough or build up in downstream lines, thereby restricting flow and increasing pumping costs. It fouls probes, flow elements, and other instruments and equipment in the wastestream. Plants sometimes use degreasers to emulsify the fats, oil, and grease (FOG) in the wastewater, making it easier to handle in the primary system. Bioaugmentation (the addition of microbes that are genetically engineered to use FOG as their food source) can also be used to reduce the amount of FOG.

Secondary scum consists of the same substances found in primary scum, but most of it tends to be floating solids (either activated sludge or biofilm, depending on the type of secondary treatment used).

Other skimmings may include those from oil and grease separators, which are similar to the primary skimmings, but without most of the floating debris. They may include solvents that require different processing and disposal.

Scum's quantity and moisture content typically is not measured, although WWTPs that treat and dispose of it separately have more impetus to do so. Scum can be pumped directly to the digestion process; concentrated and then incinerated with other residuals; dried and landfilled; or recovered by firms that specialize in recycling and reprocessing FOG. Like screenings and grit, scum is laden with pathogens and should be handled carefully.

Ash. Ash is the product of thermal reduction processes (e.g., fluidized bed reactor, multiple hearth furnace, or wet-air oxidation). Although the volume of residuals is reduced to approximately 4% of the original volume, because of the oxidation of the organic fraction, the mass is still 20 to 50% of the original mass (digested solids produce more ash content) (WEF, 1992). Ash typically is disposed via landfilling.

In Japan, heat-dried solids are mixed with crushed stone and melted in a furnace. This produces an ash–stone slag used increasingly for construction materials (e.g., road bedding, clay pipes, water-permeable concrete blocks, and paving bricks and tiles) (Kanezashi and Murakami, 1992; Sato and Watanabe, 1994). Ash can be considered a beneficial product, although large amounts of it are still landfilled in Japan. The Japanese and the Europeans are also aggressively recovering the thermal energy expended in the incineration process and augmenting it with methane gas combustion produced in the anaerobic digestion process (Kaneko, 1992; Sato and Watanabe, 1994). While incineration is declining as a disposal alternative in many countries, it is increasing in Japan, which has little usable land.

CHARACTERIZATION

Residuals can be characterized by their physical, chemical, and biological properties. The tools used to measure these residuals range from physical testing (e.g., capillary suction time and settled volume) to biological testing (e.g., biochemical oxygen demand and coliform counts). Chemical testing is used to ascertain the major and minor constituents in the residual. Most of the testing methods referred to below can be found in *Standard Methods for the Examination of Water and Wastewater* (APHA et al., 2005). Some of these methods are controversial; residuals vary and are truly unique to every plant. *Standard Methods* (APHA et al., 2005) and 40 *CFR* 503 (U.S. EPA, 2004) are exceptional references. For questions about the use of these methods for a specific application, contact the appropriate U.S. EPA representative.

PHYSICAL PROPERTIES. A substance's physical properties relate to the human senses (touch, sight, and smell). Some physical properties of residuals include temperature, viscosity, thickness (or ability to be thickened via physical means), color, settleability, volatile and nonvolatile solids, and odor. Odor is highly subjective and discussed more extensively in Chapter 13. Several other parameters are interrelated (e.g., temperature, viscosity, and settleability). Physical solids processes (e.g., gravity, gravity belt, or dissolved-air-flotation (DAF) thickening; and belt filter presses, pressure filters, and centrifugation) focus on primary and secondary sludges. Their goal is to increase the sludge's solids content by removing water. For example, settling primary sludge can increase the total solids to between 4 and 7%; additional gravity thickening can further increase the total solids. Secondary sludge can be thickened via rotary drum concentrators, gravity thickeners, DAF units, or centrifuges. These units typically use polymers to increase both the total solids concentration and solids capture rates. Secondary sludge can be thickened from 0.5 to 1.5% total solids to approximately 4 to 7%, assuming it is not septic or has some other thickening-inhibiting characteristics. Recently, membranes have begun to be used to thicken sludge, achieving mixed-liquor suspended solids (MLSS) concentrations in excess of 20 000 mg/L.

Tests have been established to determine the degree to which sludge has been thickened or dewatered. Septicity can greatly affect the sludge's dewatering characteristics. Improperly conditioned sludge also can give undesirable results. So, the following tests are offered as indicators of sludge quality.

Settled sludge volume and sludge volume index are typical thickening tests used not only to evaluate an activated sludge's settling quality, but also to analyze gravity thickening. The 30-minute settling test provides the best data on the sludge's settling characteristics. A settleometer (2-L graduated cylinder) is typically used for this test. The average settled sludge volume (SSV) is 100 to 300 mg/L (the higher the result, the poorer the sludge will settle). The results of this test can be combined with the MLSS to obtain the sludge volume index (SVI):

$$SVI = SSV \times 1000/MLSS$$

This test indicates the sludge's ability to settle properly (the lower the SVI, the better the settling quality). Sludge with an SVI of 100 or less is typically considered to be a good settling sludge (WEF, 2002). Capillary suction time and the related time-to-filter tests are used to determine the rate that water is released from sludge. The procedures for conducting these tests are presented in *Standard Methods* (APHA et al., 2005).

Sludges typically are conditioned with inorganic chemicals or polymers before dewatering. Standardized methods of determining chemical and polymer doses are presented in the manual *Sludge Conditioning* (WPCF, 1988). For belt filter presses, centrifuges, pressure filters, and vacuum filters (now largely obsolete), tests measuring the sludge's specific resistance, capillary suction time, centrifuge (bench-scale), penetrometer, and floc strength are also discussed in detail in *Sludge Conditioning* (WPCF, 1988).

CHEMICAL PROPERTIES. The major chemical constituents of sludges and biosolids are total nitrogen; total Kjeldahl nitrogen (TKN); organic nitrogen; ammonia nitrogen; total phosphorus; alkalinity; volatile acids; and other typical parameters (e.g., biochemical oxygen demand, chemical oxygen demand, total organic carbon, and pH). Related analytical methods can be found in *Standard Methods* (APHA et al., 2005) and similar documents.

The moisture content in residuals can be determined by calculating the percent solids and subtracting it from 100%. Likewise, the volatile solids in content in residuals can be found by determining the percent ash and subtracting it from 100%. Tracking solids through various treatment processes can help utility staff establish solids balances and inventories, which are useful in managing treatment efficiencies. For example, an increase in inert solids may indicate a buildup of WWTP solids because of higher recycle loadings as a result of poor solids capture in the thickening and dewatering processes.

Digester controls typically focus on volatile solids, volatile acids, alkalinity, pH, and temperature. Increasing volatile acids, decreasing alkalinity and pH, or poor volatile solids reduction indicate poor digester performance (WPCF, 1987).

Residuals also contain small amounts of organic and inorganic chemicals. These chemicals are present in such small quantities that sophisticated instruments are required to measure them. The inorganic substances that typically are monitored include both metals (e.g., arsenic, cadmium, chromium, copper, lead, mercury, molybdenum, nickel, selenium, zinc, and iron) and nonmetals (e.g., calcium and potassium).

Nitrogen, phosphorus, calcium, and potassium can substantially raise crop yields, so they are important constituents in biosolids. Nitrogen typically is present in three forms: ammonia, nitrate, and organically bound. Nitrite concentrations are typically low. The organic fraction of nitrogen in biosolids is simply equal to the TKN concentration minus the ammonia. Exposure to air will convert some ammonia to nitrate, but most is in the form of ammonium bicarbonate. Much of the ammonia is lost when biosolids are dried especially if the pH is elevated to 12 or more in postliming operations. Much of the ammonia volatilizes when biosolids are surface-applied to land (rather than added via subsurface injection). The odor associated with ammonia volatilization should be a major concern when deciding which disposal method to use. Phosphate is present as orthophosphate or total phosphate (including polyphosphates). Total phosphate enhances the fertilizer value of biosolids.

Although beneficial in small dosages, metals typically are considered indicators of potential harm when using or disposing of ash and biosolids. They are particularly harmful to anaerobic digestion and are often the rate-limiting parameters in land application. For example, if the chromium concentration is higher than that of other regulated metals, it will be used to establish the residual's maximum disposal rate and maximum cumulative loading rate. There is little a typical secondary treatment plant can do to reduce metal concentrations. The best and most economical approach typically is to categorize the sources of metals and reduce or eliminate them. Conducting a pretreatment program to reduce influent sources of metals and other undesirable wastewater constituents can help meet applicable regulatory requirements for residuals use or disposal.

Residuals contain minute amounts of organic constituents. These organics originate at both industries and households. Some are refractory, like insecticides (e.g., aldrin, eldrin, heptaclor, lindane, malathion, dichlorodiphenyldichloroethane, dichlorodiphenyldichloroethylene, dichlorodiphenyltrichloroethane, and chlordane) (WEF, 1999). Polychlorinated and polybrominated hydrocarbons (e.g., hexachlorobenzene and bromodichloromethane) are potential health and environmental hazards and largely resistant to biological decomposition (WPCF, 1990). Other hazardous compounds (e.g., phenols, benzenes, and toluenes) are more easily degraded by biological systems (WEF, 1999). The U.S. Environmental Protection Agency is likely to begin regulating organics at WWTPs, starting with dioxin and polychlorinated biphenyls (WEF, 1995). The agency does not currently require annual dioxin analysis of biosolids, but regulators highly recommend it. Some states (e.g., Ohio) do require dioxin testing and reporting. A typical WWTP has neither the laboratory equipment nor the technical expertise to analyze the metals and organics in its wastestreams. So, independent laboratories analyze wastewater and residuals samples for most U.S. treatment facilities.

Chemical processing plays an important role in sludge treatment. The use of polymers in thickening and dewatering has increased in recent years, especially as gravity belt thickeners, belt filter presses, and centrifuges have gained popularity. Utility staff should periodically test polymers to select the proper type for a given sludge because what works well for one sludge may not be suitable for another. Also, as sludge characteristics change over time, so might the chemical requirements.

Lime is also important at many WWTPs. It is used to adjust alkalinity in activated sludge (especially before nitrification) and "postlime" biosolids. Lime consumption depends on the type and dryness of the solids being adjusted. The pH of the liquid slurry can be measured easily by an electrode.

BIOLOGICAL PROPERTIES. All residuals (except ash) contain appreciable numbers of biological constituents (e.g., pathogenic bacteria, protozoan cysts, helminthic organisms, and enteric viruses). Indicator organisms (e.g., fecal coliform and fecal streptococci) help wastewater treatment professionals track pathogenic organisms. U.S. regulations address fecal counts in both wastewater and residuals. [For more information on pathogenic reduction requirements for biosolids, see *Standards for the Use and Disposal of Sewage Sludge* (U.S. EPA, 2004).]

Most treatment facilities rely on biological processes. Anaerobic and aerobic digestion are the most typical forms of solids treatment. U.S. regulations require that WWTPs reduce the volatile content of solids by at least 38% unless other control parameters (e.g., specific oxygen uptake rate) are measured. Analytical procedures can be found in the current edition of *Standard Methods* (APHA et al., 2005).

Solids characteristics can affect thickening and dewatering. For example, septic sludge tends to thicken and dewater less readily than fresh sludge and cause the recycle flows from these processes to be less efficient in capturing solids, thereby increasing recycle solids loadings. Anaerobically digested solids tend to dewater more readily than waste activated sludge or aerobically digested solids. Temperature, pH, polymers, conditioning chemicals, and other parameters also affect thickening and dewatering characteristics (see References and Suggested Readings at the end of this chapter).

SAMPLING

Utilities take samples for two reasons: regulatory compliance and process control. Following is a discussion of representative sampling, sample collection, chain-of-custody procedures, sample preservation, quality assurance, and quality control.

REPRESENTATIVE SAMPLING. The goal of a sampling program is to obtain samples that best represent the flow being analyzed. Otherwise, any decision based on analysis of these samples is more likely to be erroneous.

Grab samples are taken for rapid analysis. They typically are used for process control, but because of the holding time requirements of certain analyses, may be used for reporting purposes. For example, fecal coliform samples are typically grab samples. Grab samples are also quick and effective tools to help adjust a chemical dose or study the effects of a process modification. For a grab sample to be representative, operators should ensure that the sample is taken in a well-mixed area and obtained in the same place each time it is taken.

Composite samples consist of a mixture of grab samples taken over a period of time (e.g., 24 hours) from the same sampling point. Composites of influent and effluent typically are collected by automated samplers to save labor and ensure accurate sampling intervals, but automated samplers can only be used on low-viscosity sludges. Compositing can be time-, flow-, or concentration-paced. Time-interval composites have been discussed. Regulations often require flow pacing, in which the grab samples (aliquots) are proportional to the flow. Concentration pacing uses a sampling system to try to maintain a predetermined concentration that is not directly correlated to flow. When collecting composites of higher viscosity sludges, which cannot be sampled via automated samplers, utility staff should develop a good standard operation procedure (SOP) that should be followed each time the sample is obtained.

Although less common than grab or composite samples, integrated sampling is useful in categorizing wastestreams that may significantly affect treatability (APHA et al., 2005). An example of this might be an industrial treatment system where two process streams are combined. An integrated sample would help utility staff study the effects of this combination.

SAMPLE COLLECTION. Samples should be collected methodically. A set of procedures has been developed that ensures sampling reliability. The procedures include instructions on sampling methods, container preparation, chain of custody, sample preservation, and storage. Unless utilities adhere to these procedures, results of the sample analysis could be disputed. For example, containers must always be clean and free of previous sample residues. *Standard Methods* (APHA et al., 2005) lists the type of material suitable for sample containers (which is based on the type of test required).

Typically, compliance testing will specify the location, frequency of analysis, type of sample (grab, composite, etc.), parameters to be measured, and method of measurement. It also typically requires that chain-of-custody procedures be used. Chain-of-custody procedures; which are required for any litigious actions, are similar to those used by law enforcement agencies to preserve evidence for trial. The major points of a chain-of-custody procedure are sample labeling (time, date, and identification number); sealing of the sample container; and a chain-of-custody sheet that includes the sample volume, type, location, time, date, person collecting the sample, person relinquishing the sample, person analyzing the sample, what parameters are to be tested, and the signatures of persons involved in the transaction. After arriving at the laboratory, the analyses are checked for accuracy, a signature attesting to the procedures used for each analysis is affixed to the custody sheet, and a report is issued. Completed custody and analysis reports are kept on file for inspection and reference.

Sample preservation may be required. Some samples (temperature, dissolved oxygen, etc.) require immediate analysis. Others (e.g., selected metals), if properly preserved and stored, can be analyzed up to 6 months later. Knowing the proper preservation and storage techniques can make the difference between a meaningful and meaningless analysis.

Sample storage is important; improperly stored samples are worthless from a legal standpoint and possibly misleading from a process control standpoint. Sample-handling requirements (e.g., sample volume required, container type, preservation required, and maximum time interval between storage and analysis) can be found in *Standard Methods* (APHA et al., 2005).

QUALITY ASSURANCE AND QUALITY CONTROL. Quality assurance and

quality control (QA/QC) programs are established by laboratory personnel to confirm the reliability of the results they generate and report. Recordkeeping provides the verification of QA/QC efforts. A typical QC program will include, at a minimum, files of all raw data, calculations, QC data, and all reports. Additionally, regulations may require a written manual that includes procedures for all tests performed, instrument (pH meters, analytic balances, thermometers, and spectrophotometers) calibration frequencies with certification dates attesting to them, and other internal control mechanisms. Typically, the QC records are kept for a specified period of time—at least 3 years in some U.S. jurisdictions (State of West Virginia, 1992). Computerized records are particularly helpful in maintaining organized files of all reports and QA/QC records, but paper documentation of verification must still be maintained.

The QC data will include a record of test results for all blanks, duplicate (replicate) samples, precision analysis (spike and percent recovery results), and standardization curves (e.g., atomic absorption analysis). Information on QC data is given in extensive detail in *Standard Methods* (APHA et al., 2005). Examples of typical data-reporting sheets, with simplified test procedures and standard analytical calculations, can be found in *Basic Laboratory Procedures for Wastewater Examination* (WEF, 2002).

As part of QA/QC programs, some U.S. states require certification of laboratory personnel (State of West Virginia, 1992). These requirements stipulate specific levels of education and experience for laboratory managers, supervisors, and technicians. Certificates are issued where applicable, following successful inspection and completion of all the requirements contained in the governing code of regulations.

In sum, these internal and external controls are designed to ensure that reported results are reasonably accurate and reliable. However, the analytical results cannot ensure the quality and reliability of the sample taken.

REFERENCES

American Public Health Association; American Water Works Association; Water Environment Federation (2005) *Standard Methods for the Examination of Water and Wastewater*, 21st ed.; American Public Health Association: Washington, D.C.

Kaneko, S. (1992) Centralized Sludge Treatment in Yokohama. *Sew. Works Jap.*, **39**.

Kanezashi, T.; Murakami, T. (1992) Regional Sewage Sludge Treatment in Yokohama. *Sew. Works Jap.*, **33**.

Metcalf and Eddy, Inc. (2003) *Wastewater Engineering: Treatment, Disposal, and Reuse*, 4th ed.; McGraw-Hill, Inc.: New York.

Sato, K.; Watanabe, H. (1994) Present Status of Beneficial Use of Wastewater Solids in Japan and the Future Strategy. *The Management of Water and Wastewater Solids for the 21st Century: A Global Perspective*, Proceedings of Water Environment Federation Specialty Conference, Washington, D.C., June 19–22; Water Environment Federation: Alexandria, Virginia.

State of West Virginia (1992) *Regulations Governing Environmental Laboratory Certification and Standards of Performance*, State of West Virginia Code 20-5A-1, Title 46, Series 11; Department of Environmental Protection: Charleston, West Virginia.

U.S. Environmental Protection Agency (2004) *Standards for the Use and Disposal of Sewage Sludge*, 40 *CFR* 257, 403, and 503; U.S. Environmental Protection Agency: Washington, D.C.

U.S. Environmental Protection Agency (1993a) *1992 Needs Survey Report to Congress*, EPA-832/12-93-002; U.S. Environmental Protection Agency: Washington, D.C.

U.S. Environmental Protection Agency (1993b) Guidelines for Writing Permits for the Use or Disposal of Sewage Sludge. U.S. Environmental Protection Agency: Washington, D.C.

Water Environment Federation (2002) *Basic Laboratory Procedures for Wastewater Examination,* Special Publication; Water Environment Federation: Alexandria, Virginia.

Water Environment Federation (1999) *Hazardous Waste Treatment Processes,* Manual of Practice No. FD-18; Water Environment Federation: Alexandria, Virginia.

Water Environment Federation (1995) EPA and WEF Collaborate on Radioactivity in Wastewater Solids Issue. *WEF Highlights,* **32** (1), 3.

Water Environment Federation (1992) *Sludge Incineration: Thermal Destruction of Residues,* Manual of Practice No. FD-19; Water Environment Federation: Alexandria, Virginia.

Water Pollution Control Federation (1990) *Hazardous Waste Treatment Processes,* Manual of Practice No. FD-18; Water Pollution Control Federation: Alexandria, Virginia.

Water Pollution Control Federation (1988) *Sludge Conditioning,* Manual of Practice No. FD-14; Water Pollution Control Federation: Alexandria, Virginia.

Water Pollution Control Federation (1987) *Anaerobic Sludge Digestion,* Manual of Practice No. 16: Water Pollution Control Federation: Alexandria, Virginia.

SUGGESTED READINGS

National Research Council (1996) *Use of Reclaimed Water and Sludge in Food Crop Production;* National Academy Press: Washington, D.C.

U.S. Environmental Protection Agency (1990) *Analytical Methods for the National Sewage Sludge Survey;* U.S. Environmental Protection Agency, Office of Water: Washington, D.C.

U.S. Environmental Protection Agency (1986) *Test Methods for Evaluating Solid Waste;* U.S. Environmental Protection Agency, Office of Solid Waste and Emergency Response: Washington, D.C.

Water Environment Federation (1996) *Wastewater Sampling for Process and Quality Control,* Manual of Practice No. OM-1; Water Environment Federation: Alexandria, Virginia.

Chapter 29

Thickening

INTRODUCTION

DEFINITION OF THICKENING. The removal of water from biosolids is a continuum. Although there is no technical definition of *thickening*, at around 50 000 centipoises we are at a point where most biosolids are a heavy paste and no longer pour from a cup.

PURPOSE OF SLUDGE THICKENING. Biosolids enter a plant at ±350 ppm (0.035% suspended solids). To remove these solids, they have to be concentrated and separated from the fluids. Concentrating takes place at several points during the treatment process and the basic goal is to reduce the process volume, which, in turn, will reduce the costs of downstream sludge processing. For example, in feeding an anaerobic digester, a plant may require 20 days residence time in the digester. If the plant is producing 100 000 gpd of 2.0% sludge, they will need 20 days times 100 000 gpd, or 2 million gallons of digester capacity. If the sludge can be thickened to 3%, the daily volume is only 66 000 gal and the resulting digester volume is 1.3 million gallons, which represents a considerable savings.

Similarly, a plant may wish to concentrate sludge before transport or disposal. The Deer Island Wastewater Treatment Plant in Boston, Massachusetts, thickens digested sludge before pumping it several miles under Boston Harbor to the dewatering facility. This reduces pumping costs and also reduces the volume of centrate produced at the dewatering site; both of these factors represent significant cost reductions after thickening.

THICKENING MODES

There are several points in the treatment process where thickening is commonly done. The most common are thickening before sludge stabilization or dewatering. The clarifier underflow is often quite thin and, as mentioned earlier, preconcentrating it will reduce the amount of process equipment downstream.

Recuperative thickening is also occasionally used. Recuperative thickening is where the sludge is withdrawn from a process (typically digestion), thickened, and returned to the digester.

The last stage in the treatment process is when the sludge is thickened after stabilization and before beneficial reuse, which is a common practice when the biosolids are disposed of as a liquid.

ANCILLARY EQUIPMENT

This section addresses equipment needed to support thickening equipment.

GRINDERS, PUMPS, AND FLOW METERING. Some devices are easily damaged or plugged by oversized materials. There are three sources of oversized particles: those that somehow slipped by bar screens; those that are generated within the plant, such as rag and string balls; and miscellaneous debris, such as wrenches, mop heads, and construction materials.

Grinders or screening devices are used whenever there is a likelihood of oversized particles causing damage. Although grinders are usually less expensive than screens, either device will suffice in terms of removing oversized particles. At one time, there were high-speed grinders available that caused a lot of shear and low-speed grinders that did not. However, the high wear rates of high-speed grinders drove them out of the market, leaving only the low-speed grinders.

There are two broad categories of pumps: centrifugal and positive-displacement. As a rule, positive-displacement pumps create less shear on the sludge. Because shear is presumed to be detrimental to thickening and dewatering, positive-displacement pumps are preferred. However, the cost of positive-displacement pumps poses a problem. As the volume pumped increases, the positive-displacement pump cost-per-volume pumped goes up. Therefore, centrifugal pumps usually are used to move very large flowrates. When the sludge is concentrated the volume is less, and the pump is more likely to be a positive-displacement type.

In recent years, the price of variable-frequency drives has dropped, and it is practical to install variable-speed centrifugal pumps. This allows a flow meter to control the pump speed, which greatly reduces both the power consumption and shear imparted to the sludge.

Another consideration is flow pulsing. A thickening or dewatering centrifuge has a liquid residence time of fewer than 2 seconds. Pulsations in flow are not dampened out very much in the centrifuges, so the clarified liquid will reflect the peak flow, not the average flow. Along those lines, most mechanical thickening devices require polymer to operate. Assuming the polymer flow is constant, pulsing feed sludge will result in variation in the polymer dosage. Pulsation dampeners are a valuable addition to the pumping system but must be maintained to avoid failure.

The following are positive-displacement thickened sludge pumps listed in order of increasing pulsation:

- Progressing cavity,
- Double disc,
- Rotary lobe,
- Diaphragm, and
- Plunger.

The last two pumps create very large pulsations and should not be used in instances where this will be a problem.

Although older plants used a variety of variable-speed devices, variable-frequency drives are presently the least costly. One advantage of variable-frequency drives is that they are easily programmed to provide a "soft start". This reduces the wear and tear on the pump drives. These drives also provide a readout of the motor speed at no additional cost. For a positive-displacement pump, the fluid flow is proportional to the pump's revolutions per minute. This provides a flowrate readout at no cost. Unfortunately, as the pump wears, the actual flowrate drifts off of the new pump calibration. The result is that a year or so after start-up, no one knows what the flowrate is on any of the pumps. As such, pump speed is a poor substitute for flow meters.

POLYMER ADDITION. In water and wastewater treatment, all filtration processes are dependent on polymers to work. In contrast, although they can perform much better with polymer, sedimentation processes can work without polymer. Polymer selection and application is covered in depth in Chapter 33.

GRAVITY THICKENING

Prior to nitrification requirements (circa 1980), gravity thickening was the principal method used for mixed sludges (both primary and secondary). Dissolved air floatation

(DAF) was occasionally used for thickening secondary biological sludges due to their light, diffuse nature.

With nitrification, gravity thickening of mixed sludges became difficult due to zero dissolved oxygen in the thickener, which allowed nitrates to be broken down, nitrogen to be released, and the sludges to float rather than be thickened. Larger plants then used gravity thickening for primary sludges and DAF for secondary biological sludges.

Gravity belt, rotary drum, and centrifuge thickening have virtually replaced gravity and DAF thickening due to their relatively low capital cost, ease of operation, relatively low odor generation, and small size. The major problem with these three methods is thickening the sludge to too high a concentration, which can cause downstream pump failure. Six to eight percent solids are normal discharges from these units (with polymer) and 12 to 14% solids can also be achieved. Process control instruments are beginning to be used to control the solids load going to thickening devices and to control the thickened solids from the operation.

The following three methods are typically used to thicken at various points in the treatment stream:

- Gravity thickening concentrates through the use of sedimentation. The normal acceleration of gravity provides the driving force for the separation. The most typical configuration for gravity thickeners is circular, as shown in Figure 29.1; however, rectangular systems are also used. The physical processes governing gravity thickening vary according to the nature of the solid being thickened. Sedimentation and thickening occur in different modes depending on the concentration and flocculent nature of the solids being handled. Three basic types of sedimentation include discrete settling (unhindered), hindered or zone settling, and compression or compaction settling. Gravity thickening can be aided by mechanical stirrers (pickets) and by chemical addition.
- Flotation thickening concentrates through sedimentation. The attachment of microscopic air bubbles cling to the suspended solids, thereby reducing their specific gravity to less than that of water. The air-attached particles then float to the surface for removal. This method is well-suited to wastes containing high amounts of finely divided solids. The most common thickener in this category is the DAF thickener.
- Centrifugal thickening uses centrifugal force to enhance the sedimentation of suspended solids. The centrifuges generate ±2000 times the force of gravity; it is this acceleration that causes suspended solids to migrate through the suspending liquid, away from the rotation axis, and to concentrate on the bowl wall.

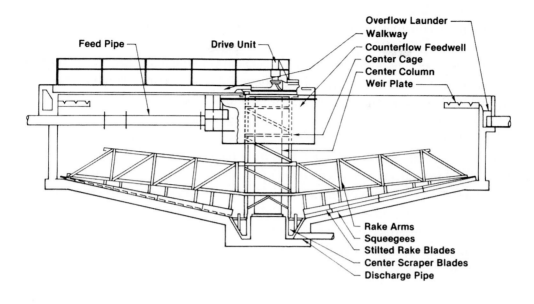

FIGURE 29.1 Schematic of typical open tank gravity thickener.

- Gravity belt thickening concentrates through filtration. Water filters through the conditioned solids, driven by gravity separation of water and solids on a continuously moving, porous, horizontal belt.
- Rotary drum thickening concentrates solids through filtration, much like the gravity belt thickener, except that instead of a belt it uses a rotating perforated drum or screen.

Conditioning agents, typically polyelectrolytes (polymers) specific to the waste stream being treated, may be required to enhance the separation of free water from the charged particles in all of the aforementioned thickening methods. Conditioning agents are discussed in Chapter 9 and Chapter 30.

DESCRIPTION OF PROCESS. Gravity thickening characteristics vary according to the nature of the solids being thickened. Settling and thickening will occur in different modes depending on the concentration and flocculent nature of the solids being handled. Three basic types of settling occur: discrete settling, hindered or zone settling, and compression or compaction settling. These are described as follows:

- Discrete settling occurs in dilute suspensions in which particles maintain their own characteristic settling rates. Although these rates may be affected by the presence of neighboring particles, the settling rate of a particle depends primarily on its size and density. Discrete settling is characteristic of very low solids loading rates and, for this reason, does not apply to most gravity thickeners.
- Zone settling occurs as the loading rates increase and individual particles become increasingly influenced by neighboring particles. Particles maintain position relative to one another as they settle, forming a porous matrix supported by the displaced fluid as the solids settle. Zone settling is characterized by an easily identified interface between the solids-laden liquor and adjacent supernatant. The solids concentration, fluid viscosity, and rate of underflow withdrawal dictate the zone settling rate.
- Compression settling occurs when flocculent particles develop a structure in which they are supported by adjacent particles rather than the displaced fluid. The compression settling rate is determined by channel formation in the floc structure that allows water to escape while the solids mass settles and compresses into itself.

Prevailing theory on gravity thickener design is based on zone settling, with the settling rate ideally depending on solids concentration alone. Downward solids transport in a gravity thickener stems from settling because of gravity and bulk movement resulting from pumping from the tank bottom. The sum of these two transport mechanisms provides the total solids flux.

Supernatant from a gravity thickener recycles through the treatment facility. Supernatant clarity is maintained to limit the recycled solids and biochemical oxygen demand (BOD) through the plant to minimums, thus limiting increased system solids and organic loading rates. Overflow from a smoothly operating gravity thickener, typically with less than 350 mg/L suspended solids, may be sent as influent to the secondary treatment process (Metcalf and Eddy, 2003). Supernatant from a gravity thickener with operational difficulties should be recycled to primary clarifiers (because it can be a source of strong odors). Diluting thickener influent sludge with "fresh" secondary effluent, chlorinated effluent, or waste activated sludge (WAS) can enhance the reliability of gravity thickening.

Gravity thickener loadings may be expressed both in terms of solids loading rates, the most useful parameter, and overflow rates. Typical gravity thickener loading rates for primary sludge range from 100 to 150 kg/m²·d (20 to 30 lb/d/sq ft), whereas typical loading rates for activated sludge range from 20 to 40 kg/m²·d (4 to 8 lb/d/sq ft). Gravity thickener overflow rates typically range between 16 and 32 m³/m²·d (390 and 780 gpd/sq ft). The overflow rate reduces as the proportion of secondary sludge increases.

The following list includes various sludge feed characteristics and their expected thickening effects:

- Primary sludge
  - Readily gravity-thickened,
  - Tends to settle quickly, and
  - Forms thick sludge without chemical aids;
- Waste activated sludge
  - Low settling rates,
  - Resists compaction, and
  - Tends to stratify causing flotation;
- Temperature
  - 15 to 20 °C (59 to 69 °F) acceptable if secondary-to-primary sludge ratio ranges from 4:1 to 6:1 and
  - Higher temperatures may require additional dilution of sludge feed;
- Upstream grinder pump
  - Breakup of oversized material to enhance settling;
- Chemical addition
  - Improves thickener efficiency; and
- Separate scum-handling procedures
  - Reduces odors and improves aesthetics of thickeners.

DESCRIPTION OF EQUIPMENT. Gravity thickeners (Figure 29.2) typically are circular tanks with side water depths of 3 to 4 m (10 to 13 ft) and diameters of up to 25 m (82 ft). Thickener bottoms are designed with a floor slope between 1:6 and 1:3. Occasionally, gravity thickeners may be rectangular units; however, their performance typically has been unsatisfactory.

Gravity thickener mechanisms tend to be of stronger construction than clarifier mechanisms because of the higher torques involved. Gravity thickeners may be designed as an integral part of a clarifier when the clarifier includes a central, deep well with a raking mechanism for thickening. If very high solids concentrations or high viscosities are anticipated, a lifting mechanism may be supplied to raise the thickener mechanism (rake) above the sludge blanket when torques are extremely high. This has

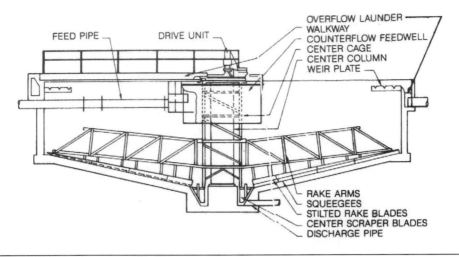

FIGURE 29.2   Typical gravity thickener.

been used with lime solids or other chemically precipitated sludges. As solids are re-moved from the tank and the torque is reduced, the lift will lower the mechanism back into the sludge blanket until it reaches its original position. Typical mechanism tip speeds are 0.08 to 0.1 m/s (15 to 20 ft/min).

Gravity thickeners may include the following equipment:

- Pickets, vertical pipes, or angles may be attached to the thickener arm to aid in releasing gas and preventing rat-holing or coning. However, pickets may cause difficulties in wastewater treatment plants (WWTPs) without effective screen-ing for removal of rags and fibrous materials.
- Variable-speed drives may be used to increase rake speed to agitate the sludge blanket and release trapped gas bubbles, and to prevent rat-holing or coning. Prolonged operation at high speeds will reduce the ultimate solids concentra-tion and reduce the life of the thickener drive mechanism.
- Scum removal equipment may include a skimmer and scum box.

The ancillary equipment should include positive-displacement pumps (plunger, rotary lobe, diaphragm, or progressing cavity pumps). Process control equipment in-cludes sludge blanket indicators (light path, sonic, or variable-height taps), online pro-cess monitors on the feed or underflow, torque readouts on the rake drive, and timers to vary the pump on–off (or speed) sequences. In cold weather areas or areas where odors are a problem, thickeners are typically covered (most commonly by a dome, al-though they can be a more traditional form as shown in Figure 29.3).

START-UP. It is important to ensure that all guards and covers are properly secured on the machine and that lockout/tagout is not in place. The following procedures apply during start-up:

- Following confined-space entry protocol, check the mechanism to ensure that there are no obstructions in the tank (e.g., ice or tools).
- Check the drive mechanism alarm switch and cutout switch to ensure they are operating properly. Do not attempt to adjust these switches. Alarm and cutout switches are set by the factory; therefore, avoid field adjustment unless performed by a manufacturer's service representative.

FIGURE 29.3 Enclosed gravity thickeners (City of Boulder, Colorado, Wastewater Treatment Plant).

- Avoid starting a thickener that contains accumulated solids. To avoid overload, the solids should be disposed of before starting the mechanism.
- Depending on your configuration, start your thickener first with fresh dilution water, then apply sludge flow. This will maintain freshness and proper overflow rates.
- Before pumping from the system to downstream processes, allow sufficient solids to accumulate in the thickener to completely seal the hopper feeding the underflow discharge pipe.
- Check and adjust the skimming mechanism to increase the amount of scum drawn into the scum box and to reduce the amount of supernatant carried with the skimmings.
- Always avoid starting a thickener with ice in the unit. The system, including the feed (stand) pipe, should be drained to prevent freezing whenever it is taken out-of-service.

SHUTDOWN. The following procedures apply during shutdown:

- If shutdown is for a short period of time (from 1 to 6 hours), continue the fresh or chlorinated dilution water. This action will keep the sludge "fresh".
- Continue the programmed withdrawal of solids until you are down to 50% of the typical sludge blanket depth (measured from the floor) or the top of the cone, whichever is greater.
- Reduce the overflow of "fresh" dilution water to approximately 25% of normal flow for the length of the shutdown. If the shutdown is only for between 1 and 6 hours, do not shut the drive off.
- If the shutdown is for longer than 6 hours, or if you need to service tank internals, remove the remaining sludge to another online thickener or slowly return it to the primary clarifier. You may leave the tank full of chlorinated effluent unless you plan to keep the thickener out-of-service for a length of time.
- Follow lockout/tagout procedures to secure the unit before conducting any maintenance or inspection. Lockout is a method of keeping equipment from being set in motion and endangering workers. The *Code of Federal Regulations* (1991) 29 *CFR* Part 1910.147 covers the servicing and maintenance of machines and equipment in which the unexpected energization or start-up of the machines or equipment or release of stored energy could cause injury to employees. Tagout occurs when the energy isolation device is placed in a safe condition and a written warning is attached to it. Typically, the employee signs and dates

the tag. Removal of the tag and lockout device is done by the same individual when the equipment is safe for restart.

- If freezing is a possibility, drain the tank and all exposed piping, including the feed stand pipe and scum wells.

PROCESS CONTROL. Ideally, feed to a gravity thickener should be screened or shredded to prevent the entry of rags or fibrous materials, and it should be pumped continuously. If continuous pumping is impractical, the feeding schedule should closely approach continuous feeding. Continuous feeding not only promotes a stable sludge blanket, but it also tends to reduce gasification and resultant floating sludge. Successful gravity thickener operation depends on low-feed-rate pumping during long periods.

When practicing combined primary and secondary thickening, the sludge should be mixed and combined with makeup water (often called *elutriation*) before introduction to the gravity thickener. Makeup water, which is added to maintain a constant hydraulic load, can also be used as an oxidizing source. This practice produces a feed with more uniform characteristics and "freshness" that will resist the tendency to stratify. Primary sludge is most easily thickened using gravity. It tends to settle quickly and form a thick sludge that is easily pumped without chemical aid. For these reasons, primary sludge is frequently thickened separately. Unfortunately, primary sludge generation varies greatly during 24 hours. Operators are now beginning to take notice of this in programming pumping rates to the thickener.

Greater attention to the thickener is required when thickening WAS because it has a large surface area per unit mass, resulting in low settling rates and a resistance to being compacted. Sludge tends to stratify in the gravity thickener while continuing biological activity, which includes the production of gases that can cause accumulated sludge to float.

Studies have shown that when the ratio of secondary to primary sludge varies between 25 and 59%, the resultant underflow concentration remains essentially unchanged. At some point above this range, an increase in the ratio of secondary solids to primary solids will cause a decrease in underflow concentration.

If activated sludge begins to gasify and rise despite wasting from the aeration tank, a bacteria side such as chlorine, potassium permanganate, or hydrogen peroxide may be applied to the feed to reduce biological activity, thereby reducing gas and odor production. Chlorinated WWTP effluent is often used as dilution water; this tends to "sweeten" the sludge.

The gravity thickening process is pH-sensitive, and treatment facilities have been found to react differently to pH shifts. Observation of the thickening process and experimentation seems to offer the only method of determining the effects of pH shifts and how they can be handled.

Gravity thickener operation responds to changes in process temperatures; therefore, loading rates should be reduced to values at the lower end of the range when temperatures exceed 15 to 20 °C (59 to 68 °F), depending on the ratio of primary to secondary solids. Higher temperatures will require additional dilution.

High-temperature recycle streams, such as anaerobic digester supernatant or decant tank supernatant, will cause thermal stratification in gravity thickeners and reduce settling rates. In addition, recycled centrate or filtrate may contain biologically active solids, causing gasification and a rising sludge blanket. Effects of recycle streams can be reduced by slowly metering recycle flows into the gravity thickener or by introducing recycle flows at another point, causing fewer negative effects.

Scum handling procedures can significantly affect thickener cleanliness. A large scum buildup on a gravity thickener is unsightly and causes odor and fly problems. There are two methods of in-house scum handling. The first is to treat skimmings immediately after they are removed from clarifiers in the facility. The second is to pump skimmings to a thickener and consolidate them for subsequent handling. The difficulty with the second alternative is that gravity thickeners are rarely equipped with skimming equipment capable of handling large amounts of scum, particularly scum that has accumulated and possibly congealed and hardened.

Because thickening is dependant on the weight of the solids, one on top of the other, a blanket of solids in the thickener is necessary to promote compaction thickening. The blanket also reduces rat-holing or coning as well as the need for intermittent feeding of the gravity thickener. The depth of the allowable sludge blanket varies with temperature. Higher temperatures require more shallow blankets. Typically, a sludge blanket between 0.3- and 1.5-m (1- and 5-ft) deep at the tank periphery should be maintained, with greater depths for cooler temperatures and deeper tanks. At one time, the blanket readings were taken by the operator. Current practice is to use a blanket detector, and display the sludge blanket on the supervisory control and data acquisition system. Either way, the location where the blanket level is measured is important. If done manually, sampling must be from the same point each time. Depending on tank access and operator preference, measurement of blanket depth is often taken at the ⅔ radius point (measured from the center on a circular tank) off the catwalk (walkway). The key here is consistency; therefore, it may be beneficial to permanently mark the railing where depths are obtained. The sludge blanket depth at the point of sludge discharge definitely affects the ultimate solids concentration. To take advantage of this phenomenon, the operator should experiment to determine the range of depths that optimize ultimate solids concentration. The results will depend on temperature. After the optimum depth is determined, pumping should be modulated to maintain the blanket within the optimal range. Sludge depth can be determined a variety of ways—with an electronic depth finder; a "sludge judge," or circular plastic tube called a *coretaker*; or with an airlift pump. Increased retention times may cause deterioration of other aspects of the gravity thickening operation. Under no circumstances should the blanket reach the bottom of the thickener feedwell.

Retention time for sludge in the gravity thickener should range between 1 and 2 days for primary sludge, depending on temperature. The retention time for mixtures of primary and secondary sludge should range from 18 to 30 hours, depending on temperature. Longer retention times may cause thickener upsets, such as gasified sludge, and increase chemical requirements in downstream dewatering operations.

Like sludge feeding, thickened sludge pumping must be as continuous as possible. Pumping rates may be adjusted to conform to the concentration of the solids being pumped. The traditional controls are usually on–off pumping, which, of course, are not continuous. Recently, the low cost of variable-speed drives has offered excellent control of pumping rates, representing a significant improvement over the earlier operation. The best means of developing a regular, optimum pumping schedule requires varying the duration of pumping and cycle times while checking solids concentration until the best schedule is found. As this requires a good deal of time and effort, it is commonly not done, much to the detriment of operations. As a result, the operations manager should schedule time to do this on a regular basis or consider automating the process. Underflow pumping velocities of at least 0.75 m/s (2.5 ft/sec) are needed to prevent solids deposition. When pumping highly concentrated sludge, the increased viscosity will increase operating head. Small increases in concentration above an 8% level produce large increases in operating head requirements.

For thickening, makeup water consisting of plant effluent (mixed-liquor suspended solids [MLSS] direct from aeration or secondary effluent) can offer several benefits. The dissolved oxygen in the recycled liquid helps reduce anaerobic biological activity that would otherwise cause the sludge to gasify, float, and produce odors. In addition, the elutriation process tends to remove soluble carbohydrates, ammonia nitrogen, phosphates, fats, oils, and fine particulates. Chlorination of elutriation water may be used to reduce the biological activity that causes gasification of sludge. A disadvantage, however, is that elutriation will remove salts and thus may increase conditioning requirements. Too much elutriation flow will be detrimental by expanding the sludge (thinning) and recycling wastewater unnecessarily. The key is moderate use. With anaerobic digestion, wide fluctuations in underflow concentration caused by elutriation should be avoided to maintain digester stability. As with all dilution, the added water in effect dilutes the overflow. If the incoming feed is 3000 mg/L and you measure 300 mg/L suspended solids in the overflow in round numbers, you are achieving 90% capture. If you change operations, and begin to add elutriation water equal to the feed rate, you would be doubling the overflow rate and, as a consequence, the overflow rate would have to contain about 150 mg/L to achieve the same 90% capture.

Viscous drag between the thickened sludge and the rakes creates a load on the rake drive, usually measured as amperage on the rake drive motor. Should the sludge become thicker and more viscous or something jam the rake, the amperage will rise until it sounds an overload alarm. Ideally, this will not happen with modern controls, where increasing rake amperage results in automatic increasing of the underflow pumping rate. Increasing underflow pumping reduces the cake blanket, thinning out the underflow solids and thus reducing the torque on the rake.

If the rake drive does shut off in overload, discontinue the feed to the thickener to keep the problem from worsening. Do not turn the drive mechanism on and off or "jog" it because that could overload and possibly damage the drive before the drive cutout can operate. In addition, do not wire around the drive overload protection because that would likely damage the mechanism. Unfortunately, the thickener may have to be drained to remove the cause of the torque overload (possibly a foreign object or very viscous sludge). If for some other reason the thickener is out-of-service for more than 1 hour, flow to the device should be discontinued to prevent a torque overload. If the thickener is to remain out-of-service for 1 day or more, sludge should be removed and replaced with chlorinated plant effluent.

SAMPLING. At a minimum, gravity thickeners should be sampled and analyzed as follows:

- Influent for dissolved oxygen (if thickening activated sludge), total suspended solids (TSS), total solids, and flow.
- Effluent for TSS, total solids, and flow.
- Underflow (bottom sludge withdrawal) for total solids and flow.
- Depth of blanket (DOB) by way of sampler, either "bomb" or other thief method, sonic or light path, or with variable-height taps (using an air-lift arrangement). The DOB is reported as the height above the bottom of the tank at the ⅔ radius point (sample point measured from the center of the tank outward ⅔ of the distance on the catwalk for circular tanks).

TROUBLESHOOTING. This section describes typical operational problems and methods of handling these problems. Table 29.1 presents a troubleshooting guide that identifies problems and presents possible solutions. Each of several pervasive prob-

TABLE 29.1   Gravity thickener troubleshooting guide.

Indicators/ observations	Probable cause	Check or monitor	Solutions
Septic odor, rising sludge	Thickened sludge pumping rate is too low	Check thickened sludge pumping system for proper operation	Increase pumping rate of thickened sludge
		Check thickener collection mechanism for proper operation	Increase collection speed, repair mechanism
	Thickener overflow rate is too low	Check overflow rate	Increase influent flow to thickener; a portion of the secondary effluent of MLSS may be pumped to the thickener (if necessary) to bring overflow rate to 400 to 600 gpd/sq ft[a]
	Septicity in thickener	Check thickener DO[b]	Add oxidizing agents to influent sludge (0.5–1.0 mg/L)
Thickened sludge not thick enough	Overflow rate is too high	Check overflow rate	Decrease influent sludge flow rate
	Thickened sludge pumping rate is too high	Check sludge concentration	Decrease pumping rate of thickened sludge
	Short-circuiting of flow through tank	Use dye or other tracer to check for circuiting	Check effluent weirs, repair or relevel; check influent baffles, repair or relocate
Torque overload of sludge-collecting mechanism	Heavy accumulation of sludge	Probe along front of collector arms	Agitate sludge blanket in front of collector arms with water jets; increase sludge removal rate
	Foreign object jammed in mechanism	Probe with big magnet on rope	Attempt to remove foreign object with grappling device; if problem persists, drain thickener and check mechanism for free operation
	Improper alignment of mechanism	Alignment	Realign mechanism
Surging flow	Poor influent pump programming	Pump cycling	Modify pump cycling; reduce flow and increase time

TABLE 29.1   Gravity thickener troubleshooting guide (*continued*).

Indicators/observations	Probable cause	Check or monitor	Solutions
Excessive biological growths on surfaces and weirs (slimes)	Inadequate cleaning program		Frequent, thorough cleaning of surfaces Apply chlorination
Oil leak	Oil seal failure	Oil seal	Replace seal
Noisy or hot bearing or universal joint	Improper alignment	Alignment	Align joint or bearing as required
	Lack of lubrication	Lubrication	Lubricate
Pump overload	Improper adjustment of packing	Check packing	Adjust packing
	Clogged pump	Check for trash in pump	Clean pump
Fine sludge particles in effluent	Waste activated sludge (WAS)	Portion of WAS in thickener influent	Better conditioning of the WAS portion of the sludge Thicken WAS in a flotation thickener

[a]gpd/sq ft × 40.74 = L/m^2·d.
[b]Dissolved oxygen.

lems often experienced with gravity thickeners and the likely remedies for each problem are described herein.

Excessive scum is frequently caused by prolonged sludge retention times in the thickener. Wasting sludge directly from an aeration tank to keep it fresh may reduce scum. Maintaining lower sludge blanket levels or increasing elutriation may also remedy the problem. In some cases, scum removed from clarifier surfaces enters a gravity thickener for accumulation. Because the thickener's scum disposal system is often not equipped to handle large amounts of scum, this practice should be discouraged. One scum disposal aid includes a spray system covering a portion of the tank near the scum box to keep the scum soft and suitable for pumping. Spraying with chlorinated water is particularly effective for scum caused by *Nocardia* foaming. Substantial scum accumulations may be broken up with high-pressure hosing so that the skimming mechanism can remove the scum. In severe cases, scum accumulations may be vacuumed into a hauling truck and disposed of properly.

Grease buildup in underflow lines may disrupt sludge handling. High-pressure water should be used to backflush underflow lines and dislodge grease accumulations.

One means of combatting grease accumulations requires filling underflow lines with digested sludge and allowing the sludge to remain undisturbed for at least 1 day before removing the digested sludge and dislodged grease. If the line can be isolated and cleanouts are provided, steam cleaning can reduce stoppages. In severe cases, a self-propelled hydraulic rodding device known as a *jet rodder* or hydro may be used to clear blockages. As a last resort, a specially designed plug or "pig" may be inserted in the line and propelled by the sludge pump or water pressure to clean the line.

Sludge septicity is characterized by floating solids, solids carryover in the overflow, foul odors, and reduced underflow concentrations. Storing sludge too long in either the gravity thickener or in upstream clarifiers causes septicity. This condition can be resisted temporarily by chlorine or hydrogen peroxide. Aeration of thickener feed or elutriation to provide some dissolved oxygen may reduce the problem. If secondary effluent is recycled through the gravity thickener to elutriate, the overflow rate must not exceed approximately 32 $m^3/m^2 \cdot d$ (780 gpd/sq ft).

Sludge bulking is indicated by a rising sludge blanket. The blanket disperses, resulting in poor supernatant and dilute underflow. Sludge bulking provides evidence of problems elsewhere in the WWTP; therefore, only the symptoms of sludge bulking may be dealt with in the gravity thickener. The problem must be solved at its source. Control of bulking should receive high priority to avoid excessive chemical costs and reduce other process upsets throughout the facility. Any attempt to deal with bulking at the gravity thickener should be undertaken only on a short-term basis because of the expense involved. Chlorination or the addition of hydrogen peroxide to incoming sludge may reduce the gelatinous extracellular materials characteristic of some bacteria that cause bulking. Weighting agents such as bentonite may be used to capture sludge that would not otherwise settle. Polymers are sometimes used against bulking. In worst cases, the thickener contents may be pumped to a drying bed to rid the facility of poor-quality sludge if improved sludge characteristics are expected.

Rat-holing or coning occurs when clear supernatant is pumped through the sludge blanket to produce a low underflow concentration. An increase in the sludge blanket depth will reduce the chance of rat-holing. Sludge pumping at a high flowrate for short durations leads to rat-holing. There are two solutions: replacing the pump motors with variable-speed drives or replacing the pumps and drives. Both solutions will increase the duration of sludge pumping.

PERFORMANCE. Gravity thickeners typically produce underflow sludge ranging from 4 to 8% when thickening primary sludge alone. The percentage of sludge is reduced to between 3 and 6% when a combination of primary and secondary sludge is gravity-thickened. Table 29.2 indicates various sludge sources with their corresponding

TABLE 29.2   Sludge source and total solids characteristics.

Sludge source	Low-viscosity TS, %	High-viscosity TS, %
Raw primary sludges	<6	6–12
Raw secondary sludges	<2	2–6
Raw primary and secondary sludges	<3	3–8
Digested sludges	<4	4–10
Chemical sludges		
Lime	<15	15–40
Alum and ferric sludges	<2	2–6
Chemical slurries	1–30	—
Incinerator ash slurries	5–20	—

anticipated solids characteristics. The literature shows little correlation between the percentage of solids in the sludge fed to a gravity thickener and the concentration of the thickener underflow. Anticipated solids capture ranges between 90 and 95%. Thickeners are often a major source of solids recycle within the plant (Figure 29.4). Recycle is often ignored as a cost on the notion that "the place is full of sludge, so what's a little more?" However, this is misguided thinking. The cost of recycle is, at a minimum, the cost the wastewater authority charges a customer to dump solids into the plant.

PREVENTIVE MAINTENANCE. Regular maintenance is a must. Daily maintenance items include

- Checking the drive units for overheating, unusual noise, or excessive vibration; and
- Checking for unusual conditions, such as loose weir bolts.

Weekly maintenance items include

- Checking all oil levels and ensuring that the oil fill cap vent is open,
- Checking all condensation drains and removing any accumulated moisture,
- Examining drive control limit switches, and
- Visually examining the skimmer to ensure that it properly contacts the scum baffle and scum box.

Monthly maintenance items include

- Inspecting skimmer wipers for wear and
- Adjusting drive chains or belts.

FIGURE 29.4   Thickeners as a significant source of solids recycle within the plant.

Yearly maintenance items include

- Disassembling the drive and examining all gears, oil seals, and bearings. Gears and bearing surfaces should be examined for abrasive wear, scoring, pitting, and galling. Balls should be inspected and those that are scored or cracked should be replaced.
- Checking the oil for the presence of any metal that might be a precursor to future problems. The thickener should be drained and all submerged portions of the mechanism should be checked. Particular attention should be given to the adequacy of the coating system, flexible piping connections, bolted connections, and "squeegees," if used.

- Replacing any part that has an expected useful life of less than 1 year. It is far easier to replace these items during scheduled maintenance than during a breakdown (WPCF, 1987).

Other routine maintenance items include the following:

- Use corrosion-resistant materials when possible to resist the thickener's hostile environments.
- Install kick plates on the gravity thickener bridge to prevent objects from falling into the tank. An object lodging in the underflow discharge pipe or under the mechanism will quickly halt operation of the thickener.
- If objects fall into the tank, immediately halt thickener operation to prevent torque overload.
- During plant rounds, regularly observe and record the drive torque indicator, the best signal of mechanical problems.
- Regularly check the underflow pump capacity because pumps wear rapidly in a thickened sludge operation.
- Follow the manufacturer's recommended lubrication schedule and use recommended lubricant types. Oil should typically be changed after the first 250 hours of operation and every 6 months thereafter.

SAFETY. Good safety practices include the following procedures:

- When working on the thickener, disconnect the power by following lockout/tagout procedures. Lockout is a method of keeping equipment from being set in motion and endangering workers. The *Code of Federal Regulations* (1991) 29 *CFR* Part 1910.147 describes servicing and maintenance of machines and equipment in which the unexpected energization or start-up of the machines or equipment or release of stored energy could cause injury to employees. Tagout is when the energy isolation device is placed in a safe condition and a written warning is attached to it. The employee typically signs and dates the tag. Removal of the tag and lockout device is done by the same individual when the equipment is safe for restart.
- Do not disconnect the drive if the torque indicator registers any torque. If energy is stored in the springs that activate the torque indicator, releasing the springs could cause injury. This must be part of the lockout/tagout procedure.
- When working on the thickener, follow the safety procedures of Chapter 5, including those for confined space entry to areas where the toxic gases hydrogen

sulfide and methane may be present as well as areas where oxygen deficiency or an explosive atmosphere may exist.

DISSOLVED AIR FLOTATION THICKENING

PROCESS DESCRIPTION. A DAF system essentially consists of a flotation unit and a saturator (Figure 29.5). The flotation unit serves to separate the solid phase from

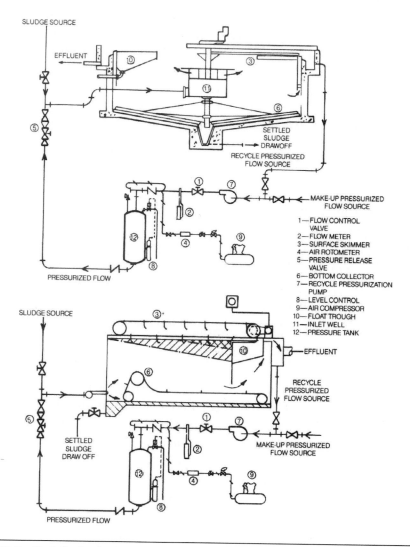

FIGURE 29.5   Detailed schematics of typical circular and rectangular DAF thickeners.

the liquid phase and the saturator dissolves air into water under pressure. Pressure-saturated water (from the saturator) is introduced to the flotation unit through a reducing valve. Just downstream of the valve, the pressure is virtually atmospheric and the saturator feed becomes highly supersaturated with air. The air precipitates out of solution in the form of very small bubbles. Just downstream from the reducing valve, the saturator feed is blended with the wastewater feed. The precipitated bubbles become attached to wastewater particles, forming bubble-particle agglomerates with a density lower than water.

If chemical aids such as polymers are used, they typically enter at the mixing point of the saturator feed and the sludge feed. Proper mixing to ensure chemical dispersion, while avoiding excessive shearing forces, produces the best aggregation of sludge particles and air bubbles. Experience has shown that introduction of polymer just downstream of the point of blending of the sludge feed and the precipitating saturator feed produces the best results.

Particular advantages of DAF systems include the following (Bratby et al., 2004):

- Ability to successfully co-thicken primary and secondary sludges;
- Amenability of thickening scum from both the primary and secondary scum collection systems;
- Allowing scum and sludge to be transported to the thickening process with maximum water flow;
- Allowing the separation and capture of grit from a continuous bottom sludge removal system when co-thickening primary and secondary sludges;
- Continuously producing a homogeneously mixed thickened sludge product that is of ideal quality for feeding digesters;
- Allowing all solids processing recycle loads to be concentrated into one stream; and
- Achieving a significant soluble BOD reduction in the DAF liquid stream.

In the flotation unit, buoyancy causes the bubble-particle agglomerates to rise to the water level and accumulate as a float. In this fashion, separation of the solid phase from the liquid phase is achieved. Because of the difference in density between the float and the water, a stage is reached where the top of the float, continuously rising above the water level, reaches the top of the unit where solids are continuously removed by scraping. Drainage of interstitial water from the float above the water level increases the solids concentration. This stage of the process is termed *thickening*.

Flotation, then, has basically two functions: (1) clarification or separation of the solid phase from the liquid phase and (2) thickening or dewatering of the separated (accumulated) solids. Figure 29.6 is a schematic of a DAF thickener.

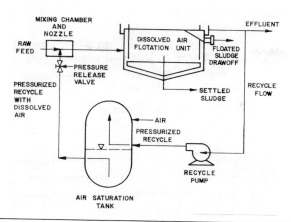

FIGURE 29.6   Dissolved air flotation thickener schematic.

Principal process variables influencing the clarification aspect of DAF are the downflow rate and the air-to-solids ratio. The downflow rate is analogous to the overflow rate used for sedimentation tanks and is defined as the total flow into the unit divided by the effective plan area of the flotation unit. The air-to-solids ratio is defined as the mass of air precipitated per unit time divided by the mass of solids introduced per unit time. The downflow rate and the air-to-solids ratio are related by the following empirical expression (Bratby, 1978):

$$v_L = K_1 \cdot (a_S)^{K2} - K_3 \tag{29.1}$$

where

     K_1 through K_3 = empirical constants for a particular waste or sludge,
             v_L = the downflow rate (m/d or gpd/sq ft), and
            a_S = the air/solids fraction.

Principal process variables influencing thickening are the solids loading rate and the height at which the float extends above the water level. Thickening of float solids occurs principally by drainage of interstitial water from the float above the water level.

Both the solids loading rate and the height of float above the water level directly influence the time allowed for drainage of the float. The solids loading rate, defined as the mass of solids introduced per unit area of flotation unit per unit time, governs the rise rate of the float above the water level. The height of float above the water level governs the distance through which the float travels before being removed from the unit.

The following empirical relationships describe thickening by DAF (Bratby, 1978; Bratby and Ambrose, 1995):

$$C_F = K_4 \cdot (d_W)^{K5} \cdot (Q_S)^{-K6} \qquad (29.2)$$

$$d_T = (d_B + d_W) = d_W[(a_S)^{K7} + K_8] \cdot a_S^{-K7} \qquad (29.3)$$

where
K_4 through K_8 = constants for a particular waste or sludge;
C_F = float solids concentration (percent) at the top of the unit (i.e., where float solids are removed from the unit);
Q_S = solids loading rate (kg/m^2·d);
d_W = height of float above water level (m);
d_B = depth of float below the water level (m); and
d_T = total float depth (m).

Note that for a given set of parameters, adjusting the air-to-solids ratio does not influence the float solids concentration. However, the air-to-solids ratio does affect the total depth of float solids.

The accuracy of the expression for total float depth ($d_B + d_W$) becomes uncertain at air-to-solids ratios below a certain value. At lower air-to-solids ratios, the float depth, in practice, tends to be greater than predicted by Equation 29.3. With activated sludge, with or without the use of polymers, the limiting air-to-solids ratio is about 0.02; with raw wastewater, using metal coagulants such as alum, the limiting air-to-solids ratio is about 0.06; with stabilization pond algae separation, using metal coagulants, the limiting air-to-solids ratio is about 0.03; and with flotation applied to thickening the sludge from highly colored waters, using polymer, the limiting air-to-solids ratio is about 0.03.

The aforementioned expressions were extended by Bratby and Ambrose (1995) to take into account the actual time spent by the float above the water level, before it is removed to the sludge hopper by the sludge scraper mechanism. Parameters that are controllable by the operator in this regard are the speed of the scraper and its periodicity of operation.

An effective time, t_E, spent by the float above the water level is defined as follows (Bratby and Ambrose, 1995):

$$t_E = (t_C/t_{ON}) \cdot (L/v) \qquad (29.4)$$

where

t_C = float scraper cycle time = t_{ON} + t_{OFF};

t_{ON} = time that the scraper is on during a given cycle (minutes);

t_{OFF} = time that the scraper is off during a given cycle (minutes);

L = effective length of rectangular dissolved air flotation thickening (DAFT), or periphery of circular DAFT (m or ft); and

v = velocity of travel of scrapers (peripheral speed for circular units) (m/d or ft/d).

Therefore, the following relationships incorporate the effects of t_E (Bratby and Ambrose, 1995):

$$C_F = K^* \cdot t_E^{[K5/(1+K5)]} \qquad (29.5)$$

$$d_T = (d_B + d_W) = Q_S \cdot t_E^{[1/(1+K5)]} \cdot (1 + K_8/a_S^{K7})/(10 \cdot K^*) \qquad (29.6)$$

where

$$K^* = [K_4/(Q_S^{(K6-K5)} \cdot 10^{K5})]^{[1/(1+K5)]}$$

At long effective times, t_E, more opportunity is provided for interstitial water in the float solids to drain and, consequently, the float solids concentration will be higher. However, there is a practical upper limit to the value of t_E. For a given d_W, there is a greater value of d_B required to supply the buoyancy to support the height of float above.

Therefore, just as increasing t_E increases d_W, at the same time it also increases the total float depth. The limitation in t_E is to avoid increasing the float depth to the extent that float solids are dragged through with the underflow, causing a deterioration of effluent quality.

From the aforementioned discussion, the parameter that is manipulated by the operator to control the performance of a thickener is the effective drainage time, t_E. This parameter is controlled by

- Adjustment of scraper on-time,
- Adjustment of scraper off-time, and
- Adjustment of scraper speed.

The preceding discussions have pointed out that the principal mechanism affecting thickening during flotation is drainage of float layers above the water level. For this reason, float solids concentration profiles through the float depth will exhibit a maximum at the top of the float, decreasing markedly with depth down to the position of the water level, then at a reduced rate through subsequent float layers until a minimum is reached at the bottom of the float. This is illustrated in Figure 29.7(a) and (b) (Bratby, 1978).

Figure 29.7(a) shows results obtained in a pilot flotation unit where float solids concentrations were taken at various depths through the float. In this case, the solids loading rate was set at 55 kg/m²·d (11.2 lb/d/sq ft) and the air-to-solids ratio at 0.036. The pilot-scale unit was fitted with a horizontal action scraper at the top of the unit. Very minimal disturbance of the float occurred and concentration profiles presented in Figure 29.7(a) are those obtained under ideal conditions. In contrast, results obtained with a full-scale proprietary flotation unit with the same sludge are presented in Figure 29.7(b). In this case, the solids loading rate was approximately 60 kg/m³·d (12.3 lb/d/sq ft) and the air-to-solids ratio was 0.036 to 0.10. A marked difference is evident in the profile. There is no evidence in the full-scale unit of the sharp decrease in concentration achieved in the pilot-scale unit operating under similar conditions.

The float solids concentration achieved in the pilot unit was about 6.5%, whereas in the full-scale unit it was about 5%. The difference in results is due to differences in

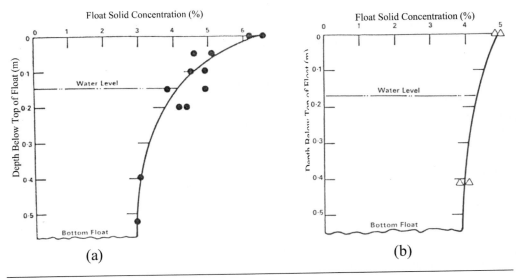

FIGURE 29.7   Float solids concentration profiles.

the methods by which float solids are removed from the two units. In the case of the pilot unit, the float scraper removes only the top-most slivers of float solids that reach the top. In the full-scale unit, the scraper blades dig into the float to a depth of about 178 mm (7 in.). This arrangement typifies most DAFT units.

In the case of the full-scale unit, the scraper arrangement had two effects. First, it maintained the top of the float at the level of the float take-off weir (i.e., it fulfilled its function of removing float solids from the unit). Second, however, it had the undesirable function of mixing the float. The influence of this disturbance will extend beyond the scraper depth. The effect of this mixing action is to replace the upper layers of the float that have attained maximum drainage by lower, wetter layers.

From this discussion it is evident that a float scraper should disturb the float as little as possible and remove only the topmost layer of the float because float solids concentration decreases markedly with depth. This requires that the bottom of the scraper be situated above the water level as far as practicable, but also that float solids do not build up excessively in front of the scraper as it travels around or along the top of the unit.

EQUIPMENT DESCRIPTION. *Flotation Thickener Tanks.* Flotation thickeners constructed of steel or concrete are either rectangular or circular tanks (Figure 29.8). Typically, the smaller rectangular units up to 2- to 3-m (8- to 10-ft) wide are steel and the larger units are concrete. Only very small circular units in municipal plants are constructed of steel (WPCF, 1980).

Flotation thickeners are equipped with both surface skimmers and floor rakes. The surface skimmers remove floating material from the thickening tank to maintain a constant average float blanket depth. Floor rakes are essential for removing the nonfloatable heavier solids that settle to the bottom of the flotation thickener.

Most units are baffled and equipped with an overflow weir. Clarified effluent passes under an end baffle (rectangular units) or peripheral baffle (circular units) and then flows over the weir to an effluent launder. The weir controls the liquid level within the flotation tank with respect to the float collection box and helps regulate the capacity and performance of the flotation unit.

A complete flotation thickener system includes a number of appurtenances to the thickening tank. These appurtenances include sludge and conditioning chemical feeding equipment and various control elements.

Saturation System. The heart of a DAFT system is the saturation system, where air is dissolved into water under pressure then released at essentially atmospheric pressure in the form of microscopic bubbles.

FIGURE 29.8 Circular DAF thickeners (Littleton/Englewood Wastewater Treatment Plant, Colorado).

The saturation system typically includes a recycle pressurization pump, an air compressor, an air saturation tank, and a pressure release valve. Although the flow through the pressurization pump typically is recycle flow, it can be makeup water. The pressure release valve controls pressure loss and distributes the gas-saturated pressurized flow into the feed sludge as dissolved air emerges from solution. The rapid reduction in pressure causes dissolved air (under pressure) to emerge from solution or "effervesce" into minute bubbles.

An alternative system that lost favor several years ago but appears to have renewed interest recently is an aspirating pumped system. Recent systems appear to have overcome the air binding problem common with older systems. To date, performance data is lacking. However, an apparent advantage is that the saturation vessel is not required, and the system is generally simpler.

The following principles are important to the saturation system (Bratby et al., 2004):

- Dissolution of a gas into a solute is a simple and straightforward process that is dependent on pressure, temperature, the solubility of the gas in question in the solute (in this instance, water), and the surface area of the liquid available for gas transfer.

- The solubility of nitrogen in water is roughly half that of oxygen.
- At 78% nitrogen and 21% oxygen, the water will more readily absorb oxygen, leaving the volume represented by the oxygen left for addition of more gas in the unvented headspace commonly provided in standard saturation tank designs, plus the volume represented by the nitrogen absorbed by the water during the critical contact period in the headspace.
- With an unvented headspace, the saturation efficiency, as measured by the volume of gas absorbed in the pressurization flow stream, rapidly deteriorates to approximately ⅔ that if the headspace gas is at or near atmospheric concentrations. Essentially, this represents a 33% loss of process capacity. It follows, then, that it is necessary to vent excess nitrogen continuously to maintain process efficiency. Because more than 80% of the energy consumed by the DAFT process is the power required for pressurizing the saturation system pumps, this also represents an opportunity to fine-tune the process to reduce energy consumption.

Other important equipment that forms part of a DAFT system is the float handling and pumping system. After removal of the float from the DAFT unit, float solids are deposited in a hopper and then pumped for further processing. This aspect of the operation requires special considerations because of the following characteristics of float solids:

- They float on the surface of the liquid; and
- They do not flow easily on flat or gently sloping surfaces, especially if the surfaces lack a smooth surface that reduces skin friction.

Adding to these characteristics, conventional sump level controls make the situation that much more difficult because pumps are typically controlled on a high/low-level basis, with the former used to start the pumps and the latter used to stop the pumps. The effect of this control scheme is to concentrate the floating material by decantation rather than pumping the solids away. If left on the surface of a sump for a protracted period and exposed to air, the solids dry out and become a dry cake, making the task of trying to move the solids toward the pump intake that much more difficult. Operators are frequently forced to spend considerable time directing high-pressure hose streams to the floating mass trying to force the material under, thereby wetting it (and decreasing its concentration) until the material can find its way into the pump.

A preferable arrangement, however, is to position the pump intake connection in a confined sump located well below the nominal floor of the sump to concentrate the floating solids in a small surface area right over the pump inlet. The pump is then controlled to shut down not on level, but rather when the pump breaks suction, guaranteeing that the floating material has reached the pump inlet (Bratby et al., 2004).

PERFORMANCE. Through experience with specific flotation thickening units, operators may determine the optimum ranges of performance for their systems. Tables 29.3 and 29.4 provide operating data from several WWTPs to compare system performance.

A float total solids content of 4% represents a typical minimum for flotation thickeners handling solids without primary sludge. Under optimum conditions, however, a 5 to 6% solids content can be expected. At least one plant routinely and successfully operates at loadings as high as 580 kg/m²·d (120 lb/d/sq ft), thickening a mixture of conventional and high-purity-oxygen WAS, and achieves more than 4% thickened solids concentration.

One factor influencing the float solids concentration is the sludge volume index (SVI) of the activated sludge. However, this is probably not the principal factor influencing the thickened concentration achieved. Other factors such as design and operation of the float scraper and proper drainage of the float layers are probably more important.

A float total solids content of 6% approximates a typical minimum for flotation thickeners handling a 50:50 mixture of secondary and primary sludge. Under optimum conditions, minimum solids contents ranging from 6 to 8% can be expected from thickening such mixtures (Bratby et al., 2004).

PROCESS CONTROL. Proper operation requires reducing variations of the feed rate and concentration. A feed sludge holding and mixing tank helps with intermittent operation. Most units are operated continuously. Some are operated with a short period shut-down during weekends, while others are operated only during certain hours of the day.

The hydraulic throughput for flotation thickening includes feed plus recycle. Flotation units are designed hydraulically to reduce entrance velocities, thereby reducing potential shearing of flocculated sludge particles.

Process control considerations include

- Operating the skimmer so that float is only skimmed above the water level, as described previously (Bratby et al., 2004).
- The speed of the float skimmer arms and the number of arms controls the rate at which floated sludge is pushed into the sludge trough, as explained earlier.

TABLE 29.3 Operating data for plant-scale dissolved air flotation units.

Location	Feed[a]	Influent SS,[b] mg/L	Subnatant SS, mg/L	Removal, %	Floatable solids, %	Loading, lb/hr/sq ft[c]	Flow area, gpm/sq ft[d]	Remarks[e]
Bernardsville, N. J.	ML	3 600	200	94.5	3.8	2.16	1.2	Standard
Bernardsville, N. J.	RS	17 000	196	98.8	4.3	4.25	0.5	Standard
Abington, Pa.	RS	5 000	188	96.2	2.8	3.0	1.2	Flotation aid after 12-hour holding
Hatboro, Pa.	RS	7 300	300	96.0	4.0	2.95	0.8	Flotation aid
Morristown, N. J.	RS	6 800	200	97.0	3.5	1.70	0.5	Standard
Omaha, Nebr.	RS	19 660	118	99.8	5.9	7.66	0.8	Flotation aid after 24-hour holding
Omaha, Nebr.	ML	7 910	50	99.4	6.8	3.1	0.8	Flotation aid
Belleville, Ill.	RS	18 372	233	98.7	5.7	3.83	0.4	Flotation aid
Indianapolis, Ind.	RS	2 960	144	95.0	5.0	2.1	1.47	Flotation aid after 12-hour holding
Warren, Mich.	RS	6 000	350	95.0	6.9	5.2	1.75	Flotation aid
Frankenmuth, Mich.	ML	9 000	80	99.1	6.8	6.5	1.3	Flotation aid
Oakmont, Pa.	RS	6 250	80	98.7	8.0	3.0	1.0	Flotation aid
Columbus, Ohio	RS	6 800	40	99.5	5.0	3.3	1.0	Flotation aid
Levittown, Pa.	RS	5 700	31	99.4	5.5	2.9	1.0	Flotation aid
Nassau Co., N. Y.								
Bay Park Sewage Treatment Plant	RS	8 100	36	99.6	4.4	4.9	1.2	Flotation aid
Bay Park Sewage Treatment Plant	RS	7 600	460	94.0	3.3	1.3	0.33	Standard
Nashville, Tenn.	RS	15 400	44	99.6	12.4	5.1	0.66	Flotation aid

[a]ML = mixed liquor from aeration tanks; and RS = return sludge.
[b]Suspended solids.
[c]lb/hr/sq ft × 4.88 = k/m²·h.
[d]gpm/sq ft × 0.679 1 = L/m²·s.
[e]Standard = no flotation aid and no holding before sampling; flottation aid = use of coagulation-flotation aid.

TABLE 29.4   Additional operating data for plant-scale dissolved air flotation units.

Location	Sludge	Loading lb/hr/sq ft[a]	Thickened sludge concentration, %	SS[b] capture, %	Air: solids ratio
Albany, Ga.	Primary and WAS[c]	0.82	6–9	88	0.013
Mankato, Minn.	Primary and WAS	0.50	5–8	90–95	0.013
Menomonie, Wis	Primary and WAS	0.67	4.8–6.7	97	0.013
Waukesha, Wis.	Primary and TF[d]	1.3	6–8	85	0.013
Hagerstown, Md.	Primary and O₂ WAS	0.96	6	97	0.016
Scranton, Pa.	Primary and WAS	1.3	5–6	85	0.016
Oconomowoc, Wis.	WAS	0.4–0.6	4.0	95	0.150
Marshalltown, Iowa	WAS	0.41	4.0	98	0.030
	WAS with chemical	0.8	4.3	99	0.018
York, Pa.	O₂ WAS	0.56	5.8	91	0.010
San Jose, Calif.	WAS	0.42–0.62	3.5–5.5	99	0.015
	WAS with chemical	1.0	3.5–5.5	99	0.015
Huntington, Ind.	WAS	0.38	4.5–5.0		
Glenco, Minn	WAS	0.22	4.0	9.5	0.015
Ocean City, Md.	O₂ WAS with chemical	0.8–1.2	4.5	99	0.013
Erie, Pa.	WAS with chemical	2.25	6.5–7.5	99	0.013
Albany, N. Y.	WAS with chemical	0.52	4.5	99	0.034
Brewer, Maine	WAS with chemical	3.25	3.5	97	0.010
Middletown, N.J.	WAS with chemical	1.25	4.5	99	0.030

[a]lb/hr/sq ft × 4.88 = k/m²·h.
[b]Suspended solids.
[c]Waste activated sludge.
[d]Trickling filter.

- The speed and the on–off times of the float skimmers should be set to maximize the float solids concentration, but not too slow to cause excessive float-depth accumulation.
- Controlling pressure within the 350 to 590 kPa (50 to 85 psig) range—find the lowest pressure that allows optimum operation—but do not operate at less than 350 kPa (50 psig). Pressures much below 350 kPa cause disproportionate increases in total float depth (see Bratby [1978] and Bratby and Ambrose [1995]).
- Controlling the saturator feed rate to maintain the required air-to-solids ratio.
- Periodically removing bottom sludge is necessary to prevent septicity, gasification, and general deterioration of float density and overall DAF efficiency.

The air-to-solids ratio (pound air/pound solids) affects the sludge rise rate and, as importantly, the total float depth. The necessary air-to-solids ratio, typically between

0.02 and 0.04 (Bratby et al., 2004), depends mostly on sludge characteristics, as explained earlier.

Dilution reduces the effect of particle interference on the rate of separation. Concentration of the sludge increases and the concentration of effluent suspended solids decreases as the sludge blanket detention time increases.

START-UP. Typically, an operator should follow these steps for routine start-up of a DAF:

- Fill the tank with screened final effluent or plant nonpotable water until overflowing.
- Start the collection mechanism for both float and underflow. Ensure that they are operating properly and then place into the normal automatic mode of operation.
- Continue plant nonpotable water flow to the DAF unit and engage the recycle system, including the compressor, if applicable. Adjust to proper pressures and flow. After functioning properly, proceed.
- Ensure that the float and underflow pumps are functional by pumping some water.
- Prepare the polymer, if applicable, and engage the polymer addition at proper flow or as jar tests have indicated if starting after prolonged shutdown, start-up, or process changes.
- Phase the nonpotable plant water out and start WAS flow to the DAF. Make process adjustments as necessary to maintain float depth as described herein.

SHUTDOWN. Typically, an operator should follow these steps for routine shutdown of a DAF:

- Stop polymer flow, if applicable, and at the same time stop WAS feed.
- If only down for a short time (30 minutes or so), there is no need to shut down the recycle system, including the compressor. If down for longer than this, shut down the recycle system.
- The float rake timer can be left on until most of the float is removed into the hopper and pumped to further processing.
- If the unit is going to be down for more than 24 hours, displace the tank contents with nonpotable water or drain and clean the tank, all troughs, and pipelines.
- In a typical operation, only the recirculation pump and retention tank discharge valves are closed when stopping a unit's operation. All other valves remain open, with the exception of valves on drain lines.

SAMPLING AND TESTING. At a minimum, DAF thickeners should be sampled and analyzed as follows:

- Influent for TSS, total solids, and flow;
- Effluent for TSS, total solids, and flow;
- Float (sludge withdrawal) for total solids and flow;
- Depth of blanket by way of sampler, either "bomb" or coretaker (sludge judge). Sample location should be near the beach where sludge is raked into the float trough;
- If applicable, determine the total solids of the polymer or whatever quality assurance tests you deem necessary for the coagulant; and
- Occasionally determine the mass of precipitated air in the saturator feed to evaluate the operating air-to-solids ratio. The methodology has been documented elsewhere (Bratby and Marais, 1975, 1977).

TROUBLESHOOTING. Table 29.5 provides causes and solutions for the most common flotation thickener problems (U.S. EPA, 1978).

PREVENTIVE MAINTENANCE. An effective preventive maintenance program will reduce equipment stoppage and breakdown. Daily maintenance items include

- Checking the drive units for overheating, unusual noise, or excessive vibration, and
- Checking for unusual conditions, such as loose weir bolts.

Weekly maintenance items include

- Checking all oil levels and ensuring that the oil fill cap vent is open;
- Checking all condensation drains and removing any accumulated moisture;
- Examining drive control limit switches; and
- Visually examining the skimmer to ensure that it properly contacts the scum baffle and scum box.

Monthly maintenance items include

- Inspecting skimmer wipers for wear and
- Adjusting drive chains or belts.

TABLE 29.5   Troubleshooting guide for flotation thickening.

Indicators/ observations	Probable cause	Check or monitor	Solutions
Floated sludge too thin	Skimmer speed too high	Visual inspection	Adjust as required
	Unit overloaded	Rise rate	Turn off feed sludge to allow unit to clear or purge the unit. with auxiliary recycle
	Polymer dosages too low	Proper operation and calibration of polymer pumps	Adjust as required
	Excessive air:solids ratio	Float appears very frothy	Reduce air flow to pressurization system
	Low dissolved air	See "Low dissolved air"	
Low dissolved air	Reaeration pump off, clogged, or malfunctioning	Pump condition	Clean as required
	Eductor clogged	Visual inspection	Clean eductor
	Air supply malfunctioning	Compressor, lines, and control panel	Repair as required
Effluent solids too high	Unit overloaded	Rise rate	Turn off feed sludge to allow unit to clear or purge the unit with auxiliary recycle
	Polymer dosages too low	Proper operation and calibration of polymer pumps	Adjust as required
	Skimmer off or too slow	Skimmer operation	Adjust speed
	Low air:solids ratio	Poor float formation with solids settling	Increase air flow to pressurization system
	Septic sludge on bottom		Periodically remove settled sludge
Skimmer blade leaking on beaching plate	Skimmer wiper not adjusted properly	Visual inspection	Adjust as required
	Hold-down tracks too high	Visual inspection	Adjust as required
Skimmer blade binding on beaching plate	Skimmer wiper not properly adjusted	Visual inspection	Adjusted as required

TABLE 29.5   Troubleshooting guide for flotation thickening (*continued*).

Indicators/ observations	Probable cause	Check or monitor	Solutions
High water level in retention tank	Air-supply pressure low	Compressor and air lines	Repair as required
	Level control system not operating	Level control system	Repair as required
	Insufficient air injection	Compressor and air lines	Repair as required
Low water level in retention tank	Recirculation pump not operating or clogged	Pump operation	Inspect and clean as required
	Level control system not operating	Level control	Repair as required
Low recirculation pump capacity	High tank pressure	Backpressure	Adjust backpressure valve
Rise rate too slow	Unit overloaded	Rise rate	Turn off feed sludge to allow unit to clear or purge the unit with auxiliary recycle
	Low dissolved air	See "Low dissolved air"	
	Polymer dosages too low	Proper operation and calibration of polymer pumps	Adjust as required

Semiannual inspections of major elements for wear, corrosion, and proper adjustment include

- Drives and gear reducers;
- Chains, belts, and sprockets;
- Guide rails;
- Saturation systems—eductors (if used) or nozzles should be inspected for wear or cleaned whenever the efficiency begins to decline, or on a semiannual basis;
- Polymer feed system; and
- Mechanical systems, including shaft bearings and bores, bearing brackets, baffle boards, flights and skimming units, suction lines and sumps, and sludge pumps.

SAFETY. The following safety considerations apply to DAF equipment:

- Comply with the applicable general safety considerations in Chapter 5, including procedures for confined-space entry to areas where the toxic gases, hydrogen sulfide and methane, may be present;

- Require the presence of at least two people when working in areas not protected by handrails;
- Keep walkways and work areas free of grease, oil, leaves, and snow;
- Keep protective guards in place unless mechanical or electrical equipment has undergone lockout/tagout. Lockout is a method of keeping equipment from being set in motion and endangering workers. The *Code of Federal Regulations (1991) 29 CFR* Part 1910.147 covers the servicing and maintenance of machines and equipment in which the unexpected energization or start-up of the machines or equipment, or release of stored energy, could cause injury to employees. Tagout occurs when the energy isolation device is placed in a safe condition and a written warning is attached to it. Typically, the employee signs and dates the tag. Removal of the tag and lockout device is done by the same individual when the equipment is safe for restart;
- Keep the pressure of the retention tank under its working pressure rating; and
- Ensure that the tank relief valve functions; inspect it regularly for corrosion.

GRAVITY BELT THICKENING

DESCRIPTION OF PROCESS. Gravity belt thickeners (Figure 29.9) work by filtering free water from conditioned solids by gravity drainage through a porous belt. The gravity drainage area is usually horizontal but may be inclined under some circumstances. Chemical conditioning is generally required to flocculate the sludge and separate the solids from the free water. Chemical conditioning may be accomplished by injecting the chemical through an injection ring and mixing it with the sludge. After chemical injection, the sludge velocity is reduced in a retention tank and the sludge is allowed to fully flocculate before overflowing by gravity onto the moving belt. As the moving belt carries the sludge, plows (chicanes) clear portions of the belt for the filtrate to drain through and gently turn over the solids, thereby exposing more free water. Prior to the discharge, most gravity belt thickeners have some type of dam or an adjustable ramp. This forces the solids to either crest the ramp or to go under the dam. In either case, the solids will tend to roll backwards due to shear forces. This force further assists in obtaining maximum performance. Thickened solids that pass the dam are discharged from the gravity belt thickeners. Any residual solids still on the belt are scraped off by a scraper (doctor) blade. The belt then continues through a high-pressure wash water spray system to clean the belt. The filtrate and the wash water are both collected for further processing or returned to the head of the plant. In some instances, the wash water may be recycled to improve capture efficiency.

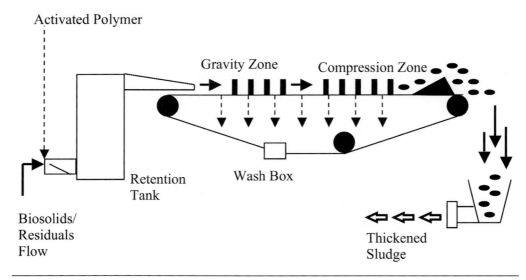

FIGURE 29.9   Gravity belt thickener process.

The hydraulic capacity of the equipment is determined by multiple factors including sludge type, sludge concentration, polymer type and polymer dosage, belt speed, belt type, as well as the obvious machine width and length.

DESCRIPTION OF EQUIPMENT. Typical gravity belt thickener systems are described herein.

Polymer Addition. Polymer addition usually occurs via injection through a multiport injection ring; it is mixed with the sludge as it flows through an inline mixing device. In some cases, the polymer may be mixed mechanically in the retention tank. However, this generally results in higher polymer consumption.

Flocculation Tank. The flocculation tank allows the incoming sludge velocity to be reduced so that flocculation can fully occur before the sludge overflows onto the moving belt. Design of the tank is critical to prevent short-circuiting within the tank.

Belt and Supports. Belts are typically woven from polyester fiber. High pH and other unusual conditions may require special materials. The belt is supported on grid strips that also serve as wipers on the bottom of the belt. This wiping action increases the drainage capacity of the belt.

Belt Tensioning. Unlike a belt press, the performance of the belt thickener is not dependent on belt tension. Once the dewatering belt on a thickener is tight enough to prevent slippage on the drive roller, additional tension is unnecessary. Additionally, the belt tension on a thickener is not dependent on the type or amount of sludge loading. As such, the requirements for belt tension are much lower on a gravity belt thickener compared to belt filter presses. Typically, 20 lb of tension force/linear inch of belt width is sufficient. All belt tension is a result of moving one roller closer to or further away from the other roller(s). This displacement may be through hydraulic, pneumatic, or mechanical actions. Once the belt is tensioned, it is not necessary to relax the tension until the belt is replaced.

Belt Drive. All belt thickeners have a variable-speed belt drive (Figure 29.10) with a typical speed range of 8 to 40 m/min. The belt speed may be either mechanically or electrically varied with speed controls at the local control panel. Typically, the belt drive is attached to a rubber-coated drive roller.

Belt Tracking. During operation of a belt thickener, the belt should more or less remain centered and not move laterally on the machine. Although the belt should not move, some type of belt tracking device will be included on most machines. The most common devices are hydraulic cylinders, pneumatic cylinders, and pneumatic bellows. Other, less common types include mechanical guides built into the machine

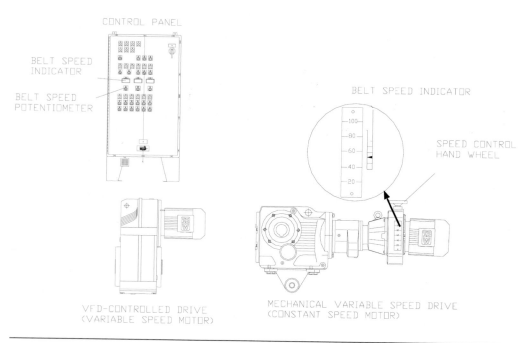

FIGURE 29.10 Basic controls for belt thickeners.

frame or rollers, jack bolts built into the bearing housing on one roller, as well as a roller that is repositionable to adjust the belt tracking. Comparatively, the belt on a belt thickener is similar to a conveyor belt; all of the tracking devices have some roots in the conveyor or papermaking industries.

Chicanes. The chicanes (Figure 29.11), or plows, on a gravity belt thickener serve several functions. As sludge first enters the gravity deck, the solids are very thin and dispersed. The chicanes clear a path on the moving belt for the filtrate to drain through. As the solids thicken, the plows start turning over the rows of solids similar to a plow making wind rows. This turning action exposes more filtrate to the moving belt and helps eliminate ponding on the gravity deck. Some plows also cause the solids to roll backwards on the deck, which creates a shearing action on the solids. This shearing action is the same force as a ramp on the machine, which greatly increases the performance of the thickener.

FIGURE 29.11   Chicane placement.

Adjustable Ramp. Most high-performance gravity belt thickeners will have some type of ramp (Figure 29.12) at the end of the gravity deck. It is an advantage for the ramp to have a method of adjustment with the solids either flowing over or under the ramp. The purpose of the ramp is to cause the solids to slightly pond on the gravity table. This ponding holds the solids longer on the moving belt, which increases the amount of filtrate removed. It also can cause the solids to roll backward against the ramp, which, again, improves the efficiency of the process.

Doctor Blade. When the system is operating properly, most of the thickened solids should drop off the end of the gravity table; however, a small amount of solids may stick to the dewatering belt. The doctor blade aids in removing these solids by gently scraping the belt to prepare it for cleaning in the high-pressure wash station.

Belt Wash. Most gravity belt thickeners will have one high-pressure wash tube. The typical requirements are 110 m^3/d (20 gpm) per meter of belt width at 585 to 620 kPa

FIGURE 29.12   Discharge ramp.

(85 to 90 psig). Although the wash water should be relatively clean, it could include final effluent or even filtrate from the gravity belt thickener. If filtrate is used, a strainer should be used to reduce the possibility of solids in the filtrate clogging the wash water nozzles. The typical installation will spray the belt on the cake side, but it is also possible to wash the belt from the non-cake side. In some applications, washing is carried out from both sides of the belt or washing is aided through the induction of a soap solution or other cleaning agent. With proper washing, the belt should last 3000 to 4000 hours in most applications.

Sludge Pumping. Sludge pumping is critical to the proper operation of any dewatering device and the gravity belt thickener is no exception. Sludge flocculation will generally rely on the head pressure from the sludge and polymer pumps for mixing of the sludge and polymer. As such, it is important to select pumps with enough head pressure to overcome the minor losses associated with this mixing. The pumps should also be such that surging or pulsing in the flow is eliminated. Although some installations will use mechanically variable speed drives on the sludge pumps, the more common approach is to have the speed of the pumps variable through a variable-speed drive controlled at the gravity belt thickener local control panel. Flow indication should also be provided at the local control panel.

Additionally, the thickened sludge may either be pumped or conveyed to the next part of the process. The thickened sludge pump will generally be controlled by the level of the sludge in the hopper and should not be controlled from a fixed-speed setpoint.

Controls. Typical controls for a gravity belt thickener provide either manual or automatic control depending on the level of plant automation. At a minimum, the following functions should be included:

- Belt drive start/stop,
- Wash water pump start/stop,
- Sludge feed pump start/stop,
- Thickened sludge pump start/stop,
- Hydraulic power unit start/stop (or air-compressor start/stop), and
- Polymer system start/stop.

Optionally, the following items are recommended:

- Belt drive speed controller,
- Sludge pump speed controller,
- Polymer pump speed controller,
- Thickened sludge pump speed controller, and
- Emergency stop push button/pull cord.

The following alarms should also be included:

- Belt misalignment,
- Belt broken,
- Low wash water pressure,
- Low hydraulic pressure/low air pressure,
- High sludge level (on the gravity deck),
- High thickened sludge level, and
- Low thickened sludge level.

PROCESS CONTROL. Numerous variables affect the overall thickening process. Listed below are most of these variables and their interrelationship with the other variables.

Sludge Feed. The sludge flowrate can be determined by the desired solids loading, the amount of feed solids per hour per meter of belt width (lb/hr/m), and the feed slurry solids concentration (%). The feed solids should be characterized to determine inorganic (ash) content, biosolids content, and solids chemistry. As a quick reference, the following types of municipal sludges are listed along with their applicable solids loading limits (lb/hr/m):

- Raw primary, 100%—2000–3000
- 50% primary, 50% waste activated—1500–2000
- Anaerobically digested blend (50% primary)—1300–1750
- Anaerobically digested blend (50% primary, 50% waste activated)—1300–1750
- Aerobically digested waste activated—1100–1500
- Waste activated, 100%—660–1200

Polymer. To accomplish optimum thickening, it is important to select the proper polymer. Regardless of the type of polymer selected, plant personnel should verify that the polymer system specified can handle the type of polymer selected for the application. Poor mixing, activation, and aging of the polymer are some of the most common causes of excessive polymer consumption. The recommended final polymer solution concentration range to condition the sludge is the following:

- Dry polymers (0.05 to 0.2% by weight),
- Emulsion/dispersion polymers (0.1 to 0.5% by volume), and
- Mannich polymers (1 to 3% by volume).

It is important to note that if the polymer solution is prepared at concentrations that are considerably higher than the solution concentrations recommended then polymer-thickening costs would increase because the polymer cannot be dispersed efficiently into the slurry. Polymer overdosing will deteriorate the sludge thickening process as well.

Polymer Dosage. For a given type of sludge, polymer dosage is generally solids-dependent (i.e., the lower the percent feed solids, the higher the chemical dosage required). Put another way, sludges that are difficult to gravity-settle will require more polymer than sludges that settle easily. Polymer should be injected into the sludge at the minimum amount required for thickening. All excess polymer is wasted and goes down the drain with filtrate. This is not cost- or process-effective.

Polymer selection and dosage are best determined by jar tests and bench-scale simulation tests, followed by a trial on the gravity belt thickeners.

There are a number of ways to automate the polymer dosage using different instruments and control systems. Information on various control options is available from most instrument manufacturers or process equipment manufacturers.

Mixing Energy. Mixing energy is the energy required to instantaneously mix the polymer with the suspended solids of the slurry. The optimum mixing energy is usually determined on-site by adjusting the throat opening inside the variable orifice mixer. For example, to increase the mixing energy, reduce the throat opening of the mixer by increasing the adjustable counterweight and turn up the adjustable bolt on the valve stop handle to allow the weight arm to move further down. Too little or too much mixing energy results in less than optimum floc formation that adversely affects thickening action.

Retention Time. Retention time is the time required for the polymer reacting with the biosolids/residuals suspended in the slurry to complete the flocculation process. Most thickening applications require 15 to 20 seconds to complete the flocculation process. With too little time, the flocs will be very small and, with too much time, the result is large, clumpy flocs. Both of these conditions lead to reduced thickening. For ideal thickening, large, uniform-sized flocs are desired.

Belt Speed. There are two changes an operator can make to belt speed: making it faster or making it slower. The slower the belt speed, the greater the residence time in the gravity section because, as the belt slows, the sludge has more time for the free water to drain. This translates into increased cake dryness. Conversely, the faster the belt speed, the greater the process throughput (assuming the sludge feed rate is increased). The belt speed should be slowly adjusted until the optimal balance between process throughput and cake dryness is achieved. The belt speed range is 8 to 40 m/min. There are two different types of belt drives offered on most gravity belt thickeners: a variable-frequency drive (VFD)-controlled, variable-speed motor drive and a mechanical variable-speed drive with a constant speed motor. Typically, the VFD-controlled belt drive speed can be adjusted by a belt drive speed potentiometer or controller located on the control panel. A belt drive indicator shows the belt speed in percent (0 to 100%). For the mechanical variable-speed drive, the speed control hand wheel can be turned to adjust belt speed.

Belt Tension. Belt tension should be set at 20 pli initially by adjusting the pressure valve on the hydraulic power unit. Because sludges vary from plant to plant, the optimum pressure should be determined once the gravity belt thickener is operating. The belt tension needs to be only high enough to obtain good traction without slipping at the drive roller. The belt tension also helps keep the belt straight and flat. Excessive belt tension will decrease belt life.

Belt Type. The opening-size weave of the belt and belt material determine the thickening characteristics of that belt. The initial belt supplied by the manufacturer should have been selected based on their experience with similar processes.

Ramp Angle. The adjustable ramp is provided to cause additional dewatering of the sludge before it is discharged from the gravity deck. This is accomplished by creating a dam across the discharge end of the gravity deck, forcing the sludge to be pushed up the ramp, turning and tumbling on itself, while additional water is caused to separate from the solids. For drier sludge, the ramp will be raised to a steeper angle. It is also possible to retract the ramp away from the belt and allow the sludge to pass under it when no further dewatering is desired.

Upstream Variables. There are other items upstream of the gravity belt thickener that can affect its performance. The information presented here is designed to illustrate some of the variables that may affect the overall performance of the gravity belt thickener with the plant slurry

Slurry Pump Selection. Positive-displacement pumps are recommended for sludge thickening applications. The preferred pumps for these applications include progressing cavity, rotary lobe, and gear pumps. These pumps allow even flow of the slurry along the pipeline to allow good dispersion of the polymer with the suspended solids of the slurry as well as a constant pressure drop across the variable orifice mixer. Other pumps that are commonly used include self-priming centrifugal pumps; however, the pressure drop across the inline polymer mixer must be carefully considered.

Sludge Characteristics. The sludge characteristics described herein should be considered in sludge thickening processes.

Solids Concentration. Solids concentration level influences the selection of a slurry conditioning program. Increasing feed solids concentration will cause lower polymer conditioning requirements. It is extremely important that the characteristics of the sludge being thickened remain relatively constant to maintain good process control of the gravity belt thickener. For example, if the feed solids concentration increases by 30% (i.e., from 2 to 2.6%), one of the following variables have to be adjusted to keep the gravity belt thickener running satisfactorily:

- Polymer dosage,
- Belt speed, and
- Slurry flowrate (to maintain constant solids loading).

Biological Sludge Content. Biological sludges usually have high cationic polymer requirements. When thickening waste activated biosolids, it is critical to understand that the specific resistance for activated sludge increases when the biological process is experiencing short mean cell residence time, low dissolved oxygen, low temperature, and high food-to-microorganism ratios to control the population of filamentous bacteria in the aeration basin(s) to prevent poor gravity belt thickener performance. Blooms of filamentous bacteria increase the polymer dosage and reduce the solids loading and cake solids during the thickening process because water is stored inside the cells of the bacteria. Higher polymer requirements usually result from high dissolved solids in sludge.

Inorganic (Ash) Content. Higher ash content usually yields higher dry cake solids. Biological sludges have ash contents ranging from 15 to 35%. Ash contents of digested biological sludge can increase to 30 to 50%. Higher ash contents are occasionally encountered from lime-stabilized sludges or chemical treatment waste sludges. In these cases, it is not unusual to have an anionic or nonionic conditioning program work best with the sludge.

Sludge Storage Time. Extended storage of raw primary and waste biological sludges before polymer conditioning increases conditioning requirements. Aeration improves the sludge thickening characteristics.

Wash Water Characteristics. The wash water used to clean the belt needs to have the following qualities to prevent poor performance of the gravity belt thickener:

- The TSS concentration should be ≤50 mg/L,
- The total dissolved solids concentration should be 1000 mg/L,

- The pH of water should be 6 to 8,
- The temperature should be 10 to 50 °C, and
- The wash water pressure should be ≥586 kPa (85 psig).

Occasionally, if the TSS concentration reaches 200 mg/L, the unit can operate marginally if the nozzles in the spray tubes are cleaned frequently with wire brushes (these are actuated by opening and closing the manual valve). If the wash water pressure drops considerably, the solids that are embedded in the belt during the filtration process cannot be dislodged, causing belt blinding after a period of time.

START-UP. After ensuring that all of the guards and covers are in place and that the unit has not been locked out/tagged out, the unit may be started. Each manufacturer will have specific instructions that should be followed. However, the basic start-up sequence for a gravity belt thickener is as follows:

- Ensure that the belt is clear of debris and that the plows/ramp/scraper are properly positioned on the belt.
- Start the hydraulic unit (or air compressor) and allow the belt to tension.
- Start the belt drive and use an initial setting of approximately 20 m/min belt speed.
- Start the wash water pump and allow the belt to pre-wet.
- Start the polymer pump and allow the fresh polymer to reach the polymer injection point
- Start the sludge pump.
- After thickened sludge is available, start the thickened sludge pump (or thickened sludge conveyor).
- After the system is running, begin fine-tuning the process by adjusting the sludge flow, polymer dose, mixing energy, belt speed, and so on until the results are within the desired process parameters. It is critical to only adjust one item at a time and to allow time for the adjustment to take effect before making another change.

SHUTDOWN. The basic shutdown sequence is as follows:

- Shut down sludge feed pump.
- Shut down polymer feed pump.
- As the thickened sludge hopper empties, shut down thickened sludge pump (or thickened sludge conveyor).

- Drain the flocculation tank on the machine.
- Wash the machine down from top to bottom.
- Allow the belts to be completely washed (this could take 15 to 45 minutes) without sludge or polymer.
- Shut down the wash water pump.
- Shut down the belt drive.
- Shut down the hydraulic unit/air compressor.

SAMPLING AND TESTING. At a minimum, gravity belt thickeners should be sampled and analyzed as follows:

- Sample influent feed for total solids, TSS, total volatile solids, pH, and flow;
- Sample wash water for TSS and flow;
- Sample thickened solids for total solids and flow;
- Sample filtrate for TSS and flow;
- Measure flow and quantity of polymer used; and
- Measure any dilution water used to make up polymer.

Samples should be collected and sealed in airtight containers until they can be analyzed. Analysis should be conducted in accordance with the latest edition of *Standard Methods for the Examination of Water and Wastewater* (APHA et al., 2005). Table 29.6 provides a sample log sheet for tracking the performance of a gravity belt thickener.

TROUBLESHOOTING. Each gravity belt thickener process consists of a system of integrated components and each component may have an effect on another portion of the system. It is important to hold as many variables constant as possible while troubleshooting other portions of the system. Troubleshooting starts with the process and then works down into the individual components of the system (Tables 29.7 through 29.12).

PREVENTIVE MAINTENANCE. Maintaining the gravity belt thickener will not only extend the life of equipment but also decrease the operational cost through increased operational efficiency such as lower polymer doses, increased throughput, and so on. For simplicity, the preventive maintenance tasks are broken down into basic tasks depending on the frequency of the activity. In all cases, the manufacturer should be consulted for additional steps or procedures for the equipment.

TABLE 29.6 Typical log sheet for thickener operations.

Operating data for month of _____, 19 ____

Day of month	Feed sludge data				Flocculant feed data		Thickened sludge data		Belt wash data		
	Belt or drum speed, rpm	Feed rate, gpm or gal[a]	Feed solids,[b] % TSS	Solids throughput,[c] lb/hr[a]	Polymer, gpm[a]	Dosage,[d] lb/ton[a]	Dilution water flow, gpm[a]	TS, %	Wash water flow, gpm[a]	TSS in filtrate, %	Solids recovery,[e] %
1											
2											
3											
4											
.											
.											
30											
31											
Total											
Average											
Minimum											
Maximum											

[a] gpm × (6.308 × 10⁻⁵) = m³/s; gal × (3.785 × 10⁻³) = m³; lb/hr × 0.126 0 = kg/s; lb/ton × 0.500 = g/kg.

[b] Percent TSS (total suspended solids), TS (total solids) recorded as 4% = 40 000 mg/L.

[c] Solids throughput: lb/hr = feed rate × % TSS × 0.25 or total volume fed × % TSS × 8.34 = lb fed.

[d] Flocculant dosage: lb flocculant/ton of sludge solids = flocculant feed rate × concentration × 500/throughput tons/hr or flocculant volume fed × concentration/(throughput tons) × 1 000).

[e] Solids recovery: 100-(thickened sludge TS) (feed sludge TSS return flow TSS)/(feed sludge TSS) (thickened sludge TS return flow TSS).

TABLE 29.7   Process troubleshooting guide.

Problem	Probable cause	Remedy
Sludge does not flocculate	Polymer is not flowing	Verify polymer system is on
		Verify polymer is flowing through plastic hoses at mixer
	Insufficient amount of polymer	Increase polymer dosage
	Wrong polymer type	Contact manufacturer or polymer representative
	Old/nonreactive polymer	Make up new polymer
	Insufficient polymer mixing	Increase polymer mixing energy
		Increase retention time in piping
Sludge does not de-water on gravity deck	Belt is blinded	Clean belts
	Poor flocculation	Increase polymer
	Wrong polymer type	Contact manufacturer or polymer representative
	Loading rate too high	Decrease sludge feed rate
Capture is poor/filtrate is dirty	Seals are worn on restrainers	Replace seals
	Insufficient polymer	Increase polymer dosage
Low cake solids	Insufficient polymer	Increase polymer dosage
	Too much polymer	Decrease polymer dosage
	Wrong polymer type	Contact manufacturer or polymer representative
	Loading rate is too high	Decrease sludge feed rate
	Belt speed is too fast	Decrease belt speed
	Ramp is too low	Increase ramp angle
Solids adhere to belt	Polymer dosage is too high	Decrease polymer dosage
	Insufficient mixing	Increase mixing energy
Sludge buildup in pans	Upset process	Clean pans and optimize process

Daily Actions.

- Clean belt by running belt drive and wash system without sludge or polymer for a minimum of 45 minutes.
- Clean spray nozzles on wash box by rotating hand wheel to a fully open position and then rotating it back to a fully closed position.

TABLE 29.8   General troubleshooting guide.

Problem	Probable cause	Remedy
Thickener does not run	Sludge pump not operating	Check sludge pump
	Polymer system not operating	Check polymer system
	Belt drive not operating	See Table 29.10
	Hydraulic unit not operating	See Table 29.9
	Water pump not operating	See Table 29.11
Scraper blade not cleaning belt	Blade is out of adjustment or requires replacement	Adjust or replace blades
	Buildup of fibrous material, such as hair, at the knife edge of the blade	Open blade and clean
Scraper blade wears quickly	Blade tension too tight	Reduce blade tension; optimize process for good belt release
	Belt speed too fast	Reduce belt speed
Roller sticking	Bearing is worn	Replace bearing
Bearing losing excessive grease	Bearing seals are blown	Replace seals
Machine runs in a jerky motion at roller bearings	Gear train is damaged	Inspect gear reducer and replace as required

- Check the oil level in the hydraulic unit; fill as required.
- Check the oil level in the air compressor (if applicable) and fill as required.
- Manually extend and retract the tension cylinder to clean and oil the rods. This will greatly extend the life of the seals.
- Cycle the steering cylinder in both directions by holding the steering paddle first one way and then the other. This will clean and oil the rods and greatly extend the life of the seals.
- Inspect all alarms and manually trip and reset the trip cord.
- Inspect the machine for any loose or worn parts such as seals, belts, and so on that need to be replaced.

Monthly Actions.

- Clean belt with a soap/bleach mixture. To prepare the soap/bleach mixture, use 1 cup of detergent and 3 cups of bleach to mix with 5 gal of water. The soap can be any laundry-type liquid detergent and the bleach can be any generic brand bleach containing 5.25% sodium hypochlorite. The water can be tap

TABLE 29.9   Hydraulic system troubleshooting guide.

Problem	Probable cause	Remedy
Hydraulic power unit fails to energize when control push button is pressed	Control panel feeder circuit in off or "tripped" position	Set breaker to "on" position
	Motor starter overload protectors in tripped position	Depress overload reset button on motor starter
Belt steering erratically, requiring constant automatic correction	Improper roller alignment, valve alignment, valve sensitivity, or belt defects	Carry out check and adjustment procedures as appropriate
Hydraulic unit operational but fails to build pressure	Incorrect motor rotation direction	Ensure that rotation is correct; if not, have a qualified electrician revise motor wiring at motor starter
	Pressure regulator is not properly adjusted	Correct adjustment
	Pressure regulator clogged by foreign material causing fluid bypass back to reservoir	Disassemble and clean valve and check for worn or broken parts; if particles are found in valve, drain and clean reservoir, refill as recommended in lubrication schedule
	Belt steering valve bypassing fluid directly back to power unit reservoir	Remove and clean belt steering valve of any foreign material; if cleaning does not improve operation, contact a qualified hydraulic repair center or contact manufacturer
	Hydraulic pump worn or damaged	Have pump serviced by a qualified hydraulic repair center; contact manufacturer if a replacement pump or part is required
Belt tension slacks when sludge is applied	Tensioning cylinder is bypassing oil internally	Repair or replace cylinder
	Pneumatic cylinder/bellow has an air leak	Repair air line or replace cylinder/bellows

TABLE 29.10   Belt drive troubleshooting guide.

Problem	Probable cause	Remedy
Main drive fails to start and drive the belt when energized	Control panel interlocks prohibit belt drive energizing until appropriate ancillary equipment is operational	See sequence of operations and control diagrams for design interlocks; energize appropriate equipment
	Belt drive speed potentiometer (or controller) set at zero	Increase setting to desired speed
	VFD electronic controller tripping chassis mounted overload protector or feeder voltage fuses burnt out	Reset overload and check fuses for continuity, and renew or reset as required; check for dried sludge or other obstructions on the belt that would put unusual starting load on belt drive
	If check out fails to show problem and incoming voltage to VFD is proper, the VFD has failed	Take VFD controller to repair center for reconditioning or replacement; contact manufacturer for service

water. Use a power wash system to spray the soap/bleach mixture on the belt surface for cleaning. The spray pressure should be about 7000 kPa (1000 psig) and not to exceed 10 300 kPa (1500 psig).

- Inspect the belt seam wires for breaks; replace if broken.

Semiannual Actions.

- Clean (or replace) hydraulic filter screen.
- Check oil level in drive unit gear box.
- Lubricate all bearings.
- Inspect polymer mixer/injection ring assembly and clean as required.
- Replace the belt seam wires.

Annual Actions.

- Replace drive unit oil.

TABLE 29.11   Belt wash water system troubleshooting guide.

Problem	Probable cause	Remedy
Water pump does not run	Pump motor not running	Verify switch is ON
		Verify power is ON
		Motor is burned out or needs to be repaired
Low pump pressure	Bypass valve at press is open	Close hand wheel valve at wash box
	Impeller is worn	Replace impeller
	Suction pressure too low	Correct suction pressure problem
	Pressure switch fault	Restart system and verify pressure is above 340 kPa (50 psi)
		Verify pressure switch is functioning by bypassing switch and restarting system
Low pressure at wash box	Line is blocked	Check piping and remove obstruction
	Bypass valve at press is open	Close hand wheel valve at wash box
	Nozzle is missing	Replace nozzle in wash tube
No water at wash box	Valve is closed	Open valves
Water bypassing shower	Valve is open	Close hand wheel valve
	Seals on wash tube are worn	Replace seals
Appearance of dirty strips	Wash water nozzles clogged	Clean or replace nozzles
	Water pressure too low	Increase pressure
	Solids lodged under scraper blade	Clean scraper blade

SAFETY. As with any process in a WWTP, there are inherent hazards that can be minimized with proper precautions. Some general hazards are

- Polymer spills creating slip hazards;
- Lubricant spills creating slip hazards;
- Sludge spills creating slip hazards;
- Chemical exposure from caustics or acids in some processes;

TABLE 29.12   Belt tracking troubleshooting guide.

Problem	Probable cause	Remedy
Belt will not track	Poor distribution of sludge	Correct distribution on gravity section
	Sludge is built up on rollers	Clean rollers
	Steering paddle not following belt	Spring on valve is worn or broken; replace spring
		Valve is sticking or frozen; repair or replace valve
		Paddle is out of adjustment; adjust paddle
	Belt is cut out of square	Remove the belt from the machine, rotate it 180 degrees, and re-install. If the belt is bad, it will go off on the opposite side; if it steers off the same side, the belt is not at fault.
	Roller has been knocked out of alignment	Check all rollers for parallel. Run a 100-ft[a] flat tape along the belt path on both sides near the end of the roller. Each side should be the same (0.5-in. tolerance)[a]; if not, call manufacturer
	Hydraulic/air pressure is low	Increase pressure
	Cylinder is stuck	Manually move steering paddle to see if cylinder responds; if not, repair/replace cylinder

[a]ft $\times$ 0.3048 = m; in. $\times$ 25.4 = mm.

- Waterborne bacterial hazards;
- Exposure to gaseous hazards such as hydrogen sulfide;
- Material handling hazards from lifting polymer, belts, and so on; and
- Electrical hazards around the controls, motors, and so on.

Another, more specific hazard associated with a gravity belt thickener is

- Exposure to the moving dewatering belt although pinch points are guarded.

All of these hazards should be minimal with a proper health and safety plan and by following safety procedures such as the *Code of Federal Regulations* (1991) 29 *CFR* Part 1910.147 as well as applicable provisions from the Occupational Safety and

Health Administration and other relevant entities. It is essential to maintain health and safety items such as machine guards, emergency stop push buttons, emergency stop pull cords, gas monitors, and so on, as well as having a detailed lockout/tagout procedure.

THICKENING WITH A CENTRIFUGE

PROCESS DESCRIPTION. Decanter centrifuges are centrifuges that have a screw conveyor inside that transports the settled solids along the bowl and out of the centrifuge (Figure 29.13 and Figure 29.14). Decanter centrifuges are the only devices where the same piece of equipment can both thicken and dewater sludge. To the centrifuge, the process is the same, varying only in degree. Centrifuges use the principle of sedimentation to separate liquids from solids—the same principle as the clarifiers and thickeners in the wet end of the plant. With sedimentation, it is the difference in density between the solids and the surrounding liquid that drives separation. On the surface of the earth, we experience the acceleration of gravity (9.8 m/s^2). In convenient shorthand, we refer to 9.8 m/s^2 as "one g." A centrifuge might generate 2000 × g on the bowl wall, meaning the acceleration was the equivalent of 2000 times the earth's gravitational acceleration. In very real terms, a one pound weight on the bowl wall would generate a 2000-lb eccentric force and cause the centrifuge to vibrate on it's isolators. In any case, where the gravity clarifiers have only 1 × g to settle the solids, the centrifuge has thousands of times the acceleration of gravity and, therefore, carries the

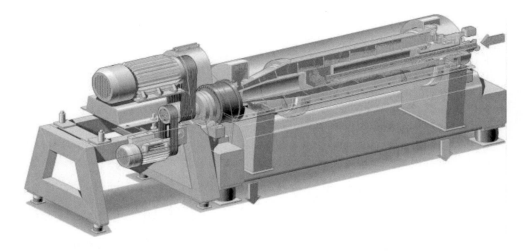

FIGURE 29.13   Thickening centrifuge (courtesy of Westfalia Separator, Inc., Northvale, New Jersey).

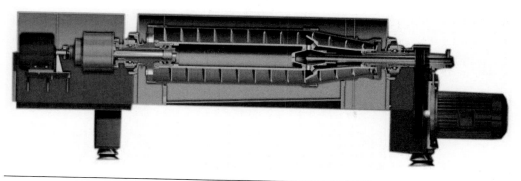

FIGURE 29.14   Cutaway of a typical centrifuge (courtesy of Alfa Laval, Inc., Richmond, Virginia).

separation farther. As it happens, the centrifuges that thicken sludges usually operate at somewhat lower speeds than dewatering centrifuges.

EQUIPMENT DESCRIPTION. Feed material enters the centrifuge through a stationary feed tube (Figure 29.15), where it then drops into the center of the conveyor that is rotating at a speed only a few revolutions per minute different from the bowl speed. This difference in speed between the bowl and the conveyor is called *differential revolutions per minute* (ΔRPM). The feed material flows along the inner surface of the pool of liquid above the compacting solids until it reaches a weir, called a *dam* (Figure 29.16). The liquid dam level is an important process adjustment on all centrifuges. The dam setting controls the liquid level in the centrifuge and is adjusted for each installation and each application. As with a clarifier, the dams must be set at the same radius to prevent short-circuiting.

Just as with a clarifier, the weir ensures the flow is evenly distributed, with no short-circuits, and controls the liquid level inside the centrifuge. As the solids travel from the feed zone to the liquid discharge end, the solids settle out of the moving layer toward the bowl wall. The screw conveyor scrapes these solids along the bowl, up the tapered beach, where they are discharged over a fixed solids weir. The difference in radius between the solids weir and the dam radius is called the *differential head* (ΔH). In most centrifuges, operators periodically adjust the ΔH by removing the dam plates and installing another set with a slightly different radius. Figure 29.15 shows how this process works. It also includes a baffle at the beach bowl intersection, sometimes called a *hydraulic lift disc*. Once a patented device, it is now commonly used on many centrifuges.

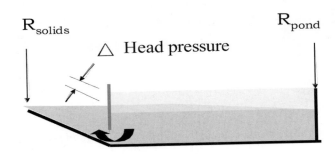

Centrifuge Operation
with a Lift Disk

R_{solids} R_{pond}

△ Head pressure

The Differential Head pressure is
proportional to:
• The solids discharge radius − Pond radius

FIGURE 29.15   Centrifuge diagram showing ΔH.

In a gravity thickener, the underflow pumping rate controls the depth of the blanket of settled solids. Increasing the pumping rate reduces the solids blanket, resulting in thinner underflow. This is the same as with the centrifuge. Increasing the rate at which solids leave the centrifuges reduces the thickness of the cake. With most thickening centrifuges, the solids discharge rate (underflow pumping rate) is a function of

$$Cake\ rate = \Delta RPM + \Delta H \tag{29.7}$$

where
     ΔRPM = the difference in speed between the bowl and the conveyor and
       ΔH = the difference in radius between the pond weir and the solids
          discharge weir.

For many centrifuges, operators periodically change the dam plates so that the differential revolutions per minute stays in mid-range (i.e., between 10 and 15 rpm). This ensures that the thickened cake can be adjusted upwards and downwards as necessary. All centrifuges control the cake rate by changing either the differential revolutions per minute, the differential head, or, for greatest flexibility, both. One centrifuge manufac-

FIGURE 29.16   Complete skid-mounted centrifuge system.

turer has perfected a dam that can be adjusted while the centrifuge is operating. This gives operators maximum flexibility to optimize performance.

Some centrifuges can thicken homogeneous sludges without polymer flocculents. Because the particles of a homogeneous sludge are all of the same composition, they have the same density and, more or less, the same particle size. The centrifuge can then treat the thickening process as a liquid/liquid separation.

Plants that have primary clarifiers followed by secondary aeration generate a WAS that is the principle example of a homogeneous sludge. The primary clarifiers remove the larger and heavier particles, and the cines and colloidal solids going to the aeration are all of a similar density. Even though the centrifuge can thicken these sludges without polymer, adding polymers increases the floc size and, therefore, the capacity of the centrifuge. The savings in buying and operating fewer centrifuges usually outweigh the cost of the polymer.

If the sludges are blends of primary solids and biological solids, they are, by definition, not homogeneous. Primary solids are denser than WAS, which means that either the capture will be poor or polymers are needed to bind the disparate bits of sludge into flocs.

Design Criteria. Centrifuges come in two general design versions: dewatering centrifuges that can also thicken sludges and thickening centrifuges that may or may not be able to dewater sludges.

Dewatering Centrifuges that Will also Thicken. The centrifuge is very useful if the plant intends to thicken sludges prior to hauling as a liquid some of the time and dewater the sludge for an alternate disposal method at other times. It costs very little to design in such versatility and the centrifuge does either at no extra cost; as such, the only added expense is a method to get thickened sludge to a storage tank or hauling vehicle (Figure 29.16). Just as the centrifuge can dewater all of the sludges generated in the plant, it can also thicken all of the sludges. Dewatering centrifuges often operate in torque control mode (sometimes called *load control* mode) when dewatering sludge. The benefit of dewatering in the torque control mode is that cake dryness is more or less proportional to the torque. With torque control, cake dryness is constant. Unfortunately, thickened sludges generate very little torque; as such, torque control does not work when thickening sludges. Thickening centrifuges normally operate in differential control mode (or in pond control mode).

Automation. There are two ways to automatically control the thickened cake, either of which uses an instrument to measure the suspended solids or the viscosity of the thickened cake. Generally, the downstream processes such as anaerobic digestion or liquid injection of the biosolids are dependant on viscosity; as such, viscosity is the more useful parameter. Figure 29.17 presents a chart of cake viscosity versus solids. Below 4 to 5%, viscosity does not change much with cake solids and control in this area is more difficult. Above 5%, the curve becomes quite steep, which gives very stable control. Alternately, there are suitable suspended solids meters that can measure the suspended solids. Whatever the method, the instrument output changes either the ΔH (pond setting) or ΔRPM to maintain the setpoint.

Specialized Centrifuges for Thickening. Centrifuge designs are driven by cost. The least costly change is to reduce the solids discharge diameter, thereby increasing the pool depth. Initially, this was done for thickening centrifuges, but now it is common for dewatering centrifuges as well. Reducing the solids diameter reduces power consumption while, at the same time, the larger volume increases the residence time in the centrifuge. Both benefits make for a more efficient centrifuge. Additionally, because thickened sludge results in very little drag on the conveyor, the back-drive device (electric gear box or hydraulic) can be smaller and, therefore, less expensive. Some centrifuges have internal vanes or inclined plates that enhance thickening. Because thickened sludges are soft, plugging the vanes is not much of a problem; however, their sus-

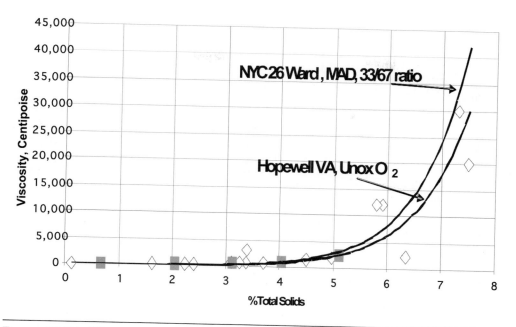

FIGURE 29.17   Viscosity versus concentration of thickened sludge.

ceptibility to plugging also prevents the use of this type of centrifuge for dewatering. Centrifuges designed just for thickening are more often chosen for larger installations where centrifuges are also used for dewatering. Although equipment selection is based on economic analysis, one noneconomic reason to favor centrifuge thickening is that the operation and maintenance procedures are the same for thickening and dewatering centrifuges and using the same device for both applications simplifies operations.

EQUIPMENT DESCRIPTION. As with other decanter centrifuges, the stationary feed tube carries the feed material into the centrifuge, depositing it at one end of the cylindrical bowl or the other. The screw conveyor, or scroll, acts as a rake to move the solids toward the solids discharge, up the inclined beach, and out of the centrifuge. The centrate flows across the surface of the pond toward the dams, or weirs, in the liquid hub.

As with dewatering centrifuges, it is a good idea to have several polymer addition points on the centrifuge. Unlike dewatering centrifuges, relatively little polymer is needed to achieve good performance. The polymer is most commonly added as close to the centrifuge as possible. Some designs, such as those depicted in Figure 29.18,

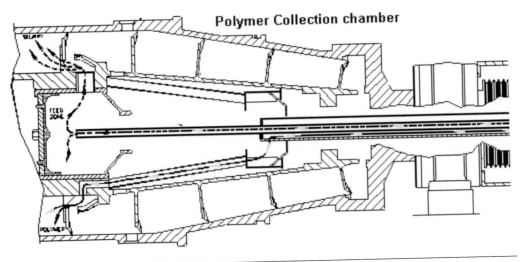

Polymer Collection chamber

FIGURE 29.18   Internal polymer addition (courtesy of Alfa Laval, Inc., Richmond, Virginia).

have an internal polymer feature where the feed tube has an annular space between the inner and outer tube. The polymer enters what looks like a tee, but in fact flows down the annular space and ultimately mixes with the sludge after it is accelerated up to bowl speed. A poor man's version is to fit a hose down the feed tube so that when the polymer is pumped into the centrifuge it first mixes with the sludge in the feed zone.

PERFORMANCE. It is difficult to honestly describe performance, as it is coupled to the sludge properties, centrifuge design, and plant operation. Generally, thickening centrifuges requires less polymer than is required to dewater the same sludge, and there are numerous installations where little or no polymer is used. Because centrifuge performance is a continuum from about 4% up to 25 to 30% solids, the centrifuge is unique in that it can produce whatever an operator needs.

PROCESS CONTROL. *Feed Characteristics.* All process devices benefit from a constant feed quality, and centrifuges are no exception. Common problems are varying rations of primary to secondary solids or feed material that has become septic. For very complicated reasons, septic sludge is more difficult to thicken than fresh sludge. Holding feed material in storage tanks under uncontrolled conditions is poor practice. When in doubt, measure the pH drop through the tankage.

Hydraulic Loading Rate. Centrifuges are available to process feed rates from 6 to 340 m³/h (25 to 1500 gpm). Although there is no intrinsic lower limit to how thin the feed can be, very viscous feedstocks are difficult to thicken, probably because the polymer does not mix well with viscous sludge.

It is also worth noting that changes in the feed rate will change the crest height over the dams and, thus, the differential head pressure. As a result, a modest change in feed rate may cause an abrupt change in process performance. It is common to run thickening centrifuges at a constant rate, and to add or drop centrifuges offline as needed to keep up with plant production. If the feed rate does change, then it is likely that the pond will have to be changed as well.

Bowl Speed. The manufacturer generally sets the bowl speed and it is rarely changed thereafter. Assuming the present speed was the correct speed several years ago is not proof that it is the best speed now. The owner should adjust the speed periodically, if only to confirm that it is correct and to remind operators that it is, indeed, a variable. It is good policy to consult the manufacturer before the bowl speed is changed.

Pond Depth (Weir Adjustment). Thickening centrifuges react to very small changes in pond height. A fair size change in a pond is 1.5 mm (1/16 in.). The goal is to set the pond so that good operation occurs with the differential revolutions per minute somewhere in its mid-range. If the differential revolutions per minute is ±5 rpm, then the pond should be reduced slightly to bring the differential revolutions per minute up to 9 to 15 rpm. Changing the feed rate substantially will require a change in pond setting. For example, increasing the feed rate will increase ΔH, which, in turn, will thin out the thickened cake. If reducing the differential revolutions per minute does not correct the problem, then the pond will have to be reduced to reduce the differential head pressure. There is one commercial design available to allow the operator to change the pond setting while the centrifuge is online. Westfalia Separator (Oelde, Germany) has a patented device called Vari-pond™. The principle is simple: Three screw motors push a nonrotating plate against the liquid discharge. As the plate nears the rotating centrifuge, the diminishing gap restricts the flow of centrate, much as restricting any weir causes the liquid level behind the weir to rise. While pond and differential revolutions per minute are, to some extent, interchangeable, in fact, changing the pond causes an immediate reaction, whereas changing the differential revolutions per minute results in a slow change in process.

Polymer Dose. Using polymer, the centrifuge can achieve very good capture of solids, which reduces the recycle of solids within the plant. It is often difficult to put a cost on recycle, but a reasonable figure is what the plant charges an industrial customer for sending suspended solids to the plant. Increasing the polymer dosage increases both the capture and the cake solids. To maintain a particular cake solids, the differential revolutions per minute or the pond must be increased at the same time. Experienced operators can change the dosage and the differential at the same time, but less experienced operators will have to go back and forth to balance everything out. As a result, it is difficult to have automatic polymer control without also having automatic cake thickness control.

START-UP. Basic start-up sequence is more or less the same as for dewatering centrifuges. Most modern centrifuges have a one-button start. Manual systems take a few minutes, but are not onerous. When the centrifuge is up to speed, the controls unlock the feed and polymer pumps, and the operator begins to put the centrifuge online. The start-up sequence is as follows:

- Turn on the feed and polymer to about one-third of the normal rate.
- Reduce the differential revolutions per minute and/or pond to minimum.
- When the cake thickness reaches normal, begin increasing the differential and the polymer feed rate. Some plants can jump directly to the normal operating conditions as soon as the cake is sealed, while others have to ramp up more slowly.

SHUTDOWN. The basic shutdown sequence is similar to that of dewatering centrifuges. Again, modern centrifuges have a one-button stop. The shutdown sequence is as follows:

- Shut off the feed and polymer and turn the flushing water on.
- When clear water exits both ends of the centrifuge, push the centrifuge stop button.
- At some point, as the centrifuge slows down, flush water will come around the feed tube or around the casing seals. Note how long it took between engaging the stop button and the water gushing out. Next time, shut the water off a minute or two sooner.
- With the flush water off, the centrifuge can usually coast to a stop without operator intervention.

SAMPLING AND TESTING. Sampling and testing should include TSS and/or total solids for the feed, total solids for thickened sludge, and TSS, ammonia, and/or phosphorus (under some conditions) for centrate.

TROUBLESHOOTING, PREVENTIVE MAINTENANCE, AND SAFETY.

There are no problems or procedures inherent to centrifuges placed in thickening applications that are different for dewatering applications. Please see sections on dewatering centrifuges in this manual for these topics.

ROTARY DRUM THICKENING

DESCRIPTION OF PROCESS. Rotary drum thickeners use rotating drums of varying sizes to capture solids with wedge wires, perforated holes, stainless steel fabric, or a combination of stainless steel and synthetic fabric. The drum has either a center shaft mounted on a steel frame or four trunnion wheels supporting its outside perimeter. A variable-speed drive unit rotates the drum at approximately 5 to 20 rpm.

Conditioned sludge enters the drum and free water (filtrate) drains through the openings into a trough underdrain. A continuous internal screw or diverted angle flights convey sludge along the drum length to the exit through the discharge chute. Wash water, periodically applied to the inside and outside of the drum, cleans solids from the screen openings. This periodic washing helps maintain high solids capture and dewatering efficiency.

DESCRIPTION OF EQUIPMENT. A rotary drum thickener, shown in Figure 29.19, Figure 29.20, and Figure 29.21, consists of an internally fed rotary drum with an internal screw, lubricated trunnion wheels, a variable- or constant-speed drive, an inlet pipe, a filtrate collection trough, a discharge chute, and an optional drum cleaning rotating brush. Figure 29.21 shows the wash water system, while Figure 29.20 shows the gear box and drive assembly.

Auxiliary equipment includes sludge feed pumps—either thickened sludge pumps or a conveyor belt—and a polymer mixing and feed system.

START-UP. Ensure that all guards and covers are properly secured on the machine and that lockout/tagout is not in place. The following sequences apply to starting up the equipment after a relatively short shutdown period:

- Inspect the unit and ensure that all of the guards and screen wash covers are in their proper place. Visually inspect the interior of the screen(s) from the discharge side without touching the equipment or getting close to any rotating assembly. Look for tools, dried sludge, or other foreign objects in the screen of influent troughs.

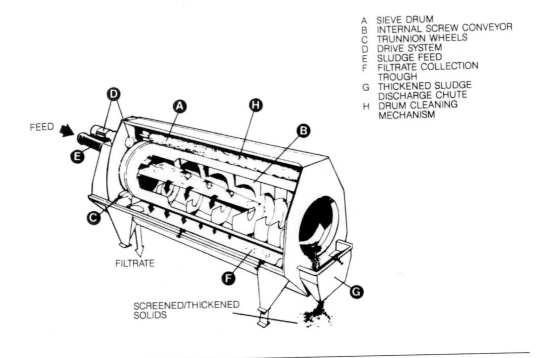

FIGURE 29.19   Drum screen thickener.

- Visually check the chain oilers and ensure that lubricant is present.
- Ensure that the floc or mixing chamber drain is open (back to the head end of plant).
- Start the spray washes.
- Start the drum rotation.
- Start the polymer pump and the dilution system after ensuring that a fresh polymer batch is prepared.
- Immediately after polymer flow begins to enter the floc tank, start the sludge flow. Operate both at 25% of normal flow until stable.
- As soon as sludge flow starts entering the floc tank, close the drain and start the mixer, if applicable. (Most floc tanks do not have mechanical mixers; they have static mixers.)
- Observe and adjust polymer and sludge flow as necessary to achieve desired results. Ramp the flowrate upward until you achieve desired results.
- Adjust drum speed as necessary.

FIGURE 29.20   Rotary drum thickener (side panel removed).

SHUTDOWN. The procedures described herein should be followed for a typical shutdown. It is important to note that the units can easily be stopped on an emergency basis and cleaned soon thereafter. Shutdown procedures are as follows:

- Shut off the sludge and polymer flow.
- Flush the sludge feed piping with nonpotable plant water and allow the polymer dilution water to continue for several minutes until flushed of polymer.
- Stop the flushing water and polymer dilution water when the screens are clear of any thickened sludge.
- Stop the flushing water and open the drain on the flocculation tanks.
- You may need to use a hose to flush down the inside of the screen from the effluent end. Be careful to stand clear of the rotating equipment.
- After the screens are clear, shut down the spray water, drum drive, and all other power to the equipment. Secure all valves to ensure that no sludge or polymers are continuing to drain the WWTP.

FIGURE 29.21   Rotary drum thickener (side panel removed).

PROCESS CONTROL. The operator may use four variables—sludge feed rate, polymer feed rate, pond depth, and drum speed—to optimize the thickening operation. Each of these variables is discussed here.

Sludge Feed Rate. Sludge feed rate, the wasting rate for the waste stream, varies from day to day. The operator determines the wasting rate each day and sets the feed system to achieve the desired results.

Polymer Feed Rate. This rate depends on the sludge feed rate and on the concentration, volatility, and settleability of the influent. Expressed as pounds of polymer per pound of dry solids, the required polymer feed rate increases as the volatile solids and the SVI increase.

Pond Depth. The incline angle of the drum controls the pond depth. The angle can be adjusted from horizontal to approximately 6 degrees above the horizontal. Increasing the angle results in a drier solids product, but also reduces the drum capacity. Decreasing the angle increases the system throughput capacity, but results in a wetter solids product. The appropriate incline angle for specific applications depends on determinations on-site. A typical starting angle for new installations ranges between 2 and 3 degrees above the horizontal.

The operator can vary drum speed in response to increasing or decreasing feed concentrations to maintain the required product thickness. The subsequent troubleshooting guide indicates how drum speed should change in response to various operational problems.

SAMPLING. At a minimum, a rotary drum thickener should be sampled and analyzed as follows:

- Sample influent for total solids and flow.
- Sample filtrate for TSS and flow.
- Sample thickened sludge for total solids and flow.
- If applicable, determine the total solids of the polymer or whatever quality assurance tests you deem necessary. Measure flow and quantity used (liquid or solid).
- Sample MLSS for SVI. Although this test is not directly with the rotary drum thickener, the SVI of the MLSS ("settling health") has a direct effect on centrifuge performance. The higher or poorer the SVI, the poorer the performance.

TROUBLESHOOTING. Table 29.13 presents a troubleshooting guide that identifies problems and presents possible solutions.

PERFORMANCE. Typical expected performance ranges are presented in Table 29.14. Table 29.15 shows typical hydraulic flow ranges for rotary drum thickeners. Performance data for rotary drum thickeners operating in the United States are contained in Table 29.16.

In each of the WWTPs, the operators advised that someone attend to the drum thickener whenever it was operating. This allows continual adjustment of the unit to handle fluctuating influent characteristics, thereby maintaining best performance. Such attention could require the addition of up to five full-time employees if the thickener operates on a 24-hour-per-day basis. Because all of the installations in Table 29.16 are fairly new, future operations may involve less time for thickener adjustment

TABLE 29.13   Troubleshooting guide for rotary drum thickeners.

Problem	Probable cause	Solutions
No floc formation in mixing tank	Not enough polymer	Increase polymer dosage or decrease sludge throughput
	Old nonactive polymer	Make up a new batch
	Wrong polymer	Complete a bench polymer selection test
	Reaction time of polymer and sludge too short	Inject polymer farther upstream before sludge pump
	Too much polymer	Increase sludge throughput and decrease polymer
Good floc formation in mixing tank but slurry exits sloppily	Drum clogged	Open wash water valve
		Turn on booster pump
		Clean wash water and brush cleaning system
	Sludge throughput too high	Decrease sludge throughput
		Increase rpm of drum
Filtrate dirty	Floc unstable	Increase or decrease polymer
		Move polymer injection point closer or into the mixing tank
	Floc too small	Same as above
		Decrease mixer speed
	Too much turbulence in drum	Sludge throughput too high
		Decrease drum speed
Sludge spills out inlet of drum	Drum rotation too slow	Increase drum rotation
	Sludge throughput too high	Decrease sludge throughput
	Dewater screens are blinded	Clean wash water system
		Adjust frequency of washing cycle

after more experience is gained in judging thickener reaction to changing influent characteristics.

PREVENTIVE MAINTENANCE. Maintenance items or considerations are as follows:

- Daily:
  - Check the drive units for overheating, unusual noise, or excessive vibration; and
  - Check for unusual conditions, such as loose screens or side covers;

TABLE 29.14   Typical rotary drum thickener performance ranges.

Type of sludge	Feed, % TS[a]	Water removed, %	Thickened sludge, %	Solids recovery, %
Primary	3.0–6.0	40–75	7–9	93–98
WAS[b]	0.5–1.0	70–90	4–9	93–99
Primary and WAS	2.0–4.0	50	5–9	93–98
Aerobically digested	0.8–2.0	70–80	4–6	90–98
Anaerobically digested	2.5–5.0	50	5–9	90–98
Paper fibers	4.0–8.0	50–60	9–15	87–99

[a]Total solids.
[b]Waste activated sludge.

TABLE 29.15   Typical flow ranges for rotary drum thickeners.

Drum diameter, m	Hydraulic loading range,[a] gpm[b]
0.6	35–90
1.5	180–240

[a]Assumes a 1% feed concentration of municipal sludge. It is important to note that variations in feed concentrations, polymer dosage, differential speed, and pond depth will act to increase or decrease the hydraulic loading rates for any given machine.
[b]gpm $\times$ (6.308 $\times$ 10^{-5}) = m^3/s.

TABLE 29.16   Rotary drum thickener performance data.

Drum size, m	Influent flowrate, gpm[a]	Influent solids concentration, %	Effluent solids concentration, %	Polymer costs, $/dry ton[b]
0.6[c]	Average = 42 Range = NA	Average = 1 Range = 1–2	Average = 6 Range = 5.5–6	Range = 11–12
1.5[d]	Average = 240 Range = 130–300	Average = 0.9 Range = 0.001–1.5	Average = 5.5 Range = 4–9	Range = 17–18

[a]gpm $\times$ (6.308 $\times$ 10^{-5}) = m^3/s.
[b]ton $\times$ 1.016 = Mg.
[c]Information based on one unit of this size that has since been taken out-of-service due to a change in plant operation.
[d]Informattion based on four units currenttly in operation.

- Weekly:
  - Flush drum screen,
  - Lubricate,
  - Check flush nozzles (clear as required), and
  - Check oil levels;
- Monthly:
  - Check brush,
  - Check reduction motor's oil level,
  - Check level in lubrication unit's grease tank,
  - Check screen casing, and
  - Check drive system; and
- Semiannually:
  - Check all connections and
  - Check brush bearings.

SAFETY. As with any process for handling sludge, hydrogen sulfide may be present with attendant odors and toxic effects. Safety precautions must be followed for proper ventilation, monitoring, and other procedures, as discussed in Chapter 5. The following general safety procedures may apply to the operation and maintenance of rotating thickeners:

- All lubrication and maintenance must be done while equipment is not in operation to prevent accidental start-up. Lockout is a method of keeping equipment from being set in motion and endangering workers. The *Code of Federal Regulations* (1991) 29 *CFR* Part 1910.147 covers the servicing and maintenance of machines and equipment in which the unexpected power-up or start-up of the equipment, or release of stored energy, could cause injury to employees. Tagout occurs when the energy isolation device is placed in a safe condition and a written warning is attached to it. Typically, the employee signs and dates the tag. Removal of the tag and lockout device is done by the same individual when the equipment is safe for restart.
- Sludge and polymers are extremely slippery when spilled; therefore, all spills must be cleaned immediately. Installation of a grated, elevated walkway can alleviate slippery conditions around polymer tanks.
- Adequate ventilation and gas testing should be provided when processing sludge.
- Consideration can be given to the addition of potassium permanganate or peroxide to the feed sludge if hydrogen sulfide (odor) levels are high.

- Never insert fingers, arms, or feet into the unit until lockout/tagout is complete and the unit is secured for maintenance.
- Be especially careful of pinch-points such as sprockets and chains on roller casters. Severe injury may occur if a finger or hand is caught in rotating equipment.
- Do not wear jewelry or loose or bulky clothing near the machine.
- Do not insert items such as tools, samplers, or probes into the drum unless lockout/tagout is complete and the unit is secure for maintenance.

REFERENCES

American Public Health Association; American Water Works Association; Water Environment Federation (2005) *Standard Methods for the Examination of Water and Wastewater*, 21st ed.; American Public Health Association: Washington, D.C.

Bratby, J.; Marais, G. v R. (1975) Saturator Performance in Dissolved-air Flotation. *Water Res.*, **9**, 929–936.

Bratby, J.; Marais, G. v R. (1977) Flotation. In *Solid/Liquid Separation Equipment Scale-Up*, Purchas, D. B., Ed.; Uplands Press Ltd.: London, England, pp. 155–198.

Bratby, J. (1978) Aspects of Sludge Thickening by Dissolved-air Flotation. *Water Pollut. Control*, **77** (3), 421–432.

Bratby, J.; Ambrose, W. (1995) Design and Control of Flotation Thickeners. *Water Sci. Technol.*, **31** (3–4), 247–261.

Bratby, J.; Jones, G.; Uhte, W. (2004) State-of-Practice of DAFT Technology—Is There Still a Place For It? Paper presented at the Water Environment Federation's 77th Annual Technical Exposition and Conference, New Orleans, Louisiana, October 2–6; Water Environment Federation: Alexandria, Virginia.

Code of Federal Regulations (1991) 29 *CFR* Part 1910.147.

Metcalf and Eddy, Inc. (2003) *Wastewater Engineering*, 4th ed.; McGraw-Hill, Inc.: New York.

U.S. Environmental Protection Agency (1978) *Field Manual for Performance Evaluation and Troubleshooting at Municipal Wastewater Treatment Facilities*; U.S. Environmental Protection Agency: Washington, D.C.

Water Pollution Control Federation (1980) *Sludge Thickening*; Manual of Practice No. FD-1; Water Pollution Control Federation: Washington, D.C.

Water Pollution Control Federation (1987) *Operation and Maintenance of Sludge Dewatering Systems*; Manual of Practice No. OM-8; Water Pollution Control Federation: Alexandria, Virginia.

Chapter 30

Anaerobic Digestion

PROCESS OVERVIEW

GENERAL THEORY. Anaerobic digestion is a multistage biochemical process that can stabilize many different types of organic material. Digestion occurs in three basic stages (Zehnder, 1978). In the first stage, extracellular enzymes (enzymes operating outside the cells) break down solid complex organic compounds, cellulose, proteins, lignins, and lipids into soluble (liquid) organic fatty acids, alcohols, carbon dioxide, and ammonia. Complex organic materials in the digester feed include primary solids, microbes grown in the aerobic stages of the liquid treatment process, and colloidal material. In the second stage, microorganisms (often referred to as *acetogenic bacteria* or *acid formers*) convert the products of the first stage into acetic acid, propionic acid, hydrogen, carbon dioxide, and other low-molecular-weight organic acids. In the third stage, two groups of methane-forming bacteria work—one group to convert hydrogen and carbon dioxide to methane and the other group to convert acetate to methane and bicarbonate (carbon dioxide in solution). Because both groups of bacteria are anaerobic, the digesters are sealed to exclude oxygen from the process. The three stages are summarized schematically in Figure 30.1.

In most cases, methane-forming bacteria control the process. Methane formers are very sensitive to environmental factors (high ammonia concentrations, low phosphorus concentrations, low pH, temperature, and the presence of toxic substances), and reproduce very slowly. Consequently, methane formers are difficult to grow and are easily inhibited. Therefore, process design and the operation of conventional anaerobic digestion are tailored to satisfy the needs of the methane-forming bacteria.

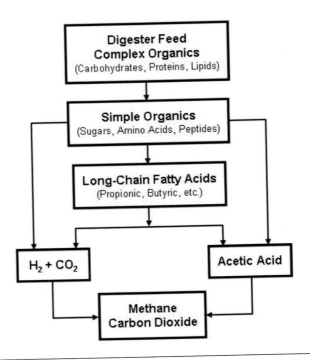

FIGURE 30.1 Biochemical processes in anaerobic digestion.

CONVENTIONAL MESOPHILIC DIGESTION. The majority of anaerobic digestion systems currently in use are configured as conventional mesophilic digesters. In these systems, all stages of the biochemical process occur in the same vessel and are operated at mesophilic temperatures [32 to 38 °C (90 to 100 °F)]. Conventional systems can be categorized as low-rate (no mixing) or high-rate processes, which include mixing and heating. The heating and mixing used in the high-rate processes produce uniform conditions throughout the tank, which results in shorter detention time and more stable conditions than low-rate processes. Consequently, most municipal digestion systems use the high-rate process.

Anaerobic digestion stabilizes solids by reducing the mass of volatile solids (VS), typically by 40 to 60%. Volatile solids reduction (VSR) can be calculated based on the mass of volatile solids in the feed and digester discharge using Equation 30.1.

$$VSR, \text{percent} = \left(\frac{VS_{\text{feed}} - VS_{\text{digested sludge}}}{VS_{\text{feed}}} \right) \times 100 \qquad (30.1)$$

Several methods of calculating VSR are presented and discussed in the U.S. Environmental Protection Agency (U.S. EPA) publication, *Control of Pathogens and Vector Attraction in Sewage Sludge* (U.S. EPA, 1999). These include the Approximate Mass Balance (AMB) method and the Van Kleeck equation. The AMB method assumes that the daily flows to the digester are steady and uniform in composition. It can be calculated using Equation 30.2:

$$\text{VSR, percent} = \left[\frac{(Q_{feed} \times VS_{feed}) - (Q_{digested\ sludge} \times VS_{digested\ sludge})}{Q_{feed} \times VS_{feed}} \right] \times 100 \qquad (30.2)$$

where

$$Q_{feed} = \text{Volumetric flowrate of the feed sludge,}$$
$$Q_{digested\ sludge} = \text{Volumetric flowrate of the digested sludge,}$$
$$VS_{feed} = \text{Volatile solids concentration of the feed sludge, and}$$
$$VS_{digested\ sludge} = \text{Volatile solids concentration of the digested sludge.}$$

The Van Kleeck method can be used when the digesters have no significant grit accumulation. It can be calculated using Equation 30.3. With the exception of the calculated VSR, all values in Equation 30.3 represent fractional volatile solids.

$$\text{VSR, percent} = \left[\frac{VS_{feed} - VS_{digested\ sludge}}{VS_{feed} - (VS_{feed} \times VS_{digested\ sludge})} \right] \times 100 \qquad (30.3)$$

Digesters are sized to provide sufficient detention time to allow stabilization. High-rate digesters are typically sized for an average solids retention time (SRT) of 15 to 20 days. Slightly shorter detention times (on the order of 12 days) are often used in European designs. Solids retention time is calculated based on the mass of solids in the digester and the mass of digested sludge removed daily. Hydraulic retention time (HRT) is calculated based on the volume of the digester and the volume of digested sludge removed daily. For systems that do not practice decanting, digester SRT and HRT are the same.

ADVANCED DIGESTION PROCESSES. Advanced digestion processes involve modifications to the conventional digestion configuration to achieve more complete digestion, promote pathogen reduction, and improve digester operation. Nearly all advanced digestion processes have reported better VSR than that achieved by conventional mesophilic digestion (Schafer et al., 2002), which increases the content of soluble nutrients (ammonia and phosphorus) in dewatering recycle streams and, thus, can increase the nutrient loads to the wastewater treatment system.

Acid/Gas Digestion. Acid/gas digestion, or two-phase digestion, is carried out in a two-reactor system to provide separate environments for the acid-forming and methane-forming bacteria so each can be optimized for the specific process. In the first phase, the feed substrates are hydrolyzed to produce volatile fatty acids (VFAs), which are converted to methane and carbon dioxide in the second phase. The acid phase is typically limited to an SRT of 1.5 to 2 days; the gas phase requires a minimum SRT of 10 to 15 days. Acid/gas systems have been used to process "difficult-to-digest" solids—including all waste activated sludge (WAS) or mostly WAS feed solids—without significant foaming.

Thermophilic Digestion. Thermophilic digestion processes include one or more stages that are operated at thermophilic temperatures [55 °C (131 °F) or higher]. The main goal of thermophilic treatment is to achieve greater pathogen destruction; however, it can also increase volatile solids destruction and decrease required detention times. Thermophilic digestion can affect the odor characteristics of the digested sludge, thereby increasing the odor potential during dewatering and cake loading. Thermophilic digestion may also affect the dewaterability of the digested sludge; however, the reported results vary. Some installations report using higher polymer dosages with thermophilic digestion, while others have reduced polymer use and increased cake solids (Haug et al., 2002). Because heating to thermophilic temperatures requires significantly more energy than needed for mesophilic digestion, heat recovery is an important step in thermophilic digestion. With the aid of heat exchangers, heat discharged from the thermophilic digester can be used to preheat the digester feed.

MULTISTAGE DIGESTION.

Multistage digestion includes various combinations of mesophilic and thermophilic treatment. Temperature-phased anaerobic digestion (TPAD) consists of at least one thermophilic stage followed by a mesophilic "polishing" stage. The objective of TPAD is to take advantage of thermophilic treatment (increased pathogen destruction and increased VSR), while including a mesophilic stage to alleviate the odor and dewaterability concerns associated with high-temperature digestion. Some multistage digestion processes also include acid/gas treatment, using the thermophilic stage with either the acid or gas phase. Several of these processes are offered as proprietary systems, including 2PAD™ by Infilco Degremont, Inc. (Richmond, Virginia) and Enzymic Hydrolysis by Monsal (United Kingdom).

PRETREATMENT PROCESSES FOR DIGESTION.

Pretreatment is an "add on" to conventional anaerobic digestion, with the objective of improving digester performance, VSR, and pathogen destruction and reducing foaming potential. Treatment typically includes application of energy in the form of ultrasound, heat, pressure, or a combination of these. A number of the pretreatment processes are proprietary.

Ultrasound Treatment. Ultrasound treatment (sonication) is used to condition WAS prior to digestion, with the goal of improving digestion and VSR and decreasing digester foaming. Ultrasound is applied to WAS by a series of probes installed in a flow-through vessel. The resulting ultrasonic waves cause cavitation, generating microbubbles that collapse and release energy. The released energy can destroy the cell structure of the microbes in the WAS. Sonication can be used to treat the entire WAS flow or only a portion of it.

Thermal Hydrolysis. Thermal hydrolysis is used to increase pathogen kill and to condition the solids for better digestion by heating raw solids under pressure for a short period. Thermal pretreatment improves the dewaterability of the digested solids by releasing the water bound in the bacterial cells. Thermal hydrolysis processes are mechanically complex and produce offgases that are a significant source of odors. Thermal hydrolysis processes are designed for temperatures of 150 to 220 °C (302 to 428 °F) and pressures of approximately 1380 kPa (200 psi). Reaction times vary from 20 to 30 minutes. A number of thermal hydrolysis processes are available that, depending on the system, can be used on thickened or dewatered sludge.

   Thermal hydrolysis processes typically include preheating sludge through a heat exchanger before it enters the reaction tank. Steam is added to the reaction tank to maintain temperature and increase pressure. On completion of the batch reaction, the solids are depressurized and pumped through heat exchangers to lower their temperature prior to mesophilic digestion. A schematic of one type of thermal hydrolysis process, the Cambi® (Norway) system, is presented in Figure 30.2.

   Wet air oxidation is a type of thermal hydrolysis in which oxygen (in the form of air or pure oxygen) is added to solids prior to heating and reacting. Wet air oxidation is available via several proprietary processes, including Zimpro® by Siemens (Warrendale, Pennsylvania).

Pasteurization. Pasteurization involves heating solids to a temperature of 70 °C (158 °F) or higher (typically for at least 30 minutes in either a batch or continuous plug-flow operating mode) to kill pathogens. It does not necessarily affect the digestibility of the sludge. Multiple heat exchangers are typically used to preheat incoming sludge and capture heat from the treated sludge. The solids concentration of the feed sludge can affect the heat exchanger's ability to reach target temperatures. A number of proprietary pasteurization systems are available, including Alpha-Biotherm, BioPasteur, Roediger™, and Eco-Therm™.

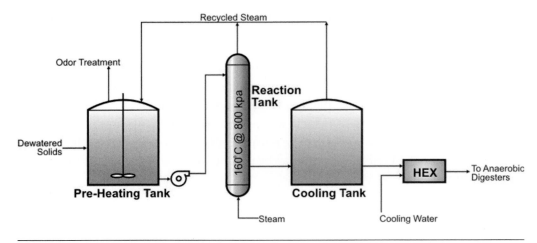

FIGURE 30.2   Thermal hydrolysis process.

Homogenization. Homogenization is used to condition WAS prior to digestion. It is a multistep process that lyses the bacterial cell walls to promote VSR during anaerobic digestion. One such process, offered by MicroSludge™, involves the use of a caustic solution to weaken the cell walls and lower the viscosity of thickened WAS. The treated WAS passes through a screen to remove large debris and enters a high-pressure homogenizer, where it is subjected to a sudden drop in pressure that lyses the cells.

HIGH SOLIDS-CONCENTRATION DIGESTION. Anaerobic digestion is typically used for feed sludge with a total solids concentrations of 3 to 5%. However, it has also been used on feed concentrations of 7 to 8% total solids at a number of facilities, mostly in Europe. The digesters in these facilities are typically equipped with either roof-mounted mechanical mixers or pumped mixing systems with internal nozzles. Digestion of high-solids feed can produce foam; therefore, antifoaming systems are usually included. The Torpey and recuperative thickening processes are types of high- solids digestion in which a portion of the digested sludge is recirculated to a sidestream or upstream thickener.

High solids-concentration digestion can reduce the needed digester volume or effectively increase the capacity of an existing digester. Thickening the feed to 6 or 7% total solids can significantly increase its viscosity, as well as change its mixing characteristics. Pumping and mixing systems designed to support digestion with feed concentrations of 3 to 5% total solids may not support digestion at significantly higher feed concentrations. Consequently, all components of the digestion system should be designed to support a high solids-concentration process. The process also effectively increases the volatile solids loading on the digester and, because high-solids digestion concentrates feed constituents, toxicity effects can be exacerbated.

Regulatory Issues. Land application of biosolids is governed by U.S. EPA's 40 *CFR* Part 503 (U.S. Code of Federal Regulations, 1993). While many of the requirements of this regulation, such as site management and application rates, are outside the scope of the anaerobic digestion process, the requirements pertaining to pathogen and vector attraction reduction can be achieved via digestion.

CLASS A AND CLASS B PATHOGEN REDUCTION.
U.S. EPA's 40 *CFR* Part 503 establishes two levels of pathogen reduction requirements: Class A and Class B. The objective of Class A, which is the more stringent requirement, is to reduce the density of *Salmonella* sp. bacteria, enteric viruses, and viable helminth ova to below detectable levels. Biosolids that meet Class A criteria are subject to few restrictions; in most localities, they can be given or sold to the public or directly land-applied by the utility. The regulations identify six alternatives to meet Class A Criteria, each of which requires that the treated biosolids meet either the following fecal-coliform or *Salmonella* sp. density at the time they are used, sold, or given away:

- Fecal coliform—fewer than 1000 most probable number (MPN) per gram of total dry solids and
- *Salmonella* sp.—fewer than 3 MPN per 4 g of total dry solids.

Of the six treatment alternatives, five may be applicable to digestion processes and are described herein; more detailed information is available in *A Plain English Guide to the EPA Part 503 Biosolids Rule* (U.S. EPA, 1994). It should be noted that to meet Class A requirements, pathogen reduction must be achieved before or concurrently with vector attraction reduction.

- *Alternative 1—Thermally Treated Biosolids.* This alternative applies to treatment systems that use heat to reduce pathogens. Several equations are used to determine the required heating time, depending on the treatment temperature and solids concentration.
- *Alternative 2—Biosolids Treated in a High-pH/High-Temperature Process.* This alternative does not apply to anaerobic digestion.
- *Alternative 3—Biosolids Treated in Other Known Processes.* This alternative is used to demonstrate that a new treatment process meets Class A pathogen reduction criteria. It requires monitoring of viruses and helminth ova until it is shown that

the treatment process consistently reduces pathogens to Class A levels. Once the process has been demonstrated to meet Class A requirements, the process parameters must be monitored to show consistent operation.

- *Alternative 4—Biosolids Treated by Unknown Processes.* This alternative differs from Alternative 3 by requiring pathogen testing (enteric viruses and helminth ova) for each batch of biosolids, rather than monitoring the process parameters. This alternative is used in situations where the actual treatment process is unknown or does not fulfill other Class A process requirements.
- *Alternative 5—Biosolids Treated by Processes to Further Reduce Pathogens (PFRP).* This alternative identifies seven PFRP that meet Class A requirements. None specifically apply to anaerobic digestion; however, pasteurization (one of the listed technologies) can be used as predigestion treatment to meet Class A criteria.
- *Alternative 6—Biosolids Treated in a PFRP Equivalent.* This alternative allows permitting authorities determine whether other treatment processes are equivalent to PFRP technologies, either for a specific installation or on a nationwide basis.

Class B requirements establish pathogen limits that are less stringent than Class A limits. Because Class B biosolids are expected to contain significantly higher concentrations of pathogens than Class A biosolids, they cannot be sold or given away and application-site restrictions apply. The following three treatment alternatives meet Class B pathogen criteria:

- *Alternative 1—Monitoring of Indicator Organisms.* The fecal coliform density must be less than 2 million MPN or 2 million colony forming units (CFU) per gram of total solids, based on the geometric mean of seven samples.
- *Alternative 2—Biosolids Treated by Processes to Significantly Reduce Pathogens (PSRP).* This alternative includes five PSRP, of which only one applies to anaerobic digestion. This process requires a minimum mean cell residence time of 15 days in the anaerobic digester at a temperature of 35 to 55 °C (95 to 131 °F).
- *Alternative 3—Biosolids Treated in a PSRP Equivalent.* This alternative allows permitting authorities to determine whether other treatment processes are equivalent to PSRP technologies, either for a specific installation or on a nationwide basis.

VECTOR ATTRACTION REDUCTION REQUIREMENTS. Both Class A and Class B biosolids must meet vector attraction reduction (VAR) requirements. These requirements are intended to reduce the putrescibility of the solids. Highly pu-

trescible solids will tend to attract vectors, such as flies and rodents. There are 11 VAR options for biosolids. The following two options apply to anaerobic digestion:

- *Option 1—Achieve 38% Reduction in Volatile Solids Content.* The VSR can be met through digestion and any additional reduction that occurs before the biosolids leave the treatment plant.
- *Option 2—Additional Digestion of Anaerobically Digested Biosolids.* In cases where the organic material in the feed solids has undergone some stabilization preceding anaerobic digestion, it may be difficult to show an additional 38% VSR. Under this option, a bench-scale anaerobic digester can be used to show compliance with VAR. The biosolids meet the VAR requirement if the VSR is less than 17% during a 40-day bench-scale test at a temperature between 30 and 37 °C (86 and 99 °F).

DIGESTER FACILITIES AND EQUIPMENT

Anaerobic digestion facilities can be constructed using a variety of tank configurations and equipment types.

TANK CONFIGURATION. The most common type of digester currently in use in North America is the cylindrical tank, also known as a "pancake" digester. Cylindrical tanks typically have cone-shaped floors, with slopes of 1:4 to 1:6 to facilitate collection and removal of heavy sludge and grit. Cylindrical digesters, which are typically constructed of concrete, can be equipped with gas-holder covers to provide storage for digester gas. Cylindrical digesters in Europe are often constructed with 1:1 aspect ratios.

Egg-shaped digesters (ESDs) have bottoms with steeper slopes than the cylindrical tanks (typically, slopes that are at least 1:1). They may be constructed of concrete or steel and are available in several variations [Figure 30.3(a) and (b)]. Common features of ESDs include a conical bottom and a domed top. The advantages of ESDs include a reduced potential for grit accumulation because of their steep bottom slopes, a geometry that allows for more efficient mixing and thus reduces energy use, and less potential for scum accumulation owing to the small liquid surface area at the top of the egg-shaped vessel. The disadvantages of ESDs include lack of capacity for gas storage within the tank and a shape that is more difficult to insulate. In addition, their relatively high profile can cause aesthetic concerns in some locations. Egg-shaped digesters are in widespread use in Europe and are gaining popularity in the United

FIGURE 30.3A   Egg-shaped digester.

FIGURE 30.3B   Egg-shaped digester.

States. While construction costs of ESDs can be significantly higher than those of cylindrical tanks, these costs can be offset by lower operating costs, which are attributable to more efficient mixing and the reduced need for cleaning.

Imhoff tanks are two-chamber digestion vessels consisting of an upper chamber for solids sedimentation and collection and a lower chamber where the settled solids are anaerobically digested. Imhoff tanks are an older technology that is now rarely seen in municipal wastewater treatment applications.

Lagoon digestion systems consist of earthen lagoons fitted with floating membrane covers that are used to contain and collect the digester gas. The lagoons are not heated and are typically not mixed. Because they are not heated, the digestion rate is variable. Consequently, lagoons are designed to provide a long sludge detention time. Lagoons are frequently used in industrial waste applications; their use in municipal wastewater treatment is less common.

SECONDARY DIGESTERS. Most medium-to-large digestion facilities include both primary and secondary digesters. Most of the stabilization and gas production occurs in the primary digester. The digestion system design loading rates and SRT are based on primary tank volume; however, if secondary digesters are heated and mixed, their active volumes can be included when calculating operating loading rates and SRT.

While primary digesters are mixed and heated to optimize digestion and meet pathogen-reduction requirements, secondary digesters may or may not be mixed and heated. Digester-heating systems are often configured to use primary digester heat-exchanger equipment to provide heat to a secondary digester when a primary digester is out of service.

An unheated secondary digester may serve the following purposes: as a storage vessel for digested solids, as a standby primary tank, and as a source of seed sludge. In addition, at installations where the digester is decanted to increase the solids concentration, the secondary digester can be used as a quiescent basin for supernatant withdrawal.

Because an important purpose of secondary digesters is their use as storage vessels, they typically have floating or membrane covers to allow drawdown. The variable storage capacity in the secondary digester is limited by the cover travel. In floating or gas-holder covers, solids withdrawal should cease before the cover rests on the corbels. Further removal will cause a vacuum to develop beneath the cover, which can cause air to be drawn into the digester and create a potentially explosive condition. Membrane covers can allow significantly greater drawdown, depending on the configuration of the installed mixing equipment.

DIGESTER COVERS. Digester covers keep oxygen out of the anaerobic environment in the digester. Covers also prevent digester gas and odors from escaping to the atmosphere, reduce the explosion hazard associated with the methane in the digester gas, and insulate the top of the digester. There are four basic types of digester covers: fixed, floating, gas-holder, and membrane (Figure 30.4). The features of the different types of covers are described in Table 30.1.

Fixed Covers. Fixed covers are available in several types of construction. Reinforced concrete covers may be flat-slab shaped or dome-shaped and may be self-spanning or column-supported with material of varying thicknesses, depending on the design.

Some fixed covers are constructed of steel plates welded to the upper chord of a truss or to an arch-rib supporting framework. The supporting framework is connected to the top of the wall by a sliding arrangement that allows the cover to expand or contract in response to changes in temperature.

Fixed-cover digesters require operator attention during solids addition and withdrawal. If the digester is full, the incoming flow must equal the outgoing flow. Adding solids without withdrawing an equal volume will cause overpressure, which can lift the cover from its wall mountings. Withdrawal of solids without a corresponding

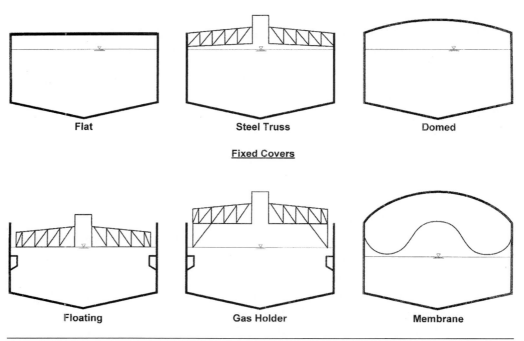

FIGURE 30.4   Digester covers.

TABLE 30.1 Digester cover features.

	Fixed cover	Floating cover	Gas holder cover	Membrane cover
Drawdown capability	Not recommended	Yes	Yes	Yes
Gas storage	No	No	Yes (limited)	Yes
Odor potential	Low	Moderate	Low to moderate	Low
Recommended use	Primary digesters	Primary or secondary digesters	Primary or secondary digesters	Primary or secondary digesters

addition will create a vacuum that can damage the covers or cause them to collapse. Both overpressurization and vacuum conditions can be alleviated by installing safety relief valves; however, even with the relief valves, fixed-cover systems often develop gas leaks at the interface of the cover and the digester wall.

Floating Covers. Floating covers float directly on the liquid surface, supported by a system of roller bearings and guide rails along the digester wall that controls both vertical and lateral movement of the cover and prevents excessive tilting. The guide rail systems are available in vertical or spiral configurations; while vertical systems are less complex, the spiral systems provide more efficient control of the cover's vertical movement. The range of vertical motion of floating covers is typically 1.5 to 3 m (5 to 10 ft), with a corresponding capacity for solids drawdown.

Floating covers offer several advantages over fixed covers, the most important of which is the ability to vary the liquid level in the digester without the danger of over- or underpressurization. Because it rises and descends with the liquid level, a floating cover compensates for the suction created when removing solids, as well as minimizes the possibility of overpressurization during solids addition.

Gas-Holder Covers. Gas holder covers are similar to floating covers; however, they are designed to accommodate gas storage as well as digester drawdown. The gas-holder cover, which floats on digester gas rather than on the liquid surface, is equipped with a skirt that extends below the liquid surface to contain gas. A concrete ballast ring below the skirt stabilizes the cover and helps control gas pressure.

The operator should check the floating cover periodically to ensure that it is level and moves freely. A tilted cover may reflect uneven loading resulting from water accumulation in the attic space, snow buildup, or binding between the wall and the cover skirt. Excessive foaming may also be obvious during the cover inspection.

Membrane Covers. Membrane gas-holder covers provide gas storage and separate the anaerobic digester contents from the atmosphere. A photo of an installed membrane cover is shown in Figure 30.5. The cover consists of an inner membrane and an outer membrane. Both membranes are attached to the tank wall, preventing escape of gas to the atmosphere. The fixed outer membrane is inflated by a blower system. The inner membrane is inflated and deflated in response to the volume of gas in the digester headspace. The inner membrane can typically travel up and down the entire depth of the digester tank, enabling it to be almost completely emptied.

CORROSION PROTECTION. Exposed surfaces of anaerobic digesters are vulnerable to corrosion caused by high concentrations of hydrogen sulfide in digester gas. If the digester is operating at a low pH, carbon dioxide can form carbonic acid, which is also corrosive. Areas affected by corrosion include any surfaces that are not submerged, such as the undersides of the covers, the tank walls above the liquid surface, and the exterior surface of the covers, which are exposed to the atmosphere. If there are any leaks in the cover that allow gas to collect in the attic space, structural elements of

FIGURE 30.5   Membrane cover (courtesy of Siemens Water Technologies).

the cover may also corrode. Corrosion can be minimized by using corrosion-resistant construction materials, protective coatings, and cathodic protection.

The exterior surfaces of covers should be inspected quarterly. Metal covers should be recoated, if necessary, using the type of coating recommended by the cover manufacturer or that described by the engineering specifications. Typically, the cover needs to be recoated every 5 to 10 years, depending on the location and weather conditions. The underside of the cover should be inspected whenever the tank is taken out of service and emptied. Repairs may include power washing, sandblasting, and recoating. Covers that show significant corrosion should be evaluated for structural integrity. Because the attic area of a digester cover can be vulnerable to digester-gas accumulation and corrosion, it should not be entered by plant staff unless proper precautions and safety codes are followed.

ODOR ISSUES. Fixed digester covers are attached directly to the digester tank wall, forming a seal that prevents the escape of digester gas and odors. Floating covers and gas-holder covers are separated from the tank wall by a small gap of approximately 8 cm (3 in.) to accommodate cover movement. The liquid in the digester forms a water seal in this area, minimizing the escape of gas. However, in the event of foaming, foam can interfere with the water seal and digester gas will escape. The most common source of digester gas odors from fixed, floating, or gas-holder covers is the pressure-relief valve. Gas can also escape from deteriorated seals and cause odors at the digester facility. Inspection and proper maintenance of the pressure-relief valves will minimize the escape of digester gas.

Membrane covers are attached directly to the digester wall, which minimizes the escape of gas; however, the gas may seep into the space between the inner and outer membranes. During pressure regulation, air can be discharged from the membrane cover. Moreover, if there are any pinhole leaks or seepage that allow digester gas into this space, the air can contain hydrogen sulfide, volatile organic compounds, and associated odors. Carbon filters can be installed on the air discharge line to remove hydrogen sulfide and minimize odors.

DIGESTER MIXING. *General Theory.* Mixing of high-rate digestion systems, along with heating, uniform feed rates, and thickening of feed sludge, is used to provide optimum environmental conditions for the microorganisms that accomplish anaerobic digestion. Effective mixing provides the following benefits:

- Distribution of influent solids throughout the digester;
- Dispersal of influent solids for maximum contact with microorganisms;

- Prevention of thermal stratification;
- Reduction of scum buildup;
- Reduction of the buildup of settled material on the digester floor;
- Dilution of digestion inhibitors, such as toxic material or unfavorable pH or temperature of feed materials; and
- Improved separation of digester gas from digester liquid.

These benefits can be grouped as process stability, scum and foam control, and prevention of solids deposition. To satisfy process requirements, the digester contents must not be subjected to large fluctuations in raw sludge substrate, solids concentration, temperature, and bacterial metabolic end products. Most anaerobic digesters contain scum, which accumulates on the liquid surface starting in locations of low turbulence. While scum accumulation is typically controlled by mixing scum into the digester contents, large-diameter tanks allow it to spread over a larger area, making it more difficult to incorporate. Appropriate configuration of digester mixing equipment can reduce or eliminate scum accumulation. Deposits of grit on the bottom of the anaerobic digester result in the loss of active volume and lead to costly and onerous cleanup procedures requiring extended downtime. Grit accumulation is most often encountered with nearly flat-bottomed conventional digesters. In digesters with sloped bottoms, grit and settleable solids slide down the sides to a single location where they can be either removed, as in the case of grit and heavy solids, or resuspended, as in the case of lighter material. If the grit and solids are allowed to remain in the digester for an extended period, they will concentrate into a solid mass that is difficult, if not impossible, to resuspend or remove through pumping.

UNMIXED DIGESTION. Unmixed digestion, which was the earliest type of anaerobic digestion, takes place in low-rate or standard-rate digesters. Because they provide no auxiliary mixing, the contents of these digesters are stratified, with a scum layer on top of the liquid surface and stabilized biosolids and grit on the digester floor. Because of the relatively small volume available for the digestion process, these digesters are sized for a typical detention time of 30 to 60 days, which is adequate to achieve stabilization under low-rate conditions. Use of low-rate digestion is typically limited to small facilities treating less than 3.8 m^3/h (1.0 mgd) and is uncommon even for such installations.

MIXED DIGESTION. Mixed digestion takes place in high-rate digesters with controlled mixing and heating, uniform feed rates, and thickening of digester feed to provide optimum conditions for the microorganisms. The stable and uniform environment created by mixing promotes high-rate digestion and allows the use of higher loading rates than low- or standard-rate digestion. Types of mixing systems available for digesters include mechanical and gas mixing systems.

Mechanical Mixing. Mechanical mixing systems include pumped systems and impeller-type mixers.

Pumped Mixing Systems. Pumped mixing systems consist of pumps, piping, and nozzles (Figure 30.6). The pumps are typically "chopper" pumps or pumps incorporating in-line grinders that prevent fibrous materials from accumulating and causing plugging problems. The pumps are installed outside the tanks to facilitate maintenance. The high-velocity nozzles are mounted inside the tank and are oriented to discharge in a flow pattern that completely mixes the tank contents. Flow patterns vary by manufacturer; however, typical patterns include a toroidal configuration (helical pattern rotating around the center) and "dual-zone" mixing, combining a uniform and a vertical flow pattern to provide a nearly uniform velocity throughout the tank.

The systems are designed so the equipment inside the digester does not need regular maintenance. Pumps require the maintenance that is typical for pumps of this type. A significant increase in pressure at the discharge end of a mixing pump could indicate a problem with the associated mixing equipment, such as clogged piping or nozzles.

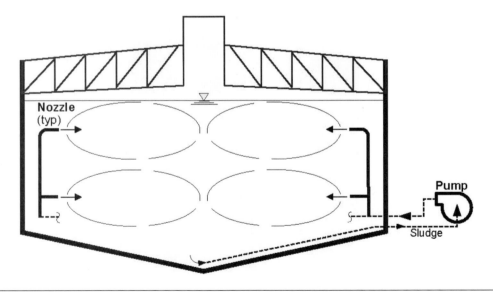

FIGURE 30.6A   Pumped mixing schematic.

FIGURE 30.6B Pumped mixing system.

FIGURE 30.6C Pumped mixing system nozzle (courtesy of Vaughan Co., Inc.).

Impeller Mixing. Impeller mixing systems consist of an impeller, a drive shaft, and a drive (Figure 30.7). Impellers can be mounted in a draft tube to direct flow within the vessel. The draft tube may include a discharge nozzle that is aimed tangentially to the wall to create a swirling action in the tank. Impeller drives are typically reversible to allow discharge at either the top or bottom of the draft tube or to reverse the impeller to clear blockages. To avoid deposits in low-velocity zones, some utilities routinely reverse impeller direction for 2 hours every 24 hours.

Various configurations are used for impeller mixing. The mixer and draft tube assembly can be mounted at the center, at the mid-radius point, or outside the tank. Impeller mixers without draft tubes are typically mounted at the center of the digester cover.

A common problem with impeller mixers is rags, hair, and other stringy materials wrapped around and entangled in the mixer blade and shaft, which interferes with their operation. Manufacturers offer "weedless" impellers to help reduce this problem. Rag buildup can be monitored by measuring the motor's current draw. A fouled mixer can sometimes be cleared by reversing its direction. To clear large accumulations, mixers can be removed via a crane or the tanks can be dewatered and the accumulated rags cut away. In addition to monitoring buildup of debris, the drive unit and bearings should be serviced in accordance with the manufacturer's recommendations. External

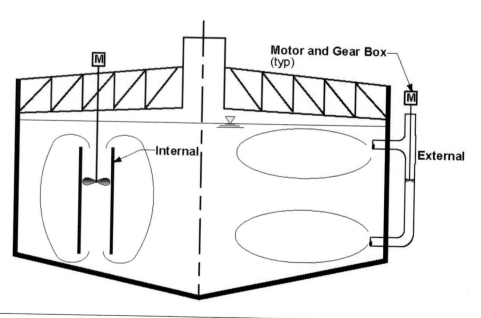

FIGURE 30.7   Mechanical draft tube mixing schematic.

draft tube mixing systems may also be subject to vibration, which can require corrective maintenance.

Gas Mixing. The four major types of gas mixing systems are bottom diffusers, lances, bubble guns, and gas lifters.

The compressor is the heart of all gas mixing systems. Compressors used include the rotary lobe, rotary vane, and liquid ring types, all of which are described in more detail in the section entitled "Gas Handling." Provisions for removing moisture and sediment from digester gas should be provided upstream from the compressors. Sediment traps should be cleaned at regular intervals. Bottom diffuser, bubble gun, and gas lifter systems include a flow-balancing manifold to distribute the flow evenly among the units inside the digesters. Condensate from the flow-balancing manifolds should be removed on a regular basis. Flow-balancing manifolds located outdoors in a cold climate should be protected from freezing.

Each gas mixing system includes means of controlling the gas pressure, such as high-pressure relief valves that allow the gas to be bypassed to the compressor suction or sent to storage when the compressor discharge pressure exceeds desired limits. If set properly, the relief valve opens only when the discharge line becomes plugged.

A low-pressure regulator bypasses gas whenever the pressure in the digester drops below a predetermined level. This prevents the development of a vacuum in the digester that could draw in air and create an explosive mixture. The low-pressure relief system should sense the gas pressure via a pipe connected directly to the digester. If the sensor is connected to the same pipe as the compressor suction, a false reading will be obtained. The sensing line must be kept clear at all times.

Pressure switches can be used to control high- and low-pressure conditions, but their use requires that the system be restarted by an operator whenever the pressure fluctuates. This is particularly troublesome during initial startup, when gas pressure varies.

Bottom Diffusers. Bottom diffuser mixing systems include diffusers or boxes on the digester floor near the center of the tank. All boxes receive and discharge equal amounts of compressed gas, creating a rising gas column.

To ensure even distribution of gas among the boxes, the gas flow must be checked periodically, because the boxes are prone to plugging. Plugging can be cleared by redirecting the entire gas flow through the affected box or by flushing it with high-pressure water. Bottom diffusers can also become covered with solids, which will affect the mixing pattern in the digester.

Lances. Lance mixing systems direct compressed gas to several discharge points in the tank (Figure 30.8). Sequencing the gas flow to various points causes the mixing action

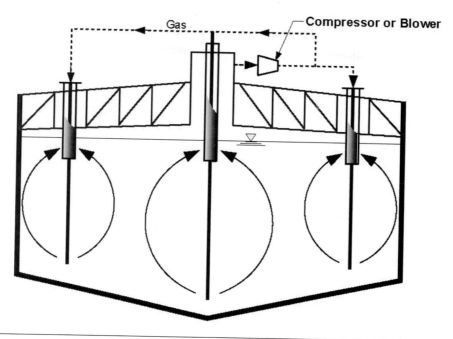

FIGURE 30.8   Lance mixing schematic.

to be distributed throughout the tank. The maximum depth of tank mixing is controlled by the depth at which the gas is discharged. Typically, the mixing will take
place within 1.8 to 2.4 m (6 to 8 ft) below the point of gas discharge. Below this depth,
materials settle and can accumulate.

Lances are typically spaced equally in a circle, at two-thirds the tank radius, with
an additional lance in the center. The lances, which are typically 5-cm-diameter (2-in.-
diameter) pipes mounted to the cover, discharge near the bottom of the tank. With
many lance configurations, the gas cycles intermittently through the lances, which can
cause them to become plugged with solids and rags. Straight pipe is used to minimize
plugging. If plugging does occur, lance caps can be removed and the lances cleaned.

The sequential flow of gas to the lances is controlled by a series of motorized
valves or by a single rotary valve. The motorized valves in the early systems allowed
moisture to collect in the valves when they were closed. These valves should be inspected regularly for moisture-related deterioration. Rotary valves allow continuous
gas flow, which reduces moisture problems. However, corrosion caused by contact
with digester gas can still be a significant problem. Therefore, the sleeves in the valves
should be greased annually.

Bubble Gun Systems. Bubble gun systems consist of bubble generators and vertical barrels. Compressed gas accumulates in a bubble generator at the base of the vertical barrel and forms a single large bubble, which rises through the pipe barrel, acting as an expandable piston. The bubble gun mixing system and bubble gun equipment are shown in Figure 30.9(a) and (b). The gas bubble forces the solids upward through the barrel and out the top, drawing solids into the barrel behind it. After leaving the barrel, the gas continues to rise to the liquid surface. Gas bubbles leave the barrel at the same rate as new bubbles are released from the generator, which results in continuous solids flow and establishes the circulation pattern. The numbers and configurations of the bubble guns depend on tank size. Because the barrels in a bubble gun system must be submerged, digesters equipped with bubble gun mixing systems have limited draw-down capability.

Even distribution of gas among the bubble guns is essential to the operation of the system. Flow-balancing controls should be checked periodically for plugged lines or clogged bubble generators. The generator siphon, with its many bends, is particularly susceptible to plugging. To clear a plug, a cleanout line is typically provided. Increasing the gas flow may also help clear a clogged generator. If these methods fail, the tank must be dewatered to gain access to the generator.

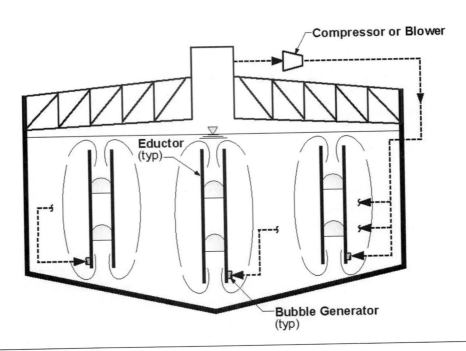

FIGURE 30.9A   Bubble gun mixing schematic.

FIGURE 30.9B   Bubble gun mixing system.

Draft Tube System. The draft tube system works much like an air lift pump (Figure 30.10). Gas is injected into the draft tube via lances installed in a vertical tube, which typically discharges the gas below the midpoint of the draft tube. As the gas is released, it carries solids upward through the draft tube, drawing in more solids at the base of the tube. The solids that leave the top of the tube flow away radially. Large tanks are equipped with multiple draft tubes; smaller vessels typically contain a single tube located in the center. Small ESDs can use a jet/draft tube system for mixing. Lances can be vulnerable to plugging, as described in the preceding "Lances" section.

Evaluation of Mixing System Performance. Because anaerobic processes are completely enclosed, it is difficult to directly observe the effectiveness of their mixing systems. Several testing procedures have been developed to measure the effectiveness of mixing, including solids profiles, temperature profiles, and tracer studies.

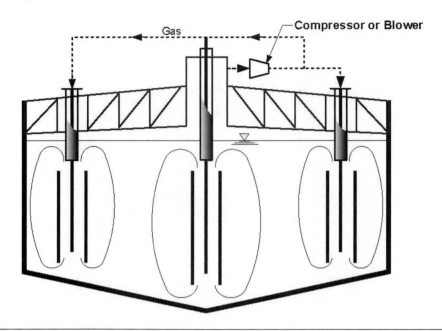

FIGURE 30.10   Draft tube mixing schematic.

The purpose of digester mixing is to produce a uniform condition in the tank. One method to evaluate solids uniformity is to perform a solids profile analysis. This test involves collection of solids samples at 0.6-m (2-ft) depth intervals, progressing from the top of the digester to the bottom. The samples can be withdrawn via sample wells (thief holes) in the cover using a spring-loaded sampling device. The samples are then analyzed for total solids. A concentration variation of more than 5000 mg/L from the average indicates poor mixing at the sample location. If the tested solids do not have a propensity to stratify, the results of the solids profile may be inconclusive. Solids profile testing should be performed every 6 months.

Another method to measure mixing efficiency is the temperature profile. The testing procedure is similar to that used for the solids profile testing, with temperature readings taken at fixed depth intervals. If any reading varies from the average by more than a specified value [typically 0.5 to 1.0 °C (1.0 to 2.0 °F)] the variation indicates poor mixing. The results of temperature profile tests may be inconclusive if the heat dispersion in the tank, even without mixing, is sufficient to produce a uniform temperature profile.

Tracer studies are yet another method to evaluate digester mixing. This method involves adding a quantity of tracer material, such as lithium, to the digester and col-

lecting samples from the digester discharge at fixed intervals. The samples are analyzed for tracer concentrations, which are then compared to theoretical concentrations
that would result from complete mixing. Although the tracer study produces the most
accurate results of the mixing evaluation methods discussed, it is also the most complicated, lengthy, and costly.

   If a digester fails to pass a mixing effectiveness test, the mixing system should be
checked for mechanical problems.

CALCULATION OF TURNOVER RATES. The turnover rate of digester contents, which is calculated by dividing the digester volume by the pumping rate, can be
used as a measure of the amount of mixing provided. Turnover rate analysis can only
be performed for mixing systems in which the pumping rate can be measured, which
limits its use to pumped mixing. Turnover rates can be calculated using the following
equation:

$$TR = DV/PR \hspace{4cm} (30.4)$$

where
   TR = turnover rate (min);
   DV = digester volume, based on sidewater depth, digester diameter, and bottom
        cone volume [L (gal)]; and
   PR = pumping rate [L/min (gpm)].

   Typical turnover rates range from 2 to 4 hours. If available, computational fluid
dynamics can be used to determine the turnover rate.

   Mixing energy evaluations can also be used to measure the amount of mixing in
a system. Typical mixing energy rates range from 7 to 13 kW/L (0.25 to 0.50 bhp/
1000 ft^3).

DIGESTER HEATING. *Heating Theory.* Each group of methane-forming microorganisms has an optimum growth temperature. If the temperature fluctuations are
too wide, methane formers cannot develop the large, stable population needed for the
digestion process. Digestion virtually ceases at temperatures below 10 °C (50 °F). Most
digesters operate in the mesophilic temperature range [32 to 38 °C (90 to 100 °F)], while
some operate in the thermophilic range [55 to 60 °C (131 to 140 °F)]. Regardless of the
operating temperature selected, the temperature of the digester contents should not

deviate from it by more than 0.6 °C (1 °F) per day. It is good practice to log the digester temperature once per shift and watch for temperature swings.

Because methane-forming microorganisms are temperature-sensitive, maintaining a constant digester temperature is one of the most important operator-controlled functions and requires a heating system that is reliable, well-maintained, and well-understood. The digestion process cannot survive more than a few days without heat.

The total heat required for digestion is based on the following:

- *Sludge heating*—the heat required to raise the temperature of the incoming raw sludge to the digester operating temperature, and
- *Transmission losses*—the heat required to make up the heat losses from the digester to its surroundings.

Solids Heating. The temperature of the incoming sludge is typically lower than the temperature of the contents of the operating digester; therefore, heat must be added. The heat needed to raise the incoming solids temperature typically represents more than 60% of the total heat demand. The heat needed to raise the temperature of the influent sludge to the digester operating temperature can be expressed by the following equation:

$$Q = (S)(C_s)(T_o - T_i) \tag{30.5}$$

where
   Q = solids heating requirement [kJ/d (Btu/hr)];
   S = solids mass flowrate [kg/d (lb/hr)];
   C_s = specific heat of solids [4.2 kJ/kg·°C (1.0 Btu/lb·°F)];
   T_o = digester operating temperature [°C (°F)]; and
   T_i = solids influent temperature [°C (°F)].

Reducing the amount of carrier water in the digester feed via thickening directly reduces the amount of heat required to raise the temperature of the digester feed to the digester operating temperature.

The digester feeding frequency affects the capacity requirements of the solids heating system. For example, a system designed to support a 24-hour-per-day feeding schedule will be overloaded if the daily volume of solids is fed in 3 hours. Such an overload may result in a temperature drop in the digester; moreover, re-establishing the proper operating temperature may take the rest of the day. Fluctuations in temperature are detrimental to the anaerobic bacteria.

Transmission Losses. The heat required to make up for heat losses from the digesters is expressed by the following equation:

$$Q = (U)(A)(T_o - T_i) \qquad (30.6)$$

where

Q = digester transmission loss [kJ/d (Btu/hr)];

U = heat transfer coefficient [kJ/d·m²·°C (Btu/hr·sq ft·°F)];

A = area of transmission loss [m² (sq ft)];

T_o = digester operating temperature [°C (°F)]; and

T_s = temperature of surroundings [°C (°F)].

Because different zones of the digester have unique heat transfer conditions, such as heat transfer coefficient or temperature of surrounding area, heat losses are calculated separately for each zone and then added together to determine the total losses for the digester. Resources for estimating heat transfer coefficients include Tables 22.12 and 22.13 of Water Environment Federation's *Design of Municipal Wastewater Treatment Plants* (WEF, 1998); *Cooling and Heating Load Calculation Manual* (McQuiston and Spitler, 1992); and information from manufacturers' catalogs. If transfer loss from one location, such as the digester covers, is particularly high, application of insulating material should be considered.

INTERNAL HEATING SYSTEMS. An internal heating arrangement transfers heat to solids in the digester vessel. Early internal heating arrangements consisted of pipes mounted to the interior face of the digester walls, and mixing tubes equipped with hot water jackets (Figure 30.11). These systems have lost popularity because much of the heating equipment and piping is inaccessible for inspection or service, unless the tank is dewatered. Also, rags and other debris collect on the pipes, reducing the heat-transfer efficiency and accelerating cleaning frequency.

EXTERNAL HEATING SYSTEMS. In external heating systems, solids are recirculated through an external heat exchanger (Figure 30.12). The recirculation pump maintains a flow velocity of 1.2 m/s (4 ft/sec), which creates turbulence across the heated surface and minimizes fouling.

The feed sludge pump is typically interlocked with the recirculation pump, causing it to operate whenever feed sludge enters the heat exchanger. The feed and the active digester sludge are blended and preheated before they enter the digester, which

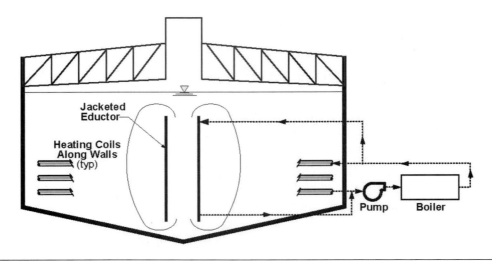

FIGURE 30.11   Internal heating schematic.

prevents the formation of cold, inactive zones of solids in the digester. The feed sludge can also be preheated by blending it with the heated sludge from the discharge of the heat exchanger.

The inlet and outlet temperatures of the heat-exchanger sludge should be monitored. A significant drop in the difference between the two indicates reduced heat transfer that may be caused by a malfunction of the solids pumps, hot water pumps,

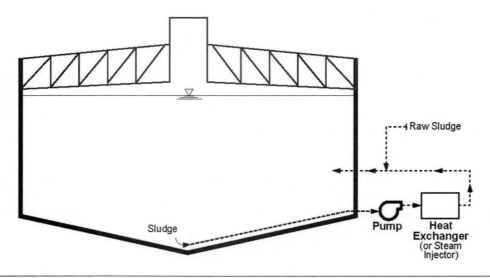

FIGURE 30.12   External heating schematic.

or the hot water supply. If these are operating properly, the heat exchanger needs to be checked for plugging, lime buildup, or fouling of the heat exchanger surfaces.

Three types of typical external heat exchangers—tube-in-tube, tube-in-bath, and spiral-plate—and their operation are described in the following sections.

Tube-in-Tube Heat Exchanger. A tube-in-tube heat exchanger consists of a serpentine arrangement of solids tubes surrounded by larger-diameter water tubes (Figure 30.13). The solids travel back and forth through the tubes via the solids-return bends, making thermal contact with the heating water on the outside of the tube. The heating water circulates through the annular space between the solids tube and the water tube countercurrent to the solids flow, which maximizes heat transfer. The temperature of heating water is typically limited to 66 °C (150 °F) to minimize solids accumulation on the inside surface of the tubes. The pressure through the heat exchanger should be monitored periodically. An increase in the drop in pressure is an indication of solids accumulation or fouling and requires cleaning of the solids tubes. The interior of the tubes can be accessed by removing the curved tubing at the end of each serpentine. If the tubes are large enough to accommodate a "pig", it is a cleaning option. If plugging

FIGURE 30.13   Tube-in-tube heat exchanger.

is excessive, grinders can be installed upstream of the heat exchangers to break up masses of fibers and rags that may have accumulated in the digester. Screening can also be used to remove fibers and rags from the solids.

Shell-and-Tube and Tube-in-Bath Heat Exchangers. Shell-and-tube and tube-in-bath heat exchangers consist of a serpentine arrangement of tubes in a hot water bath (Figure 30.14). In the shell-and-tube heat exchangers, the hot water flow is directed through baffling, which results in more efficient heat transfer. A hot water pump is used to create turbulence in the bath, thereby increasing the heat transfer. As with tube-in-tube heat exchangers, the pressure through the solids tubes should be monitored to detect caking or fouling.

Spiral-Plate Heat Exchanger. Spiral-plate heat exchangers consist of an assembly of two long strips of metal plate wrapped to form a pair of concentric spiral passages (Figure 30.15). Alternate edges of the passages are closed, forming separate channels

FIGURE 30.14A   Tube-in-shell heat exchanger.

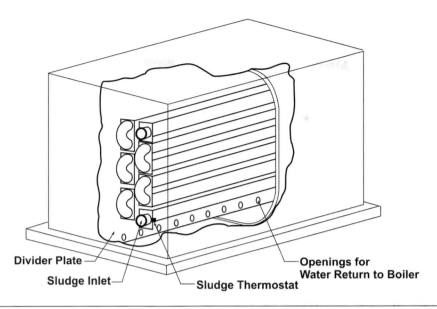

Divider Plate
Sludge Inlet
Sludge Thermostat
Openings for
Water Return to Boiler

FIGURE 30.14B   Tube-in-bath heat exchanger.

FIGURE 30.15   Spiral plate heat exchanger.

for solids and heating water. The units are typically constructed with hinged doors to provide access to the solids passages.

The solids passages of the spiral heat exchanger should be inspected regularly for plugging. Any plugging must removed from the bars separating the concentric plates. Pressure gauge readings should be taken daily to monitor the degree of plugging between cleanings. A significant increase in pressure drop indicates a plugged passage. Grinders are typically installed upstream from spiral heat exchangers to help prevent plugging.

SOLIDS-TO-SOLIDS HEAT TRANSFER SYSTEMS. Although most digestion systems do not include provisions for heat recovery from digested sludge, some new facilities are considering this option to reduce fuel use. Solids-to-solids heat transfer systems include recovery of heat from the discharge of mesophilic or thermophilic digesters to preheat feed sludge.

Solids-to-solids heat transfer systems use solids-to-solids heat exchangers, which include cube-type (Figure 30.16), tube-in-tube, and spiral-type heat exchangers. In each

FIGURE 30.16   Cube heat exchanger (courtesy of DDI).

type, hot and cold sludge are introduced to opposite sides of a heat transfer wall, through which heat is transferred from the hot sludge to the cold sludge. The cube- and spiral-type heat exchangers include provisions for access and cleaning of both sides of the heat exchanger surface.

Heat can also be transferred through a system of two solids-to-water heat exchangers and an interconnecting heating water loop, which works by first transferring heat from the hot sludge to the heating water loop in a tube-in-tube or shell-and-tube-type heat exchanger. The heated water is then pumped to another heat exchanger, which transfers heat to the cold sludge. Operation and maintenance issues associated with this system are similar to those of other tube-in-tube, shell, and tube-type heat exchangers.

STEAM INJECTION HEATING. Steam injection heating systems inject hot steam directly into solids. The systems consist of internally modulated injection heaters and solids recirculation pumps, piping, and valves. The steam is injected into the solids stream via variable area steam nozzles that are modulated to control the temperature of the solids–steam mixture. Systems can be configured with either one or two heaters. In two-heater systems, one heater heats the feed sludge and the other replaces radiant heat losses from the digester in a recirculation line. In single-heater systems, one heater is sized to meet all heat requirements.

The following conditions must be met for the steam injection system to work properly:

- Heater steam pressure [typically 103 kPa (15 psi)] must be higher than the solids pressure;
- Solids discharge piping must be kept clean to maintain a low operating pressure;
- Particles in the solids feed must be small enough to pass through the nozzle to avoid plugging; and
- The makeup water must be softened to avoid scale buildup in the boiler and steam system.

HEATING SOURCES. Digesters can be heated with either hot water or steam. Heat sources include boilers and waste heat recovery.

Hot Water Systems. Hot water systems consist of a heat source, recirculation pumps, and temperature controls. Such systems typically include two separate loops. One loop is used for the solids heat exchangers, with a recommended temperature of 66 °C

(150 °F) to minimize fouling of the heat exchanger surfaces. The second loop supports the boilers, which should be maintained at a temperature of at least 82 °C (180 °F) to avoid the formation of sulfuric acid, which leads to corrosion. A temperature control valve between the loops controls the blend of hot water from the boiler loop with re-circulated water from the heat exchanger loop to maintain the temperature upstream from the heat exchangers at 66 °C (150 °F).

The water in the heating loop should be conditioned with water treatment chemicals to inhibit corrosion and scaling; the water should be tested annually, with more chemicals added as necessary. During startup, sufficient time should be allowed for the system to reach operating temperature before making any adjustments.

Steam Systems. Steam systems consist of a conditioning makeup water system, a steam boiler, a steam-to-water heat exchanger, and a heating water loop or direct steam injection heater. The conditioning system softens the feed water to prevent buildup of scale in the system. The steam-to-water heat exchanger transfers heat from the steam produced by the boiler to the heating water loop, which operates at lower temperatures than the steam side to minimizing fouling.

The feed must be monitored regularly to ensure that it is adequately conditioned. Maintenance of the steam system consists of regular monitoring and repairing steam traps as needed.

Boilers. Fire tube, water tube, and sectional cast iron-type boilers supply heating water or steam to digester heating systems. Common fuels for heating the boilers include digester gas, natural gas, and fuel oil. Digester gas should be treated to prevent corrosion of the fuel valve train and the combustion zone. Treatment methods are discussed in the "Gas Handling" section. The boilers should be inspected as recommended by the boiler manufacturer, with particular attention to the combustion area, which is susceptible to corrosion. Codes and regulations should be followed, including those pertaining to annual boiler inspections and boiler operator certification. The facility's insurance carrier may also have special maintenance or inspection requirements that need to be identified and followed.

Heat Recovery. Digester gas is often used as fuel for engines and turbines that are used to drive equipment, such as pumps or blowers, or produce electricity. Waste heat recovered from cooling water systems or turbine exhaust gases can be used to heat digesters and buildings. Heat can also be collected as hot water from the exhaust of engines or combustion turbines and pumped to a water-to-water heat exchanger to be transferred to the heating water loop of digester sludge heat exchangers. Typically, sufficient heat can be recovered from the engines or turbines to eliminate the need for supplemental digester heat. However, a backup boiler should be available for use during combustion equipment maintenance or other outages.

GAS HANDLING. *Gas Production Theory.* Digester gas, also known as *biogas*, is formed during the last stage of digestion, when microorganisms convert organic acids and carbon dioxide (CO_2) to methane (CH_4) and water. Gas production is directly related to the quantity of volatile solids destroyed by digestion. Typical values range from 0.75 to 1.1 m^3/kg (12 to 18 cu ft/lb) of volatile solids destroyed. The total amount of volatile solids destroyed can be calculated by deducting the amount of volatile solids in the digested sludge from the total quantity fed to the digester.

Because the quantity of digester gas produced depends on destruction of volatile solids, it is used as a measure of overall digester performance. Gas quantities should be routinely measured to determine the average production for a corresponding volatile solids destruction. Gas production should then be monitored in relation to the measured volatile solids destruction to provide an indication of the condition and performance of the digestion process.

Digester Gas Characteristics. The biogas generated in the anaerobic digestion process is composed primarily of methane (60 to 65%) and carbon dioxide (35 to 40%). The physical and chemical characteristics of methane, which imparts the fuel value to biogas, are summarized in Table 30.2.

TABLE 30.2 Physical and chemical properties of methane.

Physical characteristics	Colorless, odorless
Specific gravity	0.55 at 21 °C (70 °F)
Density	0.042 kg/m^3 at 21 °C (70 °F)
Hazard	Extremely combustible
Flammability limits in air	Forms an explosive mixture with air (5 to 15% volume). Avoid naked flames or spark-producing tools when there is unburnt gas in the air.
Toxicity	Asphyxiant at high concentrations (causes insufficient intake of oxygen)
Typical heating value	37 750 kJ/m^3 (1016 Btu/cf) (natural gas)
	Biogas has a lower heating value of 22 400 kJ/m^3 (600 Btu/cf) because it typically contains only 60 to 65% methane.

Digester gas typically contains impurities, such as foam, sediment, hydrogen sulfide (H_2S), and siloxanes; it is also saturated with moisture at digester operating temperatures. The various contaminants in digester gas include the following:

- *Carbon dioxide*—Carbon dioxide in digester gas dilutes its energy content and lowers its energy value. However, removal of carbon dioxide from the gas mixture to enrich the fuel value of digester gas is not a common practice.
- *Moisture*—Moisture in digester gas will form condensate in the system piping and can combine with hydrogen sulfide to form sulfuric acid, which will accelerate deterioration of the check valves, relief valves, gas meters, and regulators. If the condensate is not drained, it can also block gas flow at low points in the piping. If moisture is not removed from the digester gas prior to combustion, it will reduce the energy available for the heating system. Gas produced by thermophilic digestion, at associated high temperatures, will contain more moisture than gas generated by mesophilic digestion.
- *Hydrogen sulfide*—Hydrogen sulfide is an extremely reactive compound that, when combined with water, forms an acidic solution that is highly corrosive to pipelines, gas storage tanks, and gas utilization equipment. If present in a digester, it will shorten the usable life of many digester gas system components. Hydrogen sulfide is also a dangerous compound that can be lethal at concentrations above 700 parts per million (ppm).
- *Siloxanes*—Siloxanes are volatile organic chemicals containing silicon that are used in many commercial, personal care, industrial, medical, and even food products. Personal care products, such as shampoos, hair conditioners, cosmetics, deodorants, detergents, and antiperspirants, are thought to be the main sources of the siloxanes found in wastewater, which are released as gas during the anaerobic digestion process. When oxidized in gas utilization equipment, they form an abrasive fine silicon dioxide sand that accumulates on moving parts or heat exchange surfaces, causing accelerated wear and loss of heat transfer efficiency.

Gas Treatment. Digester gas is a valuable resource that can be used to meet a treatment plant's energy requirements. However, the gas must be treated to remove contaminants that would otherwise damage equipment fueled by the gas and shorten its useful life. Consequently, before digester gas is used as an energy source, it may need to be treated to remove siloxanes, moisture, and sediment, as well as to lower its hydrogen sulfide concentrations.

FOAM AND SEDIMENT REMOVAL. Many systems are equipped with sediment traps and foam separators for cleaning digester gas. These devices provide a "wide spot" in the gas piping system for slowing velocities, collecting foam and particulates entrained in the gas, and removing collected condensate. The foam separator (Figure 30.17) is a large vessel with an internal plate fitted with water nozzles that provide a continuous spray. The foam- and sediment-laden gas enters the vessel near the top and travels down through the spray wash under the baffle wall and up through a second spray wash, exiting the vessel through an elevated discharge nozzle. The spray wash and the internal plate reduce foam in the gas to prevent carryover to gas utilization equipment.

HYDROGEN SULFIDE REMOVAL. Several methods are available to remove hydrogen sulfide from digester gas. The iron sponge process treats digester gas by passing it through a permeable bed of iron sponge (hydrated ferric oxide in the form of iron-dipped wood chips soaked in water). As the gas passes through the iron media, it undergoes an exothermic (heat-producing) reaction that converts the hydrogen sulfide to ferric sulfide and water.

The iron sponge can be regenerated and reused repeatedly before replacement. The need for regeneration or replacement can be determined by measuring the hydro-

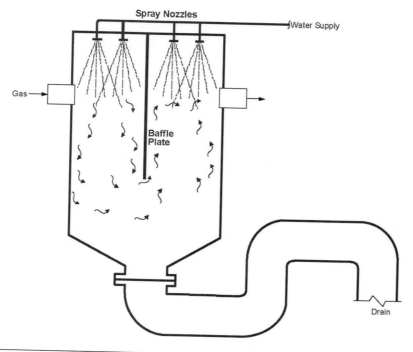

FIGURE 30.17   Foam separator.

gen sulfide concentration in the treated gas. An "exhausted" sponge typically consists of a combination of ferric sulfide and wood chips, a nonhazardous waste that can be disposed at any landfill.

Regeneration of an iron sponge is a highly exothermic reaction using water and air to release sulfur from the iron and reform the hydrated ferric hydroxide. If the iron chips are not submerged in a flowing water bath and controlled properly, they can overheat and combust spontaneously.

Several proprietary processes are available that involve passing digester gas through a bed containing media that react selectively with hydrogen sulfide (Figure 30.18). The medium used in these systems is a proprietary, free-flowing granular product that will not ignite and cannot be regenerated.

Ferric chloride injected into the digester can reduce the amount of hydrogen sulfide in digester gas by reacting with sulfur to generate a solid iron sulfide. Iron salts can be added at the following locations in the treatment process:

- The primary clarifier (helps settling and improves overall facility odor control),
- The primary clarifier and/or WAS feed to the digester,

FIGURE 30.18   Hydrogen sulfide treatment.

- The suction side of the digester sludge-recirculation pump, and
- The suction side of a mechanical mixer.

Iron salts should not be added directly upstream from the heat exchangers because this can result in deposits of vivianite on heat exchanger surfaces.

MOISTURE REDUCTION. Moisture is condensed from digester gas as it cools. Gas piping should have a slope of at least 1% toward the condensate collection point. To effectively remove the moisture, the gas flow should not exceed 3.7 m/s (12 ft/sec) countercurrent to condensate flow.

The condensate is collected in traps that should be located at the low points in long pipe runs and wherever gas is cooled. Drip taps, which can be controlled manually or automatically, provide a convenient and safe means for the removal of accumulated condensate. Manually operated drip taps are recommended for indoor applications. Float-controlled, automatic drip taps are also available, but these require frequent maintenance to keep the valves operating. Should the float stick, gas can escape to the surrounding atmosphere, which limits their use to outdoor installations (where permitted by local codes and safety considerations).

Moisture can also be removed from the gas by cooling it to approximately 4 °C (40 °F) in a refrigerant-type dryer (Figure 30.19). The dryer is typically made of stainless steel or other materials resistant to hydrogen sulfide corrosion. Corrosion can be minimized by removing hydrogen sulfide from the gas before drying. Refrigerant drying may also remove up to 20 to 40% of siloxanes from the digester gas.

CARBON DIOXIDE REMOVAL. Carbon dioxide can be removed from the digester gas via water or chemical scrubbing, carbon sieves, or membrane permeation; however, all of these technologies are expensive and their use may be cost-effective only if the gas is to be upgraded to natural gas quality and sold.

SILOXANE REMOVAL. Heating the digesters [to 35 °C (95 °F)] causes the siloxanes in the sludge to volatize into their gaseous form, which, when burned, becomes a sand-like material (silicon dioxide) that is too fine to be captured in ordinary scrubbers and collects on combustion equipment or in combustion chambers. To protect the equipment, siloxanes should be removed from the digester gas before combustion via gas dryers or activated carbon scrubbers.

Gas Drying. Siloxanes are relatively heavy volatile compounds that tend to adhere to water vapor in the digester gas stream. A significant fraction of siloxanes can be

FIGURE 30.19   Refrigerant dryer.

removed along with moisture when the gas is dried. To remove more than 90% of the siloxanes, the gas must be dried to a dew point temperature of -29 to $-23\ °C$ (-20 to $-10\ °F$) or lower, as determined by temperature sensors in the drying process. The drying equipment consists of a refrigerant-type dryer that cools the digester gas as it passes through the dryer. The dryers are typically equipped with heat recovery systems that use the heat removed from the gas to raise its exit temperature above the dew point.

While gas-drying systems are relatively simple in concept, those that operate at the temperatures required for siloxane removal are vulnerable to significant amounts of ice formation, which must be removed to protect the pipes from bursting and damaging the refrigeration system.

Activated Carbon Scrubbers. Activated carbon scrubbers used to remove siloxanes from digester gas operate according to the same principles as the carbon scrubbers used for odor control in wastewater treatment plants (WWTPs). The digester gas is passed through a vessel filled with activated carbon (Figure 30.20), which captures the organics, including siloxanes, hydrogen sulfide, and several other compounds in digester gas. With proper maintenance and replacement of the carbon, the siloxanes in the digester gas should be removed to below detection limits. However, activated carbon is not selective for siloxanes and will remove other compounds as well. Consequently, if the digester gas contains other organics, the carbon will require frequent replacement. Removal of hydrogen sulfide before it passes through the carbon scrubbers will provide better siloxane removal and extend the life of the carbon bed.

GAS COMPRESSION. Gas from the digesters is supplied at a relatively low pressure [typically, a 150- to 300-mm (6- to 12-in.) water column, or up to 450 mm (18 in.)

FIGURE 30.20   Activated carbon treatment for siloxanes.

for digesters designed for higher pressures], which makes it difficult to use the gas in some types of combustion equipment. Furthermore, in digesters with gas mixing equipment, the pressure supplied to this equipment must overcome piping losses and the static head of the liquid in the digester. In both instances, gas compression or pressure boosting is needed. Table 30.3 lists typically used gas-compression equipment and their applications.

TABLE 30.3   Gas-compression equipment.

Type	Applications	Operation and maintenance
Liquid ring compressors (centrifugal with sealing of impeller provided by water ring)	Commonly used for gas mixing at 103 kPa (15 psig) or more, or where higher pressures for gas supply are required for high-pressure storage or for engines or turbines. Gas recycle used for startup and flow control.	Able to handle dirty gas. Require rebuilding approximately every 2 years. May use plant water, or, if this causes deposits, potable water for sealing water. Lower efficiency than other compressors. Less chance of overheating when used with gas recycle for flow control. Special attention should be given to water level control in separators to prevent gas escaping. A compressor with a water seal increases system complexity, a significant sidestream flow, and additional pollutants compared to a compressor system with a seal.
Rotary positive-displacement blowers	Gas mixing [limited to about 103 kPa (15 psig)]. Flow controlled by changing speed.	Require nearly annual rebuilding if handling dirty gas. May require noise enclosure. Need thermal shutoff for recycle gas applications.
Reciprocating compressors	Generally used for lower flow and higher pressure applications. Used for higher pressure to furnish gas to engines and turbines. Usually require water-cooling system for compressor and intercoolers and after coolers. Some applications of air cooling.	Not tolerant of dirty gas. May require particulate scrubbing to clean gas. High efficiency. Higher maintenance because of close-fitting moving parts.
Hermetically sealed blowers	Low pressure [less than 14 kPa (2 psig)]. Typically used as boosters for boilers where higher flow and low pressure are required. Blower requires gas flow for cooling of sealed blower within the gas pipe. Limited flow control by gas recirculation, potential overheating.	Not tolerant of dirty gas. Can be used following gas dryer. Sealed to reduce potential gas leakage.
Multiple-stage centrifugal blowers	Used to supply gas for utilization equipment at higher flows than reciprocating compressors and at similar higher pressures. Must be designed with seals and materials that are corrosion-resistant and safe for gas use. Flow control by inlet damper.	Not tolerant of dirty gas. Require particulate scrubber to clean gas. May require lubrication oil system and oil cooling if gearbox is used.

STORAGE. Equipment fueled with digester gas typically needs a steady flow of the gas. Therefore, facilities that use digester gas for plant operation or cogeneration, rather than flaring it, should include storage provisions to dampen fluctuations in gas production.

Several types of gas storage are available. The most common means of low-pressure gas storage is the floating gas-holder cover. The storage capacity under the cover depends on the available vertical travel distance of the cover and the tank diameter. Gas storage is typically limited to 1 to 2 hours at gas production. Membrane storage, which also operates at low pressures, can provide up to twice the amount of storage provided by floating gas-holder covers. Membrane storage can be installed either on the digester, to serve both as cover and storage space, or on the ground as a standalone structure (Figure 30.21).

FIGURE 30.21   Membrane gas storage (courtesy of JDV Equipment Corporation).

Gas-storage spheres, which hold gas at pressures of 550 to 700 kN/m^2 (80 to 100 psi), are standalone structures available in a variety of sizes (Figure 30.22). They can accommodate larger quantities of gas than other storage systems, but the gas must be compressed before it is stored. The energy use associated with gas-storage spheres can be significant.

GAS SYSTEM EQUIPMENT. *Flow Measurement.* Gas production is a measure of digester performance. Reliable monitoring equipment alerts plant operators to process malfunctions and gas leaks. The flow meters used for gas monitoring can be broadly classified as positive-displacement, thermal-dispersion, and differential-pressure (Table 30.4).

Separate flow meters are recommended for each digester because digester gas-production rates vary. Separate flow meters are also recommended to monitor gas use by the utilization equipment. The gas may contain moisture and impurities, which may cause maintenance problems for the metering devices.

FIGURE 30.22   Steel gas storage sphere.

TABLE 30.4   Gas-flow indication and metering.

Type	Applications	Operation and maintenance
Positive displacement	Used in many plants. Large devices that require bypasses when servicing. Accurate gas measurement unless dirty. Do not require upstream and downstream straight-pipe sections for accuracy.	Somewhat tolerant of dirty gas. Require periodic cleaning and rebuilding.
Thermal dispersion insert	Increasingly used technology because of tolerance of dirty gas and simple installation. Accurate for low flows. Some accuracy problems with zero flow, such as for flares. Require some upstream and downstream straight-pipe distances for accuracy. Can be inserted and removed through ball valve.	Smaller than most metering devices. Require periodic cleaning of probe. Must not have liquid impinging on probe. No bypass needed, reducing need for piping and isolation valves to service.
Pressure differential orifices and Venturis	Commonly used in older plants. Upstream and downstream straight-piping sections needed.	Require frequent cleaning when used for dirty gas. Accuracy dependent on instrument cleanliness. Can significantly constrict gas flow when fouled.

Gas System Safety and Control Devices. Gas systems include safety and control equipment to limit flashbacks, measure and control system gas pressure, and burn unused gas. Gas system safety and control devices are listed in Table 30.5, along with descriptions of their use and maintenance requirements.

GAS UTILIZATION.
Digester gas is a valuable fuel that has traditionally been used to heat boilers to produce steam or hot water for process and building heating, to operate combustion turbines or engine generators to produce electric power (and hot water from heat recovery for heating), or to operate dryer facilities to remove moisture from the solids (with heat recovery to heat digesters). Digester gas collection and utilization technologies have been steadily improving over the years, and energy recovery from digester gas is now regarded as one of the more mature and successful waste-to-energy technologies.

Thermal Energy. Digesters typically produce more gas than needed for digester heating. The excess gas can be used to heat buildings or treatment processes. If digester gas is used to generate steam rather than to heat water, energy would also be available for cooling buildings via absorption chillers. Because building heating and cooling loads are seasonal, the energy requirements vary depending on ambient temperature.

TABLE 30.5   Gas system safety and control devices.

Type	Applications	Operation and maintenance
Flame traps (large pipe) and flame check valves (small piping)	Used to stop flame propagation in a pipe. Flame traps installed with thermal shutoff valves as noted below. Installed on gas supplies to engines, compressors, boilers, flares, or other sources of ignition. Also used on digester covers with pressure- and vacuum-relief valves.	Require periodic cleaning, typically monthly. May require insulation and heating in cold climates to prevent ice formation. Foam will blind the devices, rendering other devices, particularly relief valves, ineffective.
Thermal shutoff valves	Spring-loaded or pressure-operated isolation valves that trip-shut at flame temperatures. Must be installed between flame trap and source of ignition.	Replace element periodically or when tripped. Usually includes an indicator to show that the valve has tripped. May require insulation and heating in cold climates to prevent ice formation.
Pressure- and vacuum-relief valves	Provided for all digester covers to prevent overpressurization or vacuum conditions in the digester. Typically adjustable, weighted valves, built as common unit. Should be set below design limits of cover. Typically two sets provided for each cover to allow servicing of one set while keeping digester in service. Typically discharge to area around valves. Available with pipe to allow discharge away from valves.	Check seals periodically; replace if leaks develop. May require insulation and heating in cold climates to prevent ice formation. Relief valves are last safety measure for digester covers to prevent overpressure or vacuum. These valves do not provide liquid relief.
Back pressure valves	Typically provided with waste gas flares. Maintain pressure upstream by relieving overpressure to flare. Usually uses weights to set relief pressure; however, some use springs. Upstream sensing line should ideally be in gas-collection piping header, away from the valve and large enough to reduce the potential for clogging.	Frequent maintenance required to prevent sticking (open or closed). These valves regulate pressure in the gas-collection piping system and are the first level of protection for overpressurization of the digester cover.
Low-pressure check valves	Used to prevent backflow of gas. Typically constructed with leather or flexible flap to ensure operation at low pressure. Not commonly used.	Check flap and replace if not functioning.
Waste gas flare or burners	Provided to dispose of surplus gas. Usually provided with standing pilot or electronic ignition.	Provided for all plants to serve as a means of suitable disposal of gas. Untreated gas should not be freely released.
Open pipe or "candle flare"	Used for smaller plants and where stringent air-emissions control is not needed. Generally visible flame, gas-disposal device.	Good pilot-gas supply is needed (may require natural gas or propane supply). Heat and corrosion make periodic replacement necessary. Pilot system requires frequent servicing.

TABLE 30.5 Gas system safety and control devices (*continued*).

Type	Applications	Operation and maintenance
Controlled combustion flare, ground flare, or emissions controlled flare	Used at plants where stringent air emissions are required and visible flame is not desired. Control air supply to maintain sufficient temperature to completely burn gas. These flares must be operated at a specific temperature to achieve emission limits. If oversized, auxiliary fuel may be required, which increases operating costs.	Require periodic system maintenance for controls and air blowers.
Pressure indication	Manometers used for low pressure [760 mm wc (30 in. wc) or lower]. Gauges used for higher pressures, particularly for gas utilization lines.	Periodically clean manometers, replace fluid, and calibrate gauges. Pressure gauges should be provided with an isolating diaphragm to prevent gas contaminants from fouling the mechanism.

Boilers do not have to be fueled with high quality gas; however, the hydrogen sulfide concentrations in the gas should be reduced to less than 1000 ppm for use in a boiler and the water vapor should be condensed to avoid problems with the gas nozzles.

Digester gas can also be used to generate heat for such processes as thermal drying of dewatered cake. Digesters can typically produce enough gas to supply 50 to 80% of the drying energy requirement for drying, with the remainder supplied by natural gas. Heat recovered from the dryer condenser/scrubber return stream can be used to heat the digesters.

Power Generation. Digester gas can be collected and used to generate electricity with onsite power generation equipment. The heat recovered from the power-generation units in the form of hot water or steam (from combustion turbines only) can be used to heat digesters or heat or cool buildings. If all the recovered heat is used, the overall efficiency of gas use can approach 80%.

The amount of power production varies depending on digester gas production; however, most medium-to-large wastewater treatment facilities generate less than 5 MW of power. Onsite power-generation systems include traditional engine generators, combustion turbine generators (for larger facilities), microturbines, and fuel cells. Internal combustion engines are by far the most commonly used generating equipment.

Approximately 80% of the power-generation facilities at WWTPs are reciprocating engines. Depending on the scale of operation, a combustion turbine and/or a steam turbine can also be used.

The gas quality requirements for cogeneration equipment are more stringent than those for boilers. The gas should be cooled to remove moisture, and typically requires scrubbing to remove hydrogen sulfide and siloxanes.

Distribution via Natural Gas Network. Another option for digester gas utilization is to treat it to the quality of natural gas by removing all the contaminants, including carbon dioxide, and distribute the upgraded gas through a natural gas supply network. This is practical only where large quantities of digester gas are available.

SAFETY. Digester gas can be hazardous because it is composed primarily of methane and carbon dioxide, both of which are colorless and odorless and can displace air, thus causing asphyxiation. Consequently, digestion facilities should be equipped with gas detectors and analyzers, as well as a self-contained breathing apparatus for use in the event of a gas leak.

Methane is highly combustible when mixed with air at concentrations of 5 to 20% methane. To keep out air, gas-handling systems are designed to operate under positive pressure. Because air can be introduced into a digester if solids withdrawals exceed the liquid operating range provided by floating or gas-holder covers, precautions should be taken to prevent draining operating digesters below the recommended level.

Any air in the digester will mix with gas and eventually enter the gas piping system. If the air-and-gas mixture should reach a flame source (waste gas burner, boiler, or engine), it would cause the flame to flash back in the pipe. The piping is equipped with flame traps to eliminate the flashback; however, the gases can expand in the pipe and cause pressure increases sufficient to rupture the gas line before the flame trap has an opportunity to snuff the flame. To reduce the possibility of pipe rupture during a flashback, the flame trap should be located as close as possible to the potential ignition source, with a maximum distance of 9 m (30 ft).

Digester gas leaks are another explosion hazard. Digester facilities are typically equipped with monitoring instruments that measure the lower explosivity limit, as well as the concentration of hydrogen sulfide in the gas. Because of the fire and explosion hazards, areas of digester facilities can be subject to National Fire Protection Association (NFPA) standards governing electrical classifications, construction materials, and fire-protection measures. Specific information on anaerobic digestion facilities is presented in Chapter 6 of *Standards for Fire Protection in Wastewater Treatment and Collection Facilities* (National Fire Protection Association, 2003a). According to NFPA standards, most areas in anaerobic digestion facilities are considered to be susceptible to explosion hazards; consequently, safety measures include use of nonsparking tools, prohibition of smoking or open flames, provision of adequate ventilation, and explosion-proof electrical equipment.

An additional safety consideration is the toxicity of digester gas. The effects of hydrogen sulfide at different concentrations are listed in Table 30.6. Although hydrogen sulfide has a strong rotten-egg odor, after short exposure the nose becomes numb to the odor, which may lead to the false conclusion that the hazard has diminished or disappeared. Consequently, hydrogen sulfide detectors should be used and calibrated on a regular basis. Hydrogen sulfide is heavier than air; therefore, detection equipment should be installed to adequately monitor all elevations of the digester facility.

Safety equipment associated with the digester gas system, such as flame arresters, pressure-relief valves, gas burners and controls, gas compressors, condensate traps, and detection instrumentation, should be inspected regularly.

Areas in the digester facility that should be considered confined spaces include digesters that have been taken out of service and emptied, as well as the attic space under digester covers. These areas should be entered in accordance with Occupational Safety and Health Administration (OSHA) Part 1910.146, *Permit Required Confined Spaces*, or, where applicable, similar safety codes.

TABLE 30.6   Effects of hydrogen sulfide at various concentrations.

Concentration (ppm[1])	Effects/Standards
5	Easily detectable odor
10	Permissible exposure limit (PEL[2]), slight eye irritation
15	Short-term exposure limit
50	Ceiling limit, maximum concentration of exposure at any time during a work shift
100	Coughing, eye irritation, loss of sense of smell after 2 to 15 minutes
300–500	Dizziness, nausea, bronchitis, pulmonary edema; also immediately dangerous to life and health
600–700	Rapid unconsciousness, cessation of respiration, followed by death
1000–2000	Unconsciousness at once, death in a few minutes
40 000	Explosive range [approximately 4 to 44% (40 000 to 440 000 ppm)]

[1]ppm = parts per million, that is, parts of gas per million parts of air by volume. A 1% gas–air concentration is equal to 10 000 ppm.
[2]PEL = 8-hour weighted average concentration at which workers can be repeatedly exposed each work shift without adverse effects (California Occupational Safety and Hazard Agency standard).

DIGESTER OPERATIONS

STARTUP AND SHUTDOWN. *Startup.* The objective of anaerobic digester startup is to achieve steady-state operation and the required volatile solids reduction in the shortest possible time. Successful biological commissioning requires an effective equipment startup of the various systems and subsystems, including the gas-handling system, heat exchangers, solids pumps, mixing equipment, and auxiliary systems before introducing volatile solids to the digester. The three parameters that are critical to ensuring a smooth startup process are

- Maintenance of a specific operating temperature;
- Constant mixing; and
- A specific volatile solids loading rate, with a daily variation not to exceed ± 10%.

To minimize failures of the equipment or biological processes, the operator should inspect and test the digester equipment, as follows:

1. Remove all debris from the digester tank.
2. Inspect all valves for proper and smooth operation.
3. Inspect all gas control, conditioning, and safety devices, including, but not limited to, sedimentation traps, foam separators, manometers, waste gas flare, flame traps, and pressure-relief valves.
4. Inspect and lubricate the solids pumps and verify that they are free of debris and have correct drive alignments.
5. Clean heat exchangers and tighten all piping connections.
6. Adjust and lubricate the digester mixing system, whether mechanical or gas.
7. Verify that the digester boiler system is operational for approximately 24 hours before filling the digester.

Digester startup can be divided into two phases: preparation and operation. During the preparation phase—typically the first 4 days—the digester is filled to its minimum operating level, allowed to reach operating temperature, and mixing is started. Initially, the digester can be filled with any of the following liquids:

- Unthickened WAS (before polymer addition),
- Secondary clarifier effluent (unchlorinated),
- Primary clarifier effluent, and
- A combination of raw sludge and WAS.

Anaerobic seed sludge is not needed for effective biological commissioning. Digester commissioning typically requires 45 days; however, use of seed sludge can reduce this time by 7 to 10 days. The ratio of primary solids to WAS in the seed material should resemble the digester feed as closely as possible. The digester contents should be mixed and heated to the operating temperature during the initial 3-day commissioning period. The digester gas system should be returned to the main gas line when digester filling begins.

At the end of the preparation period, the digester should be operating at the desired temperature and ready to receive sludge. At this point, the digester may be loaded with the following:

- Primary clarifier effluent (preferred source),
- Secondary clarifier effluent (unchlorinated),
- Unthickened WAS (before polymer addition), and
- A combination of raw sludge and WAS.

The organic loading rate should initially be approximately 0.16 kg volatile solids/m^3·d (0.01 lb volatile solids/d/cu ft) and should be increased every 3 days, provided that the digester process-control parameters are within the desired limits. The key to the startup process is control of the volatile solids loading rate. Feed sludge composition or ratios are not important. The digester feeding should be done during a 24-hour period, if possible, to reduce the instantaneous loading rates on the digester and minimize foaming potential.

The operating parameters listed in Table 30.7 should be monitored daily during startup and analyzed using trend charts to identify both positive and negative trends and predict possible digester upsets. Within the target range, the rate of change of an individual parameter is more significant than its absolute value.

Once the operating parameters indicate that the digester is operating at steady state, organic loading should be increased every 3 days at increments of 0.16 kg of volatile solids/m^3·d (0.01 lb of volatile solids/d/cu ft) until it has reached the design volatile solids loading rate.

Shutdown. Digester shutdown should be carefully planned. Prior to shutdown, the pressure- and vacuum-relief valves should be inspected for proper operation to avoid damage to the digester cover. At least 5 weeks before shutdown, a gradual reduction in volatile solids loading should be started for the target digester, while the volatile solids loading rate to the digesters remaining on-line is incrementally increased and their liquid levels kept as high as possible to maintain the maximum HRT. Maintaining the

TABLE 30.7   Digester monitoring table.

Parameter	Units	Test method	Target	Frequency	Sample location Feed sludge	Sample location Recirculated sludge[3] or digested solids
Temperature[1]	°C (°F)	Meter	32–38 °C (90–100 °F)	Daily		X
Volatile acids	mg/L	5560C[4]	50–330	Daily		X
Alkalinity (Alk)	mg/L	2320[4]	1500–5000	Daily	X	X
VA:Alk ratio	(none)	Calculated	0.1 to 0.2	Daily	N/A	N/A
pH	pH units	Meter	6.8–7.2	Daily	X	X
Total solids[2]	%	2540B[4]	(record)	Daily	X	X
Volatile solids[2]	%	2540E[4]	(record)	Daily	X	X
Flow[2]	Liters (gallons)	Meter	(record)	Daily	X	X
Gas production	cfd	Meter	12 to 16 ft^3/lb volatile solids destroyed	Daily	Gas system	
Gas composition (CO_2)	%	Gas analyzer	Less than 35% CO_2	Daily	Gas system	

[1]Temperatures for mesophilic digesters. Thermophilic digester target temperatures may vary based on design.
[2]Measure total and volatile solids content and flow separately for all feed sludges.
[3]Sampled upstream of raw sludge introduction point.
[4]*Standard Methods* (APHA et al., 2005).

maximum HRT should minimize the biological upsets caused by the higher volatile solids loading. The on-line digester's volatile acid-to-alkalinity (VA:ALK) ratio should be monitored daily and the results posted in a trend chart format to assess the effects of increasing volatile solids loading. If the ratio changes by more than 30%, the excess volatile solids loading should be bypassed to a solids holding tank to avoid a process upset.

At least 1 week before the draining cycle is started, the target digester should be completely removed from the normal fill and draw cycle as follows:

- The digester fill valve closed.
- The digested solids withdrawal valve remains closed.
- The solids heating water pumps are secured.
- The solids-recirculation pumps remain in operation, but the heat exchanger is bypassed.
- The solids mixers remain in operation.
- All target digester-gas header valves are closed, isolating the target digester from the common gas header.

Discontinuing the feeding and heating of the digester should significantly reduce gas production.

Draining the digester tank, if necessary, includes the following additional steps:

- Discontinue mixing.
- Add nitrogen gas to the digester gas headspace to displace the digester gas through the pressure-relief valve. The digester is a confined space, and clearance for maintenance or construction activities should be accomplished in accordance with the contractor's or utility's confined-space-safety plan. The application of an inert gas for purging is effective if rapid entrance to the digester is required. Inert gas application should displace explosive gases, but will not create a safe atmosphere for entry, so forced air ventilation is mandatory.
- Monitor the gas flow using a remote monitoring sensor.

FEEDING AND WITHDRAWAL SCHEDULE. *Digester Feeding.* Uniformity and consistency are keys to digester operation. Sudden changes in feed solids volume or concentration, temperature, composition, or withdrawal rates will inhibit digester performance and may lead to foaming. The ideal feeding procedure is a continuous, 24-hour-per-day addition of a blend of different types of feed solids (primary and

WAS). Where continuous feeding is impossible, a 5- to 10-min/h feed cycle is used. Smaller WWTPs that operate a single 8-hour shift use a schedule of at least three feedings: at the beginning, middle, and end of the shift.

Two criteria may determine the digester's capacity to process solids properly: (1) the SRT and (2) the volatile solids loading rate. These criteria determine the amount of food the microorganisms must stabilize and the time available to consume that food. At feed solids concentrations below 3%, the processing capacity is typically limited by SRT. At concentrations higher than 3%, the processing capacity may be affected by the volatile solids loading rate.

For a well-mixed high-rate digester, SRT typically ranges from 15 to 20 days. With total detention times significantly shorter than 15 days, the methane-forming organisms cannot reproduce quickly enough to replace those that are withdrawn. Shorter detention times also reduce the system's buffering capacity (its ability to neutralize volatile acids). Conditions that can shorten detention time include pumping dilute sludge and excessive accumulations of grit and scum in the digester. Sending sludge with the highest possible solids concentration to the digester (within design conditions) will increase the available detention time and reduce heating requirements.

Organic loading rates for a well-mixed and heated digester typically range from 1.6 to 3.2 kg of volatile solids/m^3·d (0.1 to 0.2 lb of volatile solids/d/cu ft). The organic loading rate typically controls the anaerobic digestion process. An *organic overload* is a volatile solids loading that exceeds the daily limits by more than 10%. Typical causes of organic overloads include the following:

- Starting the digester too rapidly,
- Excessive volatile solids loadings as a result of erratic feeding or a change in feed solids composition,
- Volatile solids loadings exceeding the daily limits by more than 10%,
- Loss of active digester volume because of grit accumulation, and
- Inadequate mixing.

Solids Withdrawal. Solids should be withdrawn from the primary digester immediately prior to feeding raw sludge to prevent short-circuiting. In digesters with surface overflow, the timing and rate of solids withdrawal and feed are coordinated to occur concurrently. Solids should be withdrawn at least daily to avoid a sudden drop in the active microorganism population, as well as disruption of the digester's VA:ALK ratio, which would affect its buffering capacity. The primary digester may be regulated to simply overflow to the secondary digester or to the digested sludge storage tank as raw sludge is added. Solids may be withdrawn from the following locations:

- The bottom of the digester,
- The overflow structure, and
- Any point within a well-mixed digester.

A benefit of removing solids from the bottom of the digester is that it may also remove the grit that accumulates on the bottom of the digester. If possible, solids removal should be performed periodically.

It is important to recognize that, because digestion destroys volatile solids, the concentration of the biosolids removed from the digester will be lower than the feed concentration, unless the digester is decanted.

Decanting. Decanting is used to increase the concentration of digested solids. Because decanting requires a variable liquid level, it can only be used if the digester tank is equipped for pumped withdrawal rather than gravity flow using a static liquid level. Decanting also requires that the digester contents be settled to separate the heavier solids from the supernatant. Because settling involves discontinuing digester mixing, this method is not recommended for primary digesters. The secondary digester is typically mixed or heated until several hours or days before decanting begins.

Decanting should be controlled within a narrow range of liquid levels. The key to its success is slow withdrawal and slow filling. If the secondary digester does not include gas storage, the tank is equipped with decant valves and piping with at least four valves that are located at 12- to 18-in. intervals below the high water level.

Decanting an anaerobic digester involves the following steps:

1. Determine the secondary digester liquid level and verify that an adequate volume is available for decanting. The volume decanted should always be the same to ensure uniform detention time.
2. Collect a sample for laboratory analysis of the settleability of the digested solids using settlometer equipment. Record the results every 30, 60, or 90 minutes, depending on the solids-settling characteristics.
3. Based on the settlometer test results, determine if the secondary digester can be decanted successfully. Settlometer values will vary by installation and must be determined based on decant volume and quality.
4. Allow secondary digester contents to settle for 6 to 8 hours after mixing and solids-heating systems are turned off.

5. Slowly open the lowest decant valve and collect a sample of the decant liquid after 10 minutes of continuous flow. Close the lowest decant valve. Perform a 15-minute centrifuge analysis or microwave test of the sample to determine the approximate total solids concentration.

6. Slowly open the second-lowest decant valve and collect a sample of the decant liquid after 10 minutes of continuous flow. Close the decant valve. Perform a 15-minute centrifuge analysis or microwave test of the sample to determine the approximate total solids concentration.

7. Continue to open, collect samples, and close the various decant valves until you have identified the valve that discharges the decant liquid with the lowest total solids concentration.

8. Continue to sample the decant liquid every 20 minutes to determine its quality and minimize recycling the digester solids to the liquid stream treatment.

9. At the proper liquid level, close all decant valves and restart the heating and mixing systems.

SCUM CONTROL. Scum accumulation in digesters is common. Scum is a combination of undigested grease and oil and often contains buoyant materials, such as plastics, that are not removed at the plant's headworks. Scum floats on the digester liquid surface and can accumulate, forming a dense mat. Properly designed and operated digester mixing systems can typically blend the scum into the tank contents.

If the digester operates without mixing for longer than 8 hours, scum may rise and float on the liquid surface. After mixing is restarted, the scum is resuspended within the liquid. The primary method of scum control is to keep the digester mixing system well-maintained during operation.

DIGESTER CLEANING. The frequency of digester cleaning depends on such factors as influent screening, grit-removal efficiency, and digester mixing and configuration. Normal cleaning frequency for cylindrical digesters is every 2 to 5 years. Because of the complexity of the cleaning procedure, many utilities prefer to operate anaerobic digester systems in a "run-to-failure" mode, which means that the digesters are only cleaned after several years of steadily declining performance.

Cleaning procedures must protect against the known physical hazards of toxic gases, confined-space entry, and materials handling, but also from site-specific unknowns, including the following:

- Failure of internal or external valves and piping,
- Failure of digester roofs or cover coatings,

- Deterioration of concrete,
- Downtime and expense associated with mechanical or structural repairs, and
- Potential for overloading operational digesters.

Failure analysis of mechanical, electrical, and instrumentation systems and implementation of repairs should be considered before digester cleaning. The cleaning process should also include chemical or mechanical cleaning of digester piping systems.

Cleaning must be done using the correct lockout and tagout procedures listed in OSHA Part 1910.333, *Electrical*; confined-space-entry procedures listed in OSHA Part 1910.146, *Permit Required Confined Spaces*; or other applicable safety codes. Precautions should be taken against the following hazards:

- *Asphyxiation or suffocation*—The atmosphere should be tested for oxygen and hydrogen sulfide, and monitoring should be continued for the duration of cleaning.
- *Explosions*—Before anyone enters the digester, it should be purged with a non-combustible gas, such as carbon dioxide (CO_2) or nitrogen (N_2), to eliminate the potential for explosion.
- *Leaks*—Before cleaning is started, all of the digester's cross-connection valves to be cleaned should be secured by being closed, tagged, and locked out. The digester gas valve should also be checked to verify that it does not leak. Circuit breakers that control moving equipment in the tank should be opened and tagged.

As digested sludge is removed from the tank, air can be drawn into the tank through defective sampling ports, open hatches, or vacuum-relief valves, creating an explosion hazard. The most explosive condition is created by a methane–air mixture containing 5 to 20% methane. Monitoring for explosive conditions should be done continuously during cleaning.

After completion of cleaning and before the digester is returned to service, the digester should be refilled with any combination of the following:

- Primary clarifier effluent (preferred),
- Secondary clarifier effluent (unchlorinated),
- Unthickened WAS (before polymer addition), and
- A combination of raw sludge and WAS.

Transferring contents from a "dirty" digester to restart a clean digester is not recommended because the inert matter in the transferred material can form deposits in the newly cleaned digester.

After the liquid in the cleaned digester reaches the minimum mixing level, the heating water pumps, solids-recirculation pumps, and solids-mixing systems should be started. As the digester contents reach the proper operating temperature, the remaining liquid level should be raised using a controlled volatile solids loading program. The anaerobic digestion process should be completely stabilized and operating at the design volatile solids loading rate within 45 days from attaining the proper operating temperature. The total time for efficient and effective digester cleaning is approximately 90 days.

If a digester is to be cleaned by a contractor, the cleaning contract should define the boundaries of the work and provide the following specific information from the owner:

- Digester liquid volume;
- Analytical results of well-mixed digester contents defining the solids composition in terms of metals, nutrients, total solids, and volatile solids concentrations;
- Available disposal site (landfill, onsite lagoon, etc.) and stabilization requirements for the digester contents;
- Quality and quantity of cleaning-process sidestreams returned to the treatment plant;
- Responsibility for opening the digester and venting digester gases—this is best accomplished jointly by plant and contractor staff;
- Cleanliness criteria for tankage and piping;
- Available facilities, including work area dimensions, water supply (flowrate and pressure), and electric power supply (voltage and amperage);
- Quantities, types, and concentrations of solids from the facility that the contractor must process during the digester shutdown; and
- Coordination of lockout and tagout of power and valve isolation.

The contractor's qualifications should include the following:

- Required insurance,
- At least 5 years of actual experience,
- Company-wide and site-specific safety programs,
- Documentation verifying company-wide and site-specific employee training, and
- Appropriate references.

PRECIPITATE FORMATION AND CONTROL. The digestion process can produce crystalline precipitates that affect both the digestion system and downstream solids-handling processes. The precipitates can accumulate on pipes and dewatering equipment, causing damage and blockages and requiring costly and time-consuming maintenance (Figure 30.23). Common precipitates include struvite, vivianite, and calcium carbonate. The constituents that form these precipitates are present in undigested solids and are released during the digestion process and converted to soluble forms that can react and crystallize. Their formation varies from site to site, depending on the chemistry of the digested solids and the treatment processes. Because precipitates preferentially form on rough or irregular surfaces, glass-lined sludge piping and long-radius elbows help minimize their accumulation.

The formation of struvite {magnesium ammonium phosphate [$MgNH_4PO_4 \cdot 6(H_2O)$]} depends on the relative concentrations of these constituents, as well as the pH of the digested sludge. The turbulence caused by pumping and dewatering strips carbon dioxide from the liquid, raising its pH and creating conditions favorable for struvite formation. Struvite formation can be controlled by adding iron salts (ferric chloride or ferric sulfate) or polymeric dispersing agents. Iron salts, which can be added at such locations as the headworks, the final clarifiers, or digesters, react with the phosphate in the solids matrix to form vivianite [$Fe_3(PO_4)_2$]. While some of the available phosphate forms vivianite, the ratio of the remaining phosphate to magnesium and ammonium changes, affecting the struvite reaction. Iron requirements are site-specific, but struvite

FIGURE 30.23   Struvite deposition in pipe.

accumulation can typically be prevented via iron dosages of 7 to 14 kg (15 to 30 lb) of iron (as Fe) per dry ton of solids. Because iron preferentially reacts with hydrogen sulfide, high sulfide concentrations in the solids will increase the iron requirement.

Dispersing agents prevent struvite accumulation by interfering with crystal formation. A number of proprietary dispersing agents are available from polymer manufacturers. Required dosages vary widely.

In such locations as centrifuge centrate lines or belt filter press filtrate lines, plant effluent can be used to dilute the concentration of the limiting constituent below the precipitation threshold.

Vivianite is the desired product when iron is added to control struvite. While vivianite is a crystalline solid like struvite, it remains in the solids matrix rather than accumulating on pipes and equipment, unless iron is added directly upstream from heat exchangers; in the latter case, it can accumulate on heat exchangers, causing plugging and interfering with heat transfer.

Less common precipitates include various calcium complexes, such as calcium carbonate. Causes of calcium scaling are not as clearly understood as struvite or vivianite scaling.

Precipitates that form in pipelines or on equipment can be removed by a combination of acid-washing, pigging, and scraping.

MONITORING AND TESTING. Proper sampling is critical for effective anaerobic digester process-control monitoring. Trend charts are an effective means of tracking changes in digester feed characteristics or operation, as well as predicting possible process upsets. They can be used for monitoring volatile suspended solids, total daily solids feed, VA:ALK ratio, VSR, and gas production. According to trend charts, the rate of change of an individual constituent is more significant than its absolute value.

Anaerobic digestion control is based on samples collected from three points in the digestion process: (1) digester feed, (2) recirculated sludge, and (3) digested solids. Solids sampling points are typically located in the digester building. Each digester has one sampling point for influent and one for discharge.

Samples for performance evaluations should be collected from each active digester. Recommended sampling locations, frequencies, and analyses are listed in Table 30.7. The following guidelines for collecting and analyzing samples are essential to obtaining accurate results:

- The sample must be representative;
- pH analysis must be performed immediately to avoid a deviation in pH due to loss of carbon dioxide;

- All digesters must be sampled in rapid succession;
- All sample containers must be cleaned thoroughly before and after use; and
- If the sample is not analyzed immediately, it must be refrigerated.

After the samples are analyzed, the results should be compared with data from previous analyses to verify that the present data points are within reasonable ranges. If not, additional samples should be collected and analyzed to confirm the results. If data appear to be reasonable, the new data points should be added to the existing trend analyses to gauge overall anaerobic digester performance. If the trend charts indicate movement toward the limits, corrective actions should be taken. Early action may prevent the process parameter from moving outside the desired range. However, a single high or low data point does not constitute a trend. Corrective action should be based on multiple data points that indicate a continued change in digester performance.

Temperature. Microorganisms in the anaerobic digestion process require specific temperature ranges to live and thrive. Digestion nearly ceases at approximately 10 °C (50 °F); however, the microorganisms of concern are most effective in the mesophilic range [30 to 38 °C (85 to 100 °F)]. High-temperature digesters operate in the thermophilic range [55 to 60 °C (131 to 140 °F)]. Regardless of the selected operating temperature, the tank temperature must deviate from that value by more than 0.6 °C (1 °F) per day. Each group of methane-forming microorganisms has an optimum temperature for growth; if the temperature is permitted to fluctuate, the methane formers cannot develop a large, stable population. It is a good practice to log the digester temperature daily to eliminate temperature swings.

pH. Digesters can operate at pH ranges from 6.0 to 8.0, with the desired operating range being 6.8 to 7.2. If the pH falls below 6.0, un-ionized volatile acids become toxic to methane-forming microorganisms. At pH values above 8.0, un-ionized aqueous ammonia (dissolved ammonia) becomes toxic to methane-forming microorganisms. These key pH values reflect the ionization constants for ammonia and acetic acid. The pH of the digester contents is controlled by the amount of volatile acids and alkalinity.

Alkalinity. The digester's buffering capacity, or its ability to resist changes in pH, is indicated by its alkalinity. Calcium, magnesium, and ammonium bicarbonates are examples of buffering substances typically found in a digester. The digestion process produces ammonium bicarbonate; the other buffering substances are contained in raw sludge. The concentration of total alkalinity in a well-established heated digester typically ranges from 1500 to 5000 mg/L. As a rule, the higher the alkalinity, the more stable the digester. Alkalinity enables the digester to better handle a sudden increase in organic load without upset.

Volatile Acids. Volatile acids are intermediate digestion byproducts. Although typical volatile acid concentrations in an anaerobic digester range from 50 to 300 mg/L, higher concentrations are acceptable if sufficient alkalinity is available to buffer the acid.

Volatile Acid to Alkalinity Ratio. The VA:ALK ratio is an indicator of the progress of digestion and the balance between the acid fermentation and methane fermentation microorganisms. The VA:ALK ratio is shown in Equation 30.7:

$$\text{VA:ALK} = \frac{\text{Volatile acids (mg/L)}}{\text{Alkalinity (mg/L)}} \tag{30.7}$$

Because of the need for balance between volatile acids and alkalinity, the VA:ALK ratio is an excellent indicator of digester health. Careful monitoring of the rate of change in this ratio can indicate a problem before a pH change occurs. The VA:ALK ratio should be approximately 0.1 to 0.2. Digester pH depression and inhibition of methane production occur if the ratio exceeds 0.8; however, ratios higher than 0.3 to 0.4 indicate upset conditions and the need for corrective action (see the "Troubleshooting Guide" section of this chapter).

DIGESTER UPSETS AND CONTROL STRATEGIES. The four basic causes of digester upsets are hydraulic overload, organic overload, temperature stress, and toxic overload. Hydraulic and organic overloads occur when the design hydraulic or organic loading rates are exceeded by more than 10% per day. The overload conditions can be controlled via management of digester feeding, as well as ensuring that the effective digester volume is not diminished by grit accumulation or poor mixing. Digester feeding is controlled via proper operation of upstream headworks, clarifiers, and thickeners to ensure the feed solids concentrations.

In the event of a digester upset, an effective control strategy includes the following steps:

1. Stop or reduce solids feed,
2. Determine the cause of the imbalance,
3. Correct the cause of the imbalance, and
4. Provide pH control until treatment returns to normal.

If only one digester tank is affected, the loading on the remaining units can be carefully increased to allow the upset unit to recover. If overloading is affecting several units, reducing the feed will require a method of dealing with the excess solids by hauling them to another facility, providing temporary storage onsite, or chemically stabilizing and disposing of them.

Temperature. Temperature-related stress is caused by a change in digester temperature of more than 1 or 2 °C (2 or 3 °F) in fewer than 10 days, which would reduce the biological activity of the methane-forming microorganisms. If the methane formers are not quickly revived, the acid formers, which are unaffected by the temperature change, continue to produce volatile acids, which will eventually consume the available alkalinity and cause the pH to decline.

The most typical causes of temperature stress are overloading solids and exceeding the instantaneous capacity of the heating system. Most heating systems can eventually heat the digester contents to the operating temperature, but not without a harmful temperature variation.

Another cause of temperature stress is operating the digester outside its optimum temperature range. For example, a mesophilic digester has an optimum temperature range of 32 to 38 °C (90 to 100 °F). At temperatures lower than 32 °C (90 °F), the biological process slows. At temperatures above 38 °C (100 °F), the digester efficiency is not improved and the system is wasting energy.

Toxicity Control. The anaerobic process is sensitive to certain compounds, such as sulfides, volatile acids, heavy metals, calcium, sodium, potassium, dissolved oxygen, ammonia, and chlorinated organic compounds. The inhibitory concentration of a substance depends on many variables, including pH, organic loading, temperature, hydraulic loading, the presence of other materials, and the ratio of the toxic substance concentration to the biomass concentration. The inhibitory levels of several compounds are listed in Tables 30.8, 30.9, and 30.10.

Sodium sulfide or ferric or ferrous sulfate can be added to alleviate heavy metal toxicity. Because toxic heavy metal sulfides have low solubility and are less soluble than ferric sulfide, the toxic metals precipitate as sulfides. Ferric chloride can be used to

TABLE 30.8 Effect of ammonia nitrogen on anaerobic digestion (U.S. EPA, 1979).

Ammonia concentration, as N[a] (mg/L)	Effect
50–200	Beneficial
200–1000	No adverse effects
1500–3000	Inhibitory at pH 7.4 to 7.6
>3000	Toxic

[a]Nitrogen

TABLE 30.9   Total concentration of individual metals required to severely inhibit anaerobic digestion (U.S. EPA, 1979).

Metal	Concentration in digester contents		
	Dry solids (%)	Moles metal/kg dry solids	Soluble metal (mg/L)
Copper	0.93	150	0.5
Cadmium	1.08	100	—
Zinc	0.97	150	1.0
Iron	9.56	1710	—
Chromium			
6+	2.20	420	3.0
3+	2.60	500	—
Nickel	—	—	2.0

control sulfide concentration by precipitation as ferric sulfide. Excessive use of these chemicals can result in pH depression.

pH Control. The key to controlling the digester pH is to add bicarbonate alkalinity to react with acids and buffer the system pH to about 7.0. Bicarbonate can be added directly or indirectly as a base that reacts with dissolved carbon dioxide to produce bicarbonate. Chemicals used for pH adjustment include lime, sodium bicarbonate, sodium carbonate, sodium hydroxide, ammonium hydroxide, and gaseous ammonia. Lime addition can be messy and will produce $CaCO_3$. Although ammonia compounds can be used for pH adjustment, they may cause ammonia toxicity and increase the ammonia load on the liquid treatment processes through return streams. Consequently, their use is not recommended.

TABLE 30.10   Stimulating and inhibitory concentrations of light metal cations (U.S. EPA, 1979).

Cation	Concenration (mg/L)		
	Stimulatory	Moderately inhibitory	Strongly inhibitory
Calcium	100–200	2500–4500	8000
Magnesium	75–150	1000–1500	3000
Potassium	200–400	2500–4500	12 000
Sodium	100–200	3500–5500	8000

During a digester upset, volatile acid concentrations may begin to rise before bicarbonate alkalinity is consumed. Because pH depression does not occur until alkalinity is depleted, it may be observed only after the digester is well on its way to failure. The relationships between alkalinity, volatile acids, methane production, carbon dioxide production, and pH during an upset are indicated in Figure 30.24.

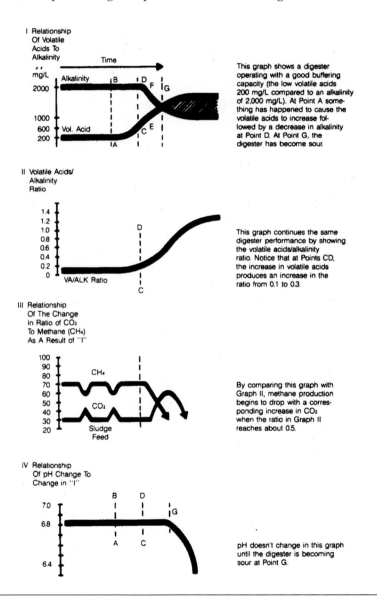

FIGURE 30.24   Change sequence in digester.

The correct chemical dosage for pH control can be calculated based on the measured concentrations of volatile acids and alkalinity. The VA:ALK ratio should be approximately 0.1 to 0.2. When the ratio exceeds 0.3 to 0.4, corrective action should be taken by decreasing the digester volatile solids loading rate from 1.6 kg/m³·d to 1.2 kg/m³·d (0.10 to 0.075 lb volatile solids/ft³·d), which will decrease solids feed and withdrawal rates by approximately 25%, and by maintaining the internal digester sludge temperature at 35 °C (95 °F), ±1 degree.

An increase in the VA:ALK ratio to 0.5 or above indicates a process upset and the need to add alkalinity. The proper alkalinity dosage can be calculated using the volatile acids concentration. The volatile solids loading rate should be decreased from 1.6 kg/m³·d to 0.8 kg/m³·d (0.10 to 0.05 lb volatile solids/ft³·d), which will reduce the solids feed and withdrawal rates by approximately 50%.

Increases in the VA:ALK ratio to 0.8 or above indicate a process upset with pH depression and inhibition of the methane formers. This requires the addition of alkalinity and reduction of volatile solids loading to 0.16 kg/m³·d (0.01 lb volatile solids/ft³·d) until the VA:ALK ratio drops to 0.5 or lower.

Alkalinity addition can be calculated using the following procedures:

1. Determine volatile acid and alkalinity concentrations (as $CaCO_3$).
2. Using a target VA:ALK ratio of 0.1 and the measured volatile acids concentration, calculate total desired alkalinity using the following equation:

$$\text{Alkalinity (mg/L)} = \frac{\text{Volatile acids (mg/L)}}{0.1} \qquad (30.8)$$

1. Subtract the measured alkalinity value from the alkalinity requirement calculated in Step 2 to determine the alkalinity addition requirement.
2. Calculate the required chemical dosage for the alkalinity addition calculated in Step 3 based on the equivalent weight ratios presented in Table 30.11.
3. Adjust chemical dosages based on the purity of delivered chemical.
4. Determine total chemical addition for the digester based on digester volume and chemical dosage calculated in Step 6 using the following equations:

$$\text{Chemical addition (kg)} = \frac{\text{Alkali addition (mg/L)} \times \text{Digester volume (L)}}{106} \qquad (30.9)$$

$$\text{Chemical addition (lb)} = \frac{\text{Alkali addition (mg/L)} \times \text{Digester volume (gal)} \times 8.34(\text{lb/gal})}{106} \qquad (30.10)$$

TABLE 30.11   Alkalinity equivalent weight ratios.

Chemical name	Formula	Ratio
Anhydrous ammonia	NH_3	0.32
Aqua ammonia	NH_4OH	0.70
Anhydrous soda ash	Na_2CO_3	1.06
Caustic soda	$NaOH$	0.80
Hydrated lime	$Ca(OH)_2$	0.74

Dose the calculated amount of chemical over an extended period to avoid scaling on the heat exchanger or pipelines. Typically, alkalinity is added over a 3- or 4-day period. Mix well, and frequently monitor volatile acids, pH, and alkalinity. Avoid toxicity from the cations associated with the alkalis. Confirm that the vacuum-relief device is operable.

Digester Foaming.   Digester foam consists of fine gas bubbles trapped in a semi-liquid matrix with a specific gravity of 0.7 to 0.95. The gas bubbles are generated below the solids layer and are trapped as they form.

While some foaming always occurs, it is considered excessive if it plugs piping or escapes from the digester. Excessive foaming can cause the loss of active digester volume, structural damage, spillage, and damage to the gas-handling system, as well as being malodorous and unsightly. The most common cause of digester foaming is organic overload, which results in the production of more VFAs than can be converted to methane. The acid formers (which release carbon dioxide) work much more quickly than the methane-forming microorganisms. The resulting increase in carbon dioxide typically increases foam formation. Factors that can contribute to organic overload include:

- Intermittent digester feeding;
- Separate feeding or inadequate blending of primary sludge and WAS;
- Insufficient or intermittent digester mixing; and
- Excessive amounts of grease or scum in digester feed (especially problematic if the digester is fed in batches).

Organic overload can be minimized by feeding the digesters continuously (or as often as possible), blending different feed sludges well before feeding, ensuring that the digester-mixing system is operable, and limiting the quantities of grease or scum in the digester feed.

TABLE 30.12 Troubleshooting guide.

Indicators	Probable cause	Check or monitor	Solutions
DIGESTER COVERS—FIXED			
1. Gas pressure higher than normal during freezing weather.	1. Pressure-relief valve stuck or closed.	1. Weights on pressure-relief valve.	1. If freezing is a problem, apply a light grease layer impregnated with rock salt to lever action and springs.
2. Gas pressure lower than normal.	2. a. Pressure-relief valve or other pressure-control devices stuck open. b. Gas line or hose leaking.	2. a. Pressure-relief valve and devices. b. Gas line and/or hose.	2. a. Manually operate pressure and vacuum relief and remove corrosion if present and interfering with operation. b. Repair broken lines as needed.
3. Leaks around metal covers.	3. Anchor bolts pulled loose and/or sealing material moved or cracking.	3. Bolts and sealing materials.	3. Repair concrete with fast-sealing concrete repair material. Apply new sealant material to leaking area using material with durable plasticity.
4. Gas leak through digester cover.	4. Fastener failure.	4. Apply soap solutions to suspected area and check for bubbles.	4. Replace defective fasteners.
DIGESTER COVERS—FLOATING			
1. Cover tilting; little or no scum around the edges.	1. a. Weight distributed unevenly. b. Water from condensation or rainwater collecting inside dome. c. Ice accumulation on surface	1. a. Location of weights. b. Check around the edges of the metal cover. (Some covers with insulating wooden roofs have inspection holes for this purpose.) c. Check for ice on cover	1. a. If moveable ballast or weights are provided, adjust location until the cover is level. If no weights are provided, use sand bags to level cover. (Note: pressure-relief valves may need to be reset if a significant amount of weights is added.) b. Use siphon or other means to remove the water. May also require significant cover repair to eliminate leaks. c. Manually remove ice accumulation
2. Cover tilting; heavy thick scum accumulating around edges.	2. a. Scum accumulation, causing excess drag on cover. b. Guides or rollers out of adjustment. c. Broken rollers.	2. a. Probe with a stick to determine scum accumulation. b. Distance between guides or rollers and the wall. c. Determine the normal position if the suspected broken part is covered by sludge. Verify correct location using manufacturer's information and/or prints if necessary.	2. a. Use chemicals or degreasing agents, such as Digest-aide or Sanfax, to soften the scum. Hose down with water. Continue on regular basis every 2 to 3 months, or more frequently if needed. b. Soften up the scum (as in "2a") and readjust rollers to prevent binding. c. Drain tank if necessary, taking care to prevent binding or uneven cover descent. Use of a crane or jacks may be required to prevent structural damage.

TABLE 30.12 Troubleshooting guide (*continued*).

Indicators	Probable cause	Check or monitor	Solutions
3. Freezing problems in PRV.	3. See fixed digester cover troubleshooting guide, Item 1.	3. See fixed digester cover troubleshooting guide, Item 1.	3. See fixed digester cover troubleshooting guide, Item 1.
DIGESTER COVERS—GAS HOLDER TYPE			
1. Cover guides and/or rollers causing cover to bind.	1. Scum accumulation restricting travel.	1. Check scum accumulation and verify amounts.	1. See floating cover troubleshooting guide, Item 1.
2. Cover binds although rollers and guides are free.	2. Damaged or binding internal guide or guy wires.	2. Lower cover to corbels. Open hatch; vent digester. Using breathing apparatus and explosion proof light, inspect internal equipment from outside digester. If cover movement is restricted, it may be necessary to secure using a crane or by other method to prevent skirt damage to sidewalls.	2. Drain and repair, holding the cover in a fixed position if necessary.
3. Cover tilting, heavy thick scum accumulating around edges.	3. a. Scum accumulation in one area causing excess drag. b. Guides or rollers out of adjustment.	3. a. Probe with a stick to determine scum accumulation. b. Distance between guides or rollers on the wall.	3. a. Use chemicals or degreasing agents, such as Digest-aide or Sanfax, to soften the scum; hose down with water. Continue on regular basis every 2 to 3 months, or more frequently if needed. b. Soften up the scum [as in 3(a)]. Readjust rollers to eliminate skirt binding.
4. Gas pressure higher than normal during freezing weather.	4. a. Water accumulation in gas line. b. Pressure-relief valve stuck or frozen in place.	4. a. Find and eliminate water accumulation point; empty all traps. b. Weights on pressure-relief valve.	4. a. Protect line from weather by covering and insulating overflow box. b. If freezing is a problem, apply light grease layer impregnated with rock salt.
5. Gas pressure lower than normal.	5. a. Pressure-relief valve or other pressure-control devices stuck open. b. Gas line or hose leaking.	5. a. Pressure-relief valve and devices. b. Gas line and/or hose.	5. a. Manually operate pressure and vacuum relief; remove corrosion if present and interfering with operation. b. Repair as needed.

TABLE 30.12   Troubleshooting guide (*continued*).

Indicators	Probable cause	Check or monitor	Solutions
SOLIDS MIXING—MECHANICAL MIXERS			
1. Corrosion of exposed parts by weather and/or corrosive wastewater gas.	1. Lack of paint or other protection.	1. Note presence of rust, corrosion, or bare exposed metal.	1. a. Construction of protective structure. b. Preparation and painting of surfaces.
2. Gear reducer wear.	2. a. Lack of proper lubrication. b. Poor alignment of equipment.	2. Excessive motor amperage; excessive noise; vibration and evidence of shaft wear.	2. a. Verify correct type and amount of lubrication from manufacturer's literature. b. Correct imbalances caused by accumulation of material on the internal moving parts.
3. Shaft seal leaking.	3. Packing dried out or worn.	3. Evidence of gas leakage when checked by soap solution or evident gas odor.	3. a. Follow manufacturer's instructions for repacking. b. Replace packing any time the tank is empty if it is not possible when unit is operating.
4. Wear on internal parts.	4. Grit or misalignment.	4. Visual observation when tank is empty. Compare with manufacturer's drawings for original size. Motor amperage will also decrease as moving parts are worn away.	4. Replace or rebuild. Experience determines the frequency of this operation.
5. Imbalance of internal parts due to accumulation of debris on the moving parts (large-diameter impellers or turbines are most affected).	5. Poor grinding and/or screening.	5. Vibration, heating of motor, excessive amperage, noise.	5. a. Reverse direction of mixer if possible. b. Stop and start alternately. c. Draw down tank and clean moving parts.
6. Power source interruption.	6. High ambient temperature.	6. Excessive amperage, corroded power connections, overheating causing circuits to kick out.	6. Protect motor by covering with ventilator housing.
SOLIDS MIXING—GAS MIXERS			
1. Compressor running hot and/or noisy.	1. a. Low lubricant level. b. High ambient temperature. c. High discharge pressure. d. Clogged suction.	1. a. Check lubricant level. b. Check for excessive ambient temperature. c. Check compressor for excessive wear. d. Check valve discharge position e. Check suction valve position	1. a. Add lubricant b. Provide ventilation if necessary c. Adjust valve position d. Backflush lines

TABLE 30.12 Troubleshooting guide (*continued*).

Indicators	Probable cause	Check or monitor	Solutions
2. Gas feed lines plugging.	2. a. Lack of flow through gas line. b. Debris in gas line.	2. Identify low temperature of gas feed pipes or low pressure in the manometer or pressure gauges.	2. a. Flush out gas lines with water. b. Clean feed lines and valves.
DIGESTER GAS SYSTEM			
1. Gas is leaking through pressure-relief valve (PRV) on roof.	1. Valve not sealing properly or is stuck open.	1. Check the manometer for normal digester gas pressure.	1. Remove PRV cover and move weight holder until it seats properly. Clean and install new ring if needed.
2. Manometer shows digester gas pressure is above normal.	2. a. Obstruction or water in main burner gas line. b. Digester PRV is stuck shut. c. Waste gas burner line pressure-control valve is closed.	2. a. If all points are operating and normal, check for a waste gas line restriction or a plugged or stuck safety device. b. Gas is not escaping properly. c. Gas meters indicate excess gas is being produced, but not going to waste gas burner.	2. a. Purge with nitrogen; drain condensate traps and check for low spots. **Caution:** Do not force air into the digester. b. Remove PRV cover and manually open valve; clean valve seat. c. Open valve
3. Manometer shows digester gas pressure below normal.	3. a. Too-fast withdrawal, causing a vacuum inside digester. b. Too much alkalinity addition.	3. a. Check vacuum breaker to verify it is operating properly. b. Sudden increase in CO_2 in digester gas.	3. a. Stop digester discharge and close off all gas outlets from digester until pressure returns to normal. b. Stop addition of alkalinity.
4. Frozen PRV valve.	4. Winter conditions.	4. Remove valve cover and inspect PRV.	4. Possible remedies: a. Place vented barrel over valve with an explosion-proof light bulb inside.
5. Pressure-regulating valve not opening as pressure increases.	5. a. Inflexible diaphragm. b. Ruptured diaphragm.	5. Isolate valve and open valve cover.	5. a. If no leaks are found (using soap solution), diaphragm may be lubricated and softened using neatsfoot oil. b. Ruptured diaphragm requires replacement.
6. Yellow flame from waste-gas burner.	6. Poor quality gas with high CO_2 content.	6. Check CO_2 content will be higher than normal.	6. Check concentration of solids feed; it may be too dilute. If so, increase solids concentration.
7. Gas flame lower than usual.	7. a. High gas usage. b. Gas leak from digester piping or safety devices. c. Low gas production due to process problems.	7. a. Check gas-production rate against gas usage. b. Check gas-collection and -distribution system, starting at digester main collection point.	7. a. This may be normal. b. Check for leaking gas; when found, isolate and repair.

TABLE 30.12 Troubleshooting guide (*continued*).

Indicators	Probable cause	Check or monitor	Solutions
8. Waste gas burner not lit.	8. a. Pilot flame not burning. b. Obstruction or water in main waste-gas line or pilot natural gas line. c. Pressure-control valve closed due to malfunction.	8. a. Pilot line pressure at waste-gas burner. b. Condensate traps. c. Waste-gas pressure-control valve.	8. a. Re-light natural gas pilot line. b. Drain condensate traps. c. Open valve and verify that setting will allow valve to open when pressure is about 0.06 kPa (¼-in. water column) above all other use point pressures.
9. Gas meter failure.	9. a. Debris in line. b. Electrical failure.	9. a. Condition of gas line. b. Fouled or worn parts.	9. a. Isolate digester; flush gas line with water, moving from digester to point of use. b. Replace worn parts.
SOLIDS TEMPERATURE CONTROL—EXTERNAL HEAT EXCHANGERS			
1. Low rate of solids fed through the exchanger.	1. High temperature override shuts pump down to prevent solids caking on the exchanger.	1. a. Pressure and water temperatures. b. Check high-temperature circuit for hot water heater to verify that the temperature setting is correct.	1. a. Open closed valve on pump. b. Check for and remove obstruction in the solids pump. c. Check for and remove obstruction in heat exchanger. d. Check temperature control valve for proper operation.
2. Recirculation pump not running; power circuits OK.	2. Temperature override in circuit to prevent pumping too-hot water through tubes.	2. Visual check; no pressure on solids line.	2. a. Allow system to cool off. b. Check temperature-control circuits.
3. Solids temperature drops and cannot be maintained at normal level.	3. a. Solids plugging heat exchanger. b. Solids recirculation line is partially or completely plugged.	3. a. Check inlet and outlet pressure and exchanger. b. Check pump inlet and outlet pressure.	3. a. Open heat exchanger and clean. b. Clean recirculation line.
4. Solids temperature rises.	4. Temperature controller is not working properly.	4. Check water temperature and controller setting.	4. If over 49 °C (120 °F), reduce temperature. Repair or replace controller.
5. Temperature readings not accurate.	5. Probes have electrical short or separation internally.	5. Compare with thermometer known to be accurate.	5. Leave probe connected to readout device; immerse in bucket of water at approximate digester temperature with thermometer in it. Compare readings.
6. Unable to maintain temperature.	6. Hydraulic overloading.	6. Feed sludge concentration.	6. Increase influent sludge concentration.

TABLE 30.12 Troubleshooting guide (*continued*).

Indicators	Probable cause	Check or monitor	Solutions
SOLIDS TEMPERATURE CONTROL—INTERNAL HEATING			
1. Low water feed rate to heat exchanger.	1. a. Air lock in line. b. Valve partially closed.	1. Inlet and outlet meter readings lower than normal but equal.	1. a. Bleed air-relief valve. b. Upstream valve may be partially closed.
2. Low heat loss between inlet and outlet water.	2. Coating on tubes.	2. Temperature gauges on inlet and outlet lines read about the same.	2. Remove coating on the water tubes.
3. Solids caked on the outside of the heat exchanger tubing.	3. Temperature too high.	3. Temperature records.	3. a. Remove solids coating. b. Control water temperature to 54 °C (130 °F) maximum.
4. Coating on inside of heat exchanger piping.	4. Temperature too high.	4. Temperature records.	4. May be controlled by adding chemicals to boiler makeup water.
SOLIDS PUMPING AND PIPELINES			
1. Pump suction and discharge pressure erratic. Pump makes unusual sounds.	1. Sand, grease, or debris plugging suction line.	1. Pump suction and discharge pressure.	1. a. Backflush the line with heated digester sludge. b. Use mechanical cleaner. c. Apply water pressure. **Caution:** Do not exceed working line pressure. d. Add approximately 3.6 g/L (3 lb/100 gal) water trisodium phosphate (TSP) or commercial degreaser. (Most convenient method is to fill scum pit to a volume equal to the line, add TSP or other chemical, then admit to the line and let stand for 1 hour.)
DIGESTED SLUDGE			
1. Gray digested sludge.	1. a. Improper digestion. b. Short-circuiting or insufficient mixing.	1. a. Layers of unmixed sludge in tank bottom.	1. a. See solids mixing troubleshooting guide.
2. Sour odor.	2. a. pH of digester is too low. b. Second stage of digestion is not performing properly. c. Overloaded digester.	2. a. Check pH at different levels. b. Check CO_2 content of gas. c. Check volatile solids loading rate.	2. a. Adjust pH using lime or other caustic. b. Let digester rest. c. Reduce volatile solids loading rate.
3. Complete lack of biological activity.	3. Highly toxic waste, such as metals or bacteriocide, in digester feed.	3. a. Gas production (or lack of). b. Analyze sample by spectrophotometer or chemical means.	3. Empty digester and re-start digester.

TABLE 30.12 Troubleshooting guide (*continued*).

Indicators	Probable cause	Check or monitor	Solutions
DECANTING			
1. Foam observed in supernatant from single-stage or primary tank.	1. a. Scum blanket breaking up. b. Excessive gas recirculation.	1. a. Check condition of scum blanket. b. Volatile solids loading.	1. a. Normal condition, but stop withdrawing supernatant if possible. b. This condition may indicate an organic overload in digester, making it necessary to slow down feeding.
2. Lumps and particles of scum in supernatant from single-stage or primary tank.	2. a. Scum layer breaking up due to excessive mixing or gas production. b. Scum layer too thick.	2. a. Observation through window in digester cover. (Unusual increase in gas production is also an indicator.) b. Depth of scum layer, measured through thief hole or through gap between floating cover and digester wall.	2. a. Decrease mixing time, adjust sludge feed. b. See scum blanket troubleshooting guide.
3. Supernatant from single-stage or primary tank is gray or brown.	3. a. Inadequate stratification; raw sludge in pockets in the tank. b. Digestion time is too short. Solids concentration is too low, or digester capacity is reduced due to grit and scum accumulation. c. Digester ecological balance is upset. d. Overloading digester.	3. a. Check mixing; under-mixing may be culprit. Take samples at various depths to detect pockets of undigested sludge. Check temperature gradient in the digester. b. Probe digester to determine grit deposits. c. CO_2 content; compare gas production to amount of volatile solids being fed. Gas production should average 0.7 to 1 m^3/kg (12 to 16 cu ft/lb) volatile solids destroyed.	3. a. Increase mixing, feeding frequency, or recirculation. b. Adjust feed concentration. Increase mixing or clean out digester. c. Reduce feed rate or increase detention time by some other means.
4. Supernatant has a sour odor.	4. a. Digester pH is too low. b. Overloaded digester (rotten egg odor). c. Toxic load (rancid butter odor).	4. pH of supernatant should be 6.8.	4. a. Adjust pH using lime or other caustic. b. Reduce volatile solids loading.

TABLE 30.12 Troubleshooting guide (*continued*).

Indicators	Probable cause	Check or monitor	Solutions
5. The supernatant solids are too high, causing plant upset.	5. a. Excessive mixing and not enough settling time. b. Supernatant draw-off point not at same level as supernatant layer. c. Raw sludge feed point too close to supernatant draw-off line. d. Not withdrawing enough digested sludge.	5. a. Observe separation pattern using 10 to 20 L of digester contents in a glass carboy. b. Locate stratum of supernatant by sampling at different depths. c. Determine volatile solids content. Should be close to value found in well-mixed sludge and much lower than raw sludge. d. Compare feed and withdrawal rates; check volatile solids to see if sludge is well digested.	5. a. Allow longer settling periods before withdrawing supernatant. b. Adjust tank operating level or draw-off pipe to get into stratum. c. Modify feed/draw-off piping configuration. d. Increase digested sludge withdrawal rates. **Caution:** Daily withdrawal should not exceed 5% of digester volume.
SCUM BLANKET			
1. Rolling movement is slight or absent.	1. a. Mixer is off. b. Inadequate mixing. c. Scum blanket is too thick.	1. a. Mixer switch or timer. b. Measure blanket thickness.	1. a. May be normal if mixers are set on a timer. If not, and mixers should be operating, check for malfunction. b. Consider increasing the mixing time. c. See Item 4 below.
2. Scum blanket is too high.	2. Supernatant overflow line is plugged.	2. a. Check gas pressure; it may be above normal or relief valve may be venting to atmosphere. b. Check supernatant line for flow.	2. a. Lower contents through bottom draw-off; rod supernatant line to clear plugging. b. Increase mixing time or break up blanket by some other physical means.
3. Scum blanket too thick.	3. Lack of mixing; high grease content.	3. Probe blanket for thickness through thief hole or through gap between floating cover and tank wall.	3. a. Break up blanket using mixers. b. Use solids-recirculation pumps and discharge above the blanket. c. Use Sanfax or Digest-aide to soften blanket. d. Break up blanket physically with pole.
4. Draft tube mixers not moving surface adequately.	4. Scum blanket too high, allowing thin sludge to travel under it.	4. Rolling movement on solids surface.	4. a. Lower solids level to 7 to 10 cm (3 to 4 in.) above top of tube, allowing thick material to be pulled into tube. Continue for 24 to 48 hours. b. Reverse mixing direction (if possible).

While intermittent or insufficient mixing can result in organic overload, it can also allow a scum layer to form on the liquid surface. Feed sludge concentrations higher than the design value can also have an adverse effect on the mixing system.

Blockages in gas piping can also contribute to foaming. If water accumulates in the gas piping, it can cause a blockage, which will increase the pressure in the digester headspace. When the blockage is removed, the sudden drop in pressure can cause the digester contents to foam. Frequent draining of water from the gas system piping will prevent this situation.

Filamentous bacteria, such as *Nocardia*, trap gas in their structure and release surface-active agents that collect on bubble surfaces, causing foaming. Filamentous bacteria can typically be controlled via proper operation of the liquid stream and digester treatment processes.

The type of solids fed to the digesters can also cause foaming problems. Digester feed consisting of 100% WAS or having a high ratio of WAS to primary solids can commonly lead to foaming.

TROUBLESHOOTING GUIDE. Possible indicators and causes of digester problems and recommendations for corrective actions are presented in Table 30.12.

REFERENCES

American Public Health Association; American Water Works Association; Water Environment Federation (2005) *Standard Methods for the Examination of Water and Wastewater*, 21st ed.; American Public Health Association: Washington, D.C.

Haug, R. T.; Harnett, W.; Ohanian, E.; Hernandez, G.; Abkian, V.; Mundine, J. (2002) Los Angeles Goes to Full-Scale Class A Using Advanced Digestion. *Proceedings of the Water Environment Federation 16th Annual Residuals and Biosolids Management Conference*; Austin, Texas, March 3–6; Water Environment Federation: Alexandria, Virginia.

McQuiston, F. C.; Spitler, J. D. (1992) *Cooling and Heating Load Calculation Manual*, 2nd ed.; American Society of Heating, Refrigerating and Air-Conditioning Engineers, Inc.: Atlanta, Georgia.

National Fire Protection Association (2003a) Chapter 6: Solids Treatment Processes. In *Standards for Fire Protection in Wastewater Treatment and Collection Facilities*; Publication 820; National Fire Protection Association: Quincy, Massachusetts.

Schafer, P.; Farrell, J.; Newman, G.; Vandenburgh, G. (2002) Advanced Anaerobic Digestion Performance Comparisons. *Proceedings of the 75th Annual Water Environment*

Federation Technical Exposition and Conference; Chicago, Illinois, Sep 28–Oct 2; Water Environment Federation: Alexandria, Virginia.

U.S. Code of Federal Regulations (1993) Title 40, 40 *CFR* Part 503.

U.S. Environmental Protection Agency (1979) *Process Design Manual—Sludge Treatment and Disposal.* U.S. Environmental Protection Agency: Washington, D.C.

U.S. Environmental Protection Agency (1994) *A Plain English Guide to the EPA Part 503 Biosolids Rule,* EPA/832R-93/003; U.S. Environmental Protection Agency, Office of Wastewater Management.

U.S. Environmental Protection Agency (1999) *Control of Pathogens and Vector Attraction in Sewage Sludge,* EPA 625/R-92-013; U.S. Environmental Protection Agency, Office of Water.

Water Environment Federation (1998) *Design of Municipal Wastewater Treatment Plants,* 4th ed.; Manual of Practice No. 8; Water Environment Federation: Alexandria, Virginia.

Zehnder, A .J. B. (1978) Ecology of Methane Formation. In *Water Pollution Microbiology.* Vol. II, R. Mitchell (Ed.), Wiley Interscience: New York; 360.

SUGGESTED READINGS

National Fire Protection Association (2003b) *National Fire Codes Supplement;* National Fire Protection Association: Quincy, Massachusetts.

Occupational Safety and Health Administration (1970) 29 Code of Federal Regulations, Part 1910, U.S. Department of Labor.

U.S. Environmental Protection Agency (1976) *Anaerobic Sludge Digestion Operations Manual,* EPA-430/9-76-001; U.S. Environmental Protection Agency, Office of Water.

Chapter 31

Aerobic Digestion

INTRODUCTION TO AEROBIC DIGESTION

Aerobic digestion is a biological treatment process that uses long-term aeration to stabilize and reduce the total mass of organic waste by biologically destroying volatile solids. This process extends decomposition of solids and regrowth of organisms to a point where available energy in active cells and storage of waste materials are sufficiently low to permit the waste sludge to be considered stable for land application (see the Rules and Regulations section of this chapter).

Aerobic Digestion may be used to treat (1) waste activated sludge (WAS) only, (2) mixtures of WAS or trickling filter sludge and primary sludge, (3) waste sludge from extended aeration plants, or (4) waste sludge from membrane bioreactors (MBRs). Aerobic digestion treats solids that are mostly a result of growth of the biological mass during the treatment process. The aerobic digestion process renders the digested sludge less likely to generate odors during disposal and reduces bacteriological hazards.

Aerobic digestion was widely used to stabilize waste solids from municipal and industrial wastewater treatment plants (WWTPs) because of their relatively simple operation, low equipment cost, and safety issues. In the past, the disadvantages associated included high energy costs, reduced exothermic-biological energy during cold weather, alkalinity depletion, poor pathogen reduction, poor volatile solids reduction, and poor standard oxygen uptake rates (SOURs). As a result of these performance difficulties, early attempts to use aerobic digestion as a solution to solids disposal and handling regulations led to relatively long solids retention time (SRT) values, which made both the capital and operating cost very unappealing.

Research was initiated by Enviroquip, Inc. (Austin, Texas) in the early 1990s in response to the new performance requirements of the Rules and Regulations for beneficial reuse (U.S. EPA, 1992). Results from their research were presented during a series of aerobic digestion workshops from 1997 to 2001 at the Water Environment Federation (Alexandria, Virginia) conferences. Each year, the work presented during the workshops was compiled in books, resulting in a five-volume series of either hardcopy books or later compact discs, which were all available to the public. Aerobic Digestion Workshops Volumes I through V include comprehensive design guidelines, operational data, and extensive research focused entirely on aerobic digestion and have been used by both the engineering community and operators as reference manuals. Please see the References section at the end of this chapter.

Research resulted in the identification of techniques that improved the process performance of aerobic digestion. These techniques are grouped into the following categories: (1) prethickening, (2) staged operation, (3) aerobic–anoxic operation, and (4) temperature control. These techniques and their benefits are summarized in Table 31.1. These techniques are explored in depth later in this chapter.

The aerobic digestion process was initially used in designs for new plants that normally treated WAS from treatment systems that did not contain a primary settling process, waste activated or trickling filter sludge only, or mixtures of waste activated

TABLE 31.1   Effects of digester improvement techniques on digester performance (Daigger et al., 1997).

Technique	Improvements in aerobic digester performance
1. Prethickening	• Increases SRT in existing facilities • Reduces overall volume requirements • Increases volatile solids destruction • Increases temperature • Accelerates digestion and pathogen destruction rate
2. Staged operation	• Reduces overall volume requirement • Improves solids stabilization • Improves pathogen destruction • Reduces capital cost of aeration equipment and tankage • Reduces oxygen requirement for process and mixing
3. Aerobic-anoxic operation	• Alkalinity recovery and pH balance • Energy savings • Nitrogen removal • Phosphorus removal
4. Temperature control	• Reduces SRT and volume • Reduces capital cost of aeration equipment and tankage • Provides reliable process all year

and trickling filter sludge. Typically, if a primary settling process was incorporated into the design, anaerobic digestion was the process of choice because reliable techniques to thicken and aerobically digest higher than 4% solids were not established at the time. Because of tighter effluent standards on both nitrogen and phosphorus being enforced in the United States in the late 1990s, primary clarifiers were slowly eliminated from the process trains. This was typically done to preserve a good carbon-to-nitrogen ratio (6:1 recommended, 4:1 minimum), which is normally required to achieve successful biological nitrogen removal. As a result of the combination of the new effluent limits and techniques, which provided the capabilities to control aerobic digestion processes and accurately predict the performance of the system, aerobic digestion has become appealing once again. A number of anaerobic digesters have been converted to aerobic digesters because of the relative easy operation and lower equipment cost and because they can produce a better quality supernatant with both lower nitrates and phosphorus, thereby protecting the liquid side upstream. Additional benefits of aerobic digestion are: achieving comparable volatile solids reduction with shorter retention periods; less hazardous cleaning and repairing tasks (California State University, 1991);

and an explosive digester gas is not produced, although the gases produced cannot be used for fuel combustion as in anaerobic digestion. Aerobic digestion has been used primarily in plants of a size less than 19 000 m^3/d (5 mgd), while, above this size, anaerobic digestion was typically the process of choice. However, in recent years, the process has been used in larger WWTPs, with capabilities up to 190 000 m^3/d (50 mgd) (Metcalf and Eddy, 2002; Water Environment Federation, 1998).

DESCRIPTION OF AEROBIC DIGESTION PROCESSES

Aerobic and facultative microorganisms use oxygen and obtain energy from the available biodegradable organic matter in the waste sludge during aerobic digestion. However, when the available food supply in the waste sludge is inadequate, the microorganisms begin to consume their own protoplasm to obtain energy for cell maintenance reactions. Eventually, the cells will undergo lysis, which will release degradable organic matter for use by other microorganisms. The end products of aerobic digestion typically are carbon dioxide, water, and "nondegradable" materials (i.e., polysaccharides, hemicellulose, and cellulose). This phenomenon, called *endogenous respiration*, is approximately described by the following equations (Metcalf and Eddy, 2002; Water Environment Federation, 1995):

$$C_5H_7O_2N + 5O_2 \rightarrow 4CO_2 + H_2O + NH_4HCO_3 \qquad (31.1)$$
Destruction of biomass in aerobic digestion

$$NH_4^+ + 2O_2 \rightarrow NO_3^- + 2H^+ + H_2O \qquad (31.2)$$
Nitrification of released ammonia-nitrogen

$$C_5H_7O_2N + 7O_2 \rightarrow 5CO_2 + 3H_2O + HNO_3 \qquad (31.3)$$
Complete nitrification

$$2C_5H_7O_2N + 12O_2 \rightarrow 10CO_2 + 5H_2O + NH_4^+ + NO_3^- \qquad (31.4)$$
With partial nitrification

$$C_5H_7O_2N + 4NO_3^- + H_2O \rightarrow NH_4^+ + 5HCO_3^- + 2N_2 \qquad (31.5)$$
Denitrification using nitrate nitrogen as electron acceptor

$$C_5H_7O_2N + 5.75O_2 \rightarrow 5CO_2 + 0.5N_2 + 3.5H_2O \qquad (31.6)$$
With complete nitrification and denitrification

The term $C_5H_7O_2N$ in eq 31.1 is the classic formula for biomass or cellular material in an activated sludge system. The cellular material is oxidized aerobically to carbon dioxide, water, and ammonia. Only approximately 75 to 80% of the cell material can be oxidized; the remaining amount is composed of inert components and organic compounds that are not biodegradable. Ammonia is released as shown in this equation, but then it is combined with some of the carbon dioxide (CO_2) to produce a form of ammonium bicarbonate (NH_4HCO_3). The main observation from this reaction is that aerobic digestion produces both ammonia and alkalinity in the presence of oxygen in the initial phase.

If additional oxygen is provided, then the second reaction will occur as shown in eq 31.2. Ammonia reduction may be required to reduce odors in aerobic digestion; however, the nitrification of the released ammonia also creates nitrate and 2 moles of acidity. The 2 moles of acidity are shown in the form of hydrogen (H^+), which can also be described as a loss of 2 moles of alkalinity.

So, in the first reaction, while the destruction of biomass produces 1 mole of alkalinity as ammonium bicarbonate, the nitrification destroys 2 moles of alkalinity. Depending on the amount of alkalinity in the feed solids, it is very easy to destroy the alkalinity and see dramatic decreases in pH, down to the 5+ range, which begins to adversely affect the digestion process.

If a digester system is capable of recovering the lost alkalinity biologically, without the need of chemical addition through denitrification, not only can nitrogen removal be accomplished, which is important to land application, but also a reduced energy requirement because process air requirements up to 18% could be realized by making use of nitrate instead of using oxygen. If a digester system is capable of complete nitrification and denitrification, then eqs 31.1 to 31.6 apply.

The techniques that enable the operators to successfully control pH, recover lost alkalinity, and optimize aerobic digestion, while saving oxygen, are described in the Design Techniques to Optimize Aerobic Digestion section in this chapter.

A number of variations of the aerobic digestion process exist, including (1) mesophilic conventional, (2) high-purity-oxygen, (3) thermophilic, and (4) cryophilic aerobic digestion. Mesophilic conventional aerobic digestion is the most typically used aerobic digestion process. As a result, the majority of this chapter reviews this process. However, a brief overview of the other aerobic digestion processes follows.

HIGH-PURITY-OXYGEN AEROBIC DIGESTION.
High-purity oxygen is used in this aerobic digestion process instead of air. Recycle flows and the resultant sludge are very similar to those obtained through conventional aerobic digestion. Typical influent sludge concentrations may vary from 2 to 4%. High-purity-oxygen aerobic

digestion is particularly applicable in cold weather climates because of its relative insensitivity to changes in ambient air temperatures due to the increased rate of biological activity and the exothermal nature of the process.

High-purity-oxygen aerobic digestion is conducted in either open or closed tanks. Because the digestion process is exothermic in nature, the use of closed tanks will result in a higher operating temperature and a significant increase in the rate of volatile suspended solids reduction. The high-purity-oxygen atmosphere in closed tanks is maintained above the liquid surface, and the oxygen is transferred to the sludge through mechanical aerators. In open tanks, the oxygen is introduced to the sludge by a special diffuser that produces minute oxygen bubbles. The bubbles dissolve before reaching the air–liquid interface (Metcalf and Eddy, 2002). High operating costs are associated with the high-purity-oxygen aerobic digestion process because of the oxygen generation requirement. As a result, high-purity-oxygen aerobic digestion is cost-effective generally only when used in conjunction with a high-purity-oxygen activated sludge system. Neutralization may be required to offset the reduced buffering capacity of the system, (Metcalf and Eddy, 2002).

AUTOTHERMAL THERMOPHILIC AEROBIC DIGESTION. Autothermal thermophilic aerobic digestion (ATAD) represents a variation of both conventional and high-purity-oxygen aerobic digestion. In this process, the feed sludge is prethickened to provide a digester feed solids concentration greater than 4%. The reactors are insulated to conserve the heat produced from the biological degradation of organic solids by thermophilic bacteria. Thermophilic operating temperatures in insulated reactors are in the range 45 to 70 °C, without external supplemental heat provided, other than the aeration and mixing devices located inside the vessels. Because of this phenomenon, the process is termed *autothermal* (Metcalf and Eddy, 2002).

Advantages. The major advantages of ATAD are as follows:

1. Decrease in retention times (smaller volume required to achieve a given suspended solids reduction) to approximately 5 to 6 days to achieve volatile solids reduction of 30 to 50%, similar to conventional aerobic digestion;
2. Greater reduction of bacteria and viruses compared with mesophilic anaerobic digestion (Metcalf and Eddy, 2002); and
3. When the reactors are well-mixed and maintained at 55 °C and above, pathogenic viruses, bacteria, viable helmith ova, and other parasites can be reduced to below detectable levels, thus meeting the pathogen reduction requirements for Class A biosolids.

Disadvantages. The major disadvantages of ATAD are as follows:

1. Poor dewatering characteristics of ATAD biosolids per John Novak (Daigger et al., 1998);
2. Objectionable odors are formed;
3. Lack of nitrification and/or denitrification per Enos Stover (Daigger et al., 1998);
4. High capital cost; and
5. Foam control is required to ensure effective oxygen transfer (Metcalf and Eddy, 2002).

This process is relatively stable, self-regulating with respect to temperature, recovers quickly from minor process upsets, and is not greatly affected by temperature variations of outside air. Because the ATAD system is capable of producing Class A biosolids, it is growing in popularity, with a total of 35 ATAD systems known to be operating in North America in 2003. In excess of 40 plants are operating in Europe (Stensel and Coleman, 2000).

Typical Design Consideration for the Autothermal Aerobic Digestion System. Note that the design parameters that follow have been adapted, in part, from Stensel and Coleman (2000) and from the Water Environment Federation (1998).

Nitrification is Inhibited in the Autothermal Aerobic Digestion Process. Because of the high operating temperatures, nitrification is inhibited, and aerobic destruction of volatile solids occurs, as described by eqs 31.1 to 31.3, without the subsequent reactions, as described in eqs 31.4 to 31.6. Additionally, many, if not all, ATAD systems may be operating under microaerobic conditions, in which oxygen demand exceeds oxygen supply (Stensel and Coleman, 2000). As described elsewhere in this chapter, ammonia is released as a result of digestion. Because nitrification is inhibited in ATAD systems, the pH in the ATAD systems will typically range from 8 to 9. Ammonia-nitrogen produced will be present in the offgas and in solution, with concentrations of several hundred milligrams per liter in each.

Effect of Liquid Sidestreams that Contain Ammonia-Nitrogen. Most of the ammonia-nitrogen will be returned to the liquid process in sidestreams from the odor control and dewatering facilities. If the effluent limits on the liquid side are very low both on nitrogen and phosphorus, the recycle stream can have a negative effect on the plant performance. If the ATAD process is chosen in conjunction with a biological nutrient removal process, both the liquid sidestreams from the odor-control and dewatering systems need to be accounted for or treated separately.

Foam. A substantial amount of foam is generated in the ATAD process as cellular proteins, lipids, and oil-and-grease materials are broken down and released into solution. The foam layer contains high concentrations of biologically active solids that provide insulation. It is important to effectively manage the foam, either with foam cutters or spray systems, to ensure effective oxygen transfer and enhanced biological activity. A freeboard of 0.5 to 1 m (1.65 to 3.3 ft) is recommended (Stensel and Coleman, 2000).

Equipment Design. Table 31.2 shows recommended design patterns for ATAD digester systems.

Prethickening. Thickening or blending facilities may be required to maintain an influent to the ATAD reactor greater than 4% solids.

Basin Configuration. Two or more enclosed insulated reactors in series should be provided and equipped with mixing, aeration, and foam-control equipment.

Continuous or batch processing is acceptable. For compliance with pathogen regulations of Class A biosolids, the withdrawal and feeding of the sludge to the reactors is performed on a batch basis. In this case, the ATAD pumping system is designed to withdraw and feed the daily amount of sludge in 1 hour or less. The reactor is then isolated for the remaining 23 hours each day, at a minimum temperature of 55 °C in stage 2 before disposal.

Figure 31.1 shows data from one of the earliest full-scale ATAD systems in Germany, operating since 1980 (U.S. EPA, 1990). This figure illustrates typical results of various operating parameters during these studies. The system achieved approximately 33% reduction in volatile suspended solids (VSS) for the WAS feed source during the study period. A "sawtooth" temperature curve is evident in both reactors and is the result of batch reactor feeding. During treatment, the pH increased from approximately 6.5 to 8.0.

Post-Autothermal Aerobic Digestion Storage/Dewatering. Post-process cooling is necessary to achieve solids consolidation and enhance dewaterability. Typically, 14 to

TABLE 31.2   Recommended design parameters for ATAD digester systems (Stensel and Coleman, 2000).

Parameters	Range	Typical
Number of reactors	2 to 3	2
Prethickened solids	4 to 6%	4%
Reactors in series		Yes
Total HRT in reactors	4 to 30 days	6 to 8 days
Temperature—stage 1	35 to 60 °C	40 °C
Temperature—stage 2	50 to 70 °C	55 °C

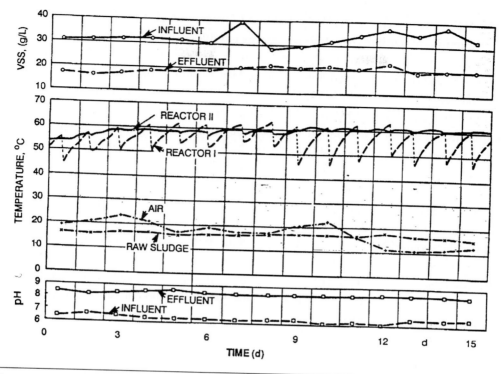

FIGURE 31.1 Data from the ATAD facility in Gemmingen, Germany (U.S. EPA, 1990).

20 days of retention time may be necessary, unless heat exchangers are used for cooling the processed biosolids.

If the reader is interested in more information on the ATAD process, the author suggests review of "Assessment of Innovative Technologies for Wastewater Treatment: Autothermal Aerobic Digestion (ATAD)" by Stensel and Coleman (2000). This document provides detailed information concerning the history, design, operation and maintenance, and performance of these systems, and can be used as an updated version of the Environmental Regulations and Technology document "Autothermal Thermophilic Aerobic Digestion of Municipal Wastewater Sludge" (U.S. EPA, 1990).

CRYOPHILIC AEROBIC DIGESTION. The operation of aerobic digesters at low temperatures (less than 20 °C) is known as *cryophilic aerobic digestion*. Although not widely used, research has been concentrating on optimizing the operation of these digesters. This research has suggested that the sludge age in the digester should be increased as the operating temperatures decrease, to maintain an acceptable level of suspended solids reduction.

CONVENTIONAL (MESOPHILIC) AEROBIC DIGESTION. Historically, the conventional aerobic digestion process was widely used to stabilize waste solids from municipal and industrial WWTPs (Grady et al., 1999). Its advantages include simple design and operation, moderate costs, and the provision of solids storage capabilities. Its disadvantages include high energy costs, reduced biological energy during cold weather, alkalinity depletion, and poor pathogen reduction. If SRT is sufficiently maintained in one or more aerated tanks, much of the biodegradable organic matter added to the digester can be stabilized. However, because of the completely mixed nature of the reactor environment, some biodegradable organic matter remains unstabilized (Grady et al., 1999). More importantly, operation in a completely mixed fashion results in bleedthrough of pathogens to the effluent. Nitrification of released ammonia-nitrogen results in consumption of alkalinity and low pH values, which inhibit digestion (Anderson and Mavinic, 1984; Daigger et al., 1997; Mavinic and Koers, 1979). As a result of these performance difficulties, relatively long SRT values were specified when solids disposal regulations were proposed and subsequently promulgated (Lue-Hing et al., 1998). Because of the above reasons, the interest in the conventional aerobic digestion process decreased.

Research was initiated in response to the new performance requirements for beneficial reuse (U.S. EPA, 1992), resulting in the identification of techniques that can improve process performance (Daigger and Bailey, 2000; Daigger et al., 1997). These techniques can be broadly grouped into the following three categories: (1) prethickening, which uses mechanical devices, such as gravity belt thickeners or gravity thickeners, to thicken before digestion; (2) staged operation, which uses either a tanks-in-series configuration or sequential batch reactor (SBR) operation to increase the plug-flow characteristic in the digester and reduce shortcircuiting; and (3) aerobic–anoxic operation, which cycles the oxygen-transfer device on and off to allow denitrification to occur (Daigger and Bailey, 2000).

DESIGN TECHNIQUES TO OPTIMIZE AEROBIC DIGESTION

PRETHICKENING. *Advantages of Prethickening.* The main advantages of this technique include the following:

1. Increased SRT and volatile solids destruction,
2. Accelerated digestion and pathogen destruction rate, and
3. Temperature elevation.

Increased Solids Retention Time and Volatile Solids Destruction. The characteristics of the waste sludge to be digested will be determined by the raw influent to the treatment system and by the treatment processes that precede aerobic digestion. The concentration of sludge to be digested is important to consider. Increasing feed sludge concentrations by thickening will result in longer SRTs with subsequent increased levels of volatile solids destruction.

Accelerated Digestion and Pathogen Destruction Rate. Because the quantity of water that must be heated is reduced, the heat released as a result of the oxidation of the biodegradable organic matter can result in elevated digester temperatures and accelerate digestion and pathogen destruction rates (Grady et al., 1999). The oxidation of biomass results in the release of its heat of combustion (approximately 3.6 kcal/g VSS destroyed) to the reactor.

Temperature Elevation. The resulting temperature increase can be significant. For example, if the feed concentration is 2% (20 000 mg/L), and 30% of the VSS is destroyed in an aerobic digester, sufficient heat is liberated to raise the temperature in the reactor by 21 °C if all of the released heat is conserved (Daigger and Bailey, 2000). The reason a temperature increase in conventional digesters is not typically observed is because, in some operations, the sludge is wasted from the activated sludge process to the digesters at 0.5 to 1.0% solids. The aeration system is shut down for 1 to 2 hours to allow the sludge to settle and the supernatant to form. The supernatant is then decanted, allowing additional sludge to be added; thus, the solids concentration typically increases to 1.5 to 2.0%. This technique reduces the volume of liquid; however, the heat released is used to heat a lot of water; thus, the temperature elevation is marginal.

    If the potential exists to capture the heat generated by the volatile solids, it can be used to control the temperature of the reactor. This could be very beneficial in cold climates. Because conventional aerobic digestion typically operates in the range 15 to 35 °C, it is classified as a mesophilic process. On the other hand, it is also important to avoid excessive temperatures, especially in the summer months.

Categories of Prethickening. Prethickened aerobic digestion is divided into five major categories, as described below, based on the thickening treatment processes used (Table 31.3) (more detailed description of these treatment processes and their capabilities is included in Chapter 29).

Category A—Thickening Treatment Process: Batch Operation or Decanting of Aerobic Digester. Batch operation involves the practice of manually decanting digested sludge. Originally, aerobic digestion was operated as a draw-and-fill process, a concept still being used at many facilities. Solids are pumped directly from the clarifiers or SBRs to

TABLE 31.3   Thickening treatment process alternatives (Daigger and Bailey, 2000).

Category	Thickening treatment process
Category A	Batch operation or decanting of aerobic digester
Solids concentration range:	1.25 to 1.75%
Polymer required?	No
Category B	Continuous feed operation using gravity sedimentation, such as gravity thickener, after digestion
Solids concentration range:	1.5 to 2%
Polymer required?	No
Category C	Using gravity thickener in loop with aerobic digestion
Solids concentration range:	2.5 to 3%
Polymer required?	No
Category D	Using membranes for thickening in-loop with aerobic digestion
Solids concentration range:	3 to 3.5%
Polymer required?	No
Category E	Using any mechanical thickener, such as gravity belt thickener, dissolved air flotation (DAF mechanisms) or centrifuge, or drum thickener before aerobic digestion
Solids concentration range:	4 to 8%
Polymer required?	Yes (note that DAF units could also be operated with polymer and achieve 3 to 5% solids. Drum thickeners could be also be operated without polymer and achieve 2 to 3% solids concentration).

the aerobic digester. The time required for filling the digester depends on the tank volume available and the volume of waste sludge. When a diffused air aeration system is used, sludge undergoing digestion is aerated continually during the filling operation. When the sludge is removed from the digester, aeration is discontinued, and the stabilized solids are allowed to settle. The clarified supernatant is then decanted and returned to the treatment process. The thickened solids are removed at a concentration between 1.25 and 1.75% solids.

For batch-feed digesters, the typical operating steps are as follows:

1. Turn off the aeration equipment and allow the solids to settle. To avoid anaerobic conditions, limit the solid–liquid separation to several hours.

2. Decant as much supernatant as possible. For quality control, analyze a sample of the supernatant for pH and suspended solids. It is also recommended to monitor parameters such as VSS, ammonia-nitrogen, nitrate-nitrogen, and alkalinity of the supernatant for the system performance.

3. Draw off the thickened, digested sludge for second-stage digestion or for disposal. For quality control, sample the drawn-off sludge. The digested solids content will be significantly higher when primary sludges are digested. Primary clarifiers can typically produce sludge in the range of 2 to 5% solids concentration. Primary to secondary sludge fed to the digester system is typically in the range of 40:60 to 20:80 primary:secondary by volume, therefore the combined feed to the digester could vary both in digester concentration and volatile reduction.

4. If WWTP flexibility permits, add the new sludge feed to the digester in short, frequent periods of time. It is suggested that the volume and concentration of the sludge added each day be somewhat uniform. At some digester installations, sludge settling and drawoff are performed once per week, while sludge addition or feed is practiced daily. Therefore, the sludge volume in the digester increases daily, until the next decanting and drawoff period.

Disadvantages of this process include the following:

1. Basins are sized based on low solids concentration and high water content. Large volumes are required.
2. High capital cost.
3. High operating cost.
4. No control of alkalinity, temperature, ammonia, nitrates, and phosphorus, as offered by categories C through E (Note: The importance of controlling these parameters is explained, in depth, in the sections to follow. This list is only to be used as a quick reference checklist).
5. Difficult to meet stringent limits on supernatant effluent.

Category B—Thickening Treatment Process: Continuous-Feed Operation Using Sedimentation After Digestion. Category B is the thickening treatment process consisting of continuous-feed operation using sedimentation, such as a gravity thickener, after digestion. This is typically a continuous aerobic digestion process that closely resembles the activated sludge process. Solids are pumped directly from the clarifiers or SBR or MBR into the aerobic digester. The digester operates at a fixed level, with the overflow going to a solid–liquid separator. Thickened and stabilized solids are removed for further processing. A lower solids concentration typically is obtained with continuous operation. The category B process can produce marginally better quality effluent than category A because the aerobic digestion basin is operated at a fixed level, and aeration transfer efficiency is optimized.

For continuous-feed digesters, the process can be improved by the following:

1. Adjusting the rate of settled return sludge to obtain the best balance between return sludge concentration and supernatant quality,
2. Adjusting the settling chamber's inlet and outlet flow characteristics to reduce shortcircuiting and unwanted turbulence that hinders solids concentration, and
3. Modifying the weir and piping arrangements.

Disadvantages of this process include the following:

1. Basins are sized based on low solids concentration and high water content. Large volumes are required.
2. High capital cost.
3. High operating cost.
4. No control of alkalinity, temperature, ammonia, nitrates, and phosphorus, as offered by categories C through E (Note: The importance of controlling these parameters is explained in depth in the following sections. This list is only to be used as a quick reference checklist).
5. Difficult to meet stringent limits on supernatant effluent.
6. If nitrification and denitrification is not controlled between the digester and thickener, it leads to anaerobic conditions and undesirable odors (Note: The importance of controlling these parameters is explained in depth in the following sections. This list is only to be used as a quick reference checklist).

Category C—Thickening Treatment Process: Using a Gravity Thickener in Loop with Aerobic Digestion. This process typically consists of two main phases—the in-loop phase and the isolation phase. This process usually includes four main basins, two digesters, one premix basin, and a gravity thickener. For feeds from SBRs and MBRs, additional basins are incorporated to the design to optimize flexibility; however, these four basins are still included as the main components of the process. During the in-loop phase, a digester, a premix, and a thickener operate in loop. The main results during the in-loop phase are the reduction of volatile solids, reduction of ammonia, and an increase in solids concentration. The in-loop thickener has two main functions—it thickens as well as denitrifies. The in-loop digester acts as a volatizer and reduces the majority of volatile solids. During the isolation phase, no contamination occurs in a digester, and the digester's main function is to complete the additional pathogen reduction to meet the sludge requirements. The digesters are fed in a batch operation, either 8, 16, or 24 times per day; however, the process is considered a "modified batch process" because one digester is fed in short-batch intervals for an extended period of time—typically 10 to 20 days (equal to the length of the in-loop phase)—before it goes to the isolation phase (10 to 20 days). This process produces 2.5 to 3% solids.

Advantages of this process include the following (Note: For points 1, 2, 3, and 4, the importance of controlling these parameters is explained in the following sections. This list is only to be used as a quick reference checklist):

1. Process provides all the benefits of aerobic–anoxic operation.
2. Process provides all the benefits of staged operation.
4. Good control of alkalinity without the need to turn air on and off.
5. Can easily meet stringent limits on supernatant effluent on ammonia, nitrates, phosphorus, and total suspended solids (TSS).
6. Process provides better SOUR and pathogen reduction when compared to categories A and B as a result of true isolation.
7. Moderate capital cost.
8. Low operating cost.
9. No polymer required, achieving 3% solids.

Category D—Thickening Treatment Process: Using Membranes for In-Loop Thickening with Aerobic Digestion. Membrane technology is considered a fairly new process in the United States and has only been used in Europe and Japan for the last 16 years. The thickening applications are even more limited, with the oldest installations operating successfully with no odor issues, typical cleaning frequency of membranes, and no membrane replacements dated back to 1998. Applications range from 3 to 5% solids concentrations, operating in continuous or batch mode, in isolation and in series. The process incorporates a wastewater membrane, which is applicable for high solids. Membranes used for these applications are either flat plate, such as those manufactured by Enviroquip Kubota Company (Texas), or hollow fiber, such as those manufactured by Zenon Membrane Solutions (Oakville, Ontario, Canada). This process can be used in lieu of any of the processes listed in categories A through C; however, it is not recommended that the design solids exceed 3.5%.

Membranes can be fitted into existing basins to provide both thickening and digestion at the same time because airflow is required to clean the membranes.

Designs include two-, three-, four-, or five-basin configurations, operating in batch or in series. Figures 31.2 and 31.3 show two possible operating scenarios using membranes in loop with digestion to thicken to 3% solids. Figure 31.2 shows a five-stage batch operation setup, and Figure 31.3 shows a five-stage, in-series operation setup.

Advantages of this process include the following (Note: For points 1, 3, 4, and 5, the importance of controlling these parameters is explained in depth in the following sections. This list is only to be used as a quick reference checklist):

1. Provides the best control of supernatant effluent because there is no danger of solid overflow resulting in poor supernatant (separate scum removal is not required).
2. Requires small footprint, which is ideal for high-rate digestion.
3. Provides good temperature control.
4. Provides all the benefits of staged operation.
5. Provides all the benefits of aerobic–anoxic operation.
6. Provides better SOUR and pathogen reduction compared with categories A and B, as a result of true isolation.
7. Moderate capital cost (smaller footprint).

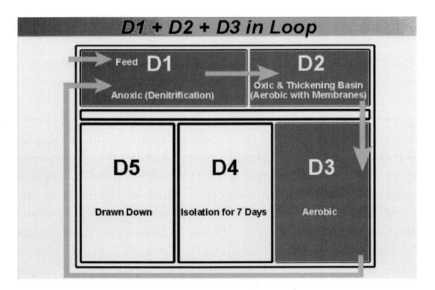

FIGURE 31.2 A five-stage batch operation setup using membranes for in-loop thickening as part of an aerobic digestion process (Daigger et al., 2001).

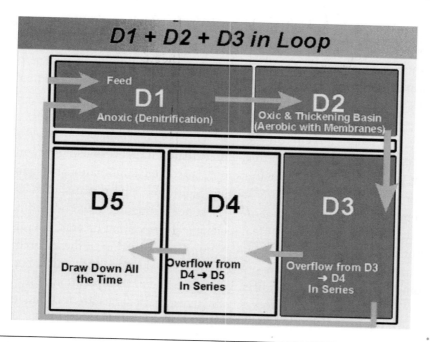

FIGURE 31.3 A five-stage, in-series operation setup using membranes for in-loop thickening as part of an aerobic digestion process (Daigger et al., 2001).

8. Low operating cost (less air is required for the process because the viscosity factor is lower; therefore, the airflow requirement is lower compared with thickened systems using polymer).

9. Following nitrification and denitrification, there is reduced phosphorus release. If a biophosphorus permit applies on the liquid sidestream, the permeate can be treated with alum or ferric chloride; therefore, the phosphorus can be fixed and removed with the sludge (please refer to the Aerobic–Anoxic Operation section of this chapter for more in-depth discussion on phosphorus).

Additional advantages that membranes offer above all other processes with respect to phosphorus removal include the following:

• Membrane plants produce zero TSS.
• It is important to note that the permeate from a membrane digester system is collected in the aerobic phase, as opposed to the supernatant or decant that is typically collected from an anoxic phase on the systems in all the other categories described above.

- Most membrane systems in thickening applications operate at approximately 0.5 to 1 mg/L dissolved oxygen, which is ideal for this application.
- Another factor that affects phosphorus, ammonia, and nitrogen is pH. Membrane systems include an anoxic zone to balance alkalinity, so pH balancing is always integral to the process.

Category E—Thickening Treatment Process: Using Any Mechanical Thickener Before Aerobic Digestion. Category E is the thickening treatment process using any mechanical thickener, such as a gravity belt thickener, dissolved air flotation mechanism, centrifuge, or drum thickener before aerobic digestion. This process uses mechanical prethickening devices that use polymers as conditioning agents to maximize thickening. The designer can choose the ideal mechanical device and desired operating solids concentration, such as 4, 5, or 6%, to maximize the reduction of the aerobic digestion basins. This process gives ultimate process control to the operator to meet performance in the summer and winter by modifying the schedule of the operation of the mechanical device as desired. For this process, two digesters in series are recommended, as a minimum (in-series operation will be addressed later in this chapter). Because of the flexibility and reliability of the process and the capital and operating cost savings, this process is also preferred in WWTP designs capable of handling more than 7600 m^3/d (2.0 mgd) in flow.

Advantages of this process include the following:

1. Provides the ultimate process control of treatment of solids.
2. Provides the capabilities to maximize reduction in volume by choosing the ideal machine and ideal solids concentration for the application.
3. Maximizes reduction in footprint, which is ideal for high-rate digestion (a deep tank is necessary).
4. Provides the best temperature control when flexibility is included in the design (cold weather not an issue with these systems, however, provisions may be required to prevent thermophilic conditions in the summer).
5. Provides all the benefits of staged operation.
6. Provides all the benefits of aerobic–anoxic operation.
7. Following nitrification and denitrification, there is reduced phosphorus release. If a biophosphorus permit applies on the liquid stream, treat the permeate with alum or ferric chloride. Phosphorus can be fixed and removed with sludge.
8. Provides excellent SOUR and pathogen reduction.
9. Moderate capital cost.

10. Low operating cost. The alpha values and transfer efficiency are lower in higher solids concentration digester of 4 to 6% compared with digesters operating in the range 2 to 3%. However, because of the reduced basin volume required for these systems, the airflow required for both process and mixing is comparable. Mixing requirements typically are higher than process air requirements for systems with lower solids concentration, so, in comparison, the overall operating horsepower for higher solids is less.

11. While in the prethickened mode, supernatant is not produced because the liquid in the digesters are in the thixotropic stage.

BASIN CONFIGURATION—STAGED OPERATION OR BATCH (MULTIPLE BASINS).
Traditional aerobic digester basins have been designed as single basins, and, if multiple basins were supplied, they were typically operated in parallel.

As described below, multiple tanks in series or in isolation operated in a batch operation have proven to improve both the pathogen destruction and SOUR. For this reason, this chapter will focus on the staged operation systems.

The Environmental Regulations and Technology manual "Control of Pathogens and Vector Attraction in Sewage Sludge" (U.S. EPA, 1999a) offers the following comments on both batch and staged operation. Sludge can be aerobically digested using a variety of process configurations, including continuously fed or fed in a batch mode, in a single- or multi-stage. Batch operation will typically produce less volatile solids reductions for primary sludge than the other options because there are lower numbers of aerobic microorganisms in it. Single-stage completely mixed reactors with continuous feed and withdrawal are the least effective of these options for bacterial and viral destruction, mainly because organisms that have been exposed to the adverse conditions of the digester for a short time can leak through to the product sludge.

Probably the most practical alternative to a single-stage completely mixed reactor is staged operation, such as the use of two or more completely mixed digesters in series. The amount of processed sludge passing from inlet to outlet would be greatly reduced compared with single-stage operation. If the kinetics of the reduction in the pathogen densities is known, it is possible to estimate how much improvement can be made by staged operation.

Farrah et al. (1986) have shown that the decline in densities of enteric bacteria and viruses follow first-order kinetics. If first-order kinetics are assumed to be correct, it can be shown that a 1-log reduction of organisms is achieved in half as much time in a two-stage reactor (of equal volume in each stage) as in a one-stage reactor. Direct experimental verification of this prediction has not been carried out, but Lee et al. (1989) have qualitatively verified the effect.

It is reasonable to give credit for an improved operating mode. However, because not all factors involved in the decay of microorganisms' densities are known, some factor of safety should be introduced. It is recommended that for staged operation using two stages of approximately equal volume, the time required be reduced to 70% of the time required for single-stage aerobic digestion in a continuously mixed reactor. This allows a 30% reduction in time instead of the 50% estimated from theoretical considerations. The same reduction is recommended for batch operation or for more than two stages in series. Thus, the time required would be reduced from 40 to 28 days at 20 °C (68 °F), and from 60 to 42 days at 15 °C (59 °F). These reduced times are also more than sufficient to achieve adequate vector attraction reduction. Approval of this process as a Process to Significantly Reduce Pathogens (PSRP) by the permitting authority may be required in specific states in the United States. For more information on this topic, please refer to the Solids Retention Time × Temperature Product section.

The benefits of a two-stage reactor system in series or in isolation include the following (Note: Please see more information about points 5 and 6 in the Dissolved Oxygen and Oxygen Requirements section of this chapter):

1. Improvement in pathogen destruction;
2. Improvement in SOUR;
3. Capital cost reduction of aeration equipment;
4. Capital cost reduction of tankage;
5. Air flow reduction, as a result of process requirements in the first digester; and
6. Air flow reduction, as a result of mixing requirements in the second digester.

AEROBIC–ANOXIC OPERATION. This technique enables operators to successfully nitrify/denitrify, control pH, recover lost alkalinity, and optimize aerobic digestion, while saving energy.

Nitrification and Denitrification. As described in the Description of Aerobic Digestion Processes section of this chapter, the oxidation of biomass produces carbon dioxide, water, and ammonia, as described in eq 31.1.

Table 31.4 includes a summary of the nitrification and denitrification eqs 31.1 to 31.6 presented earlier in this chapter to facilitate the explanation of the process.

Ammonia is combined with some of the carbon dioxide to produce a form of ammonium bicarbonate. Often, partial nitrification occurs, and a portion of the nitrogen is left as ammonia. The system will nitrify until the pH drops enough that it begins to

TABLE 31.4   Summary of the nitrification and denitrification equations (Daigger et al., 1997).

$C_5H_7O_2N + 5O_2 \rightarrow 4CO_2 + H_2O + NH_4HCO_3$ Destruction of biomass in aerobic digestion	(31.1)
$NH_4^+ + 2O_2 \rightarrow NO_3^- + 2H^+ + H_2O$ Nitrification of released ammonia-nitrogen	(31.2)
$C_5H_7O_2N + 7O_2 \rightarrow 5CO_2 + 3H_2O + HNO_3$ Complete nitrification	(31.3)
$2C_5H_7O_2N + 12O_2 \rightarrow 10CO_2 + 5H_2O + NH_4^+ + NO_3^-$ With partial nitrification	(31.4)
$C_5H_7O_2N + 4NO_3^- + H_2O \rightarrow NH_4^+ + 5HCO_3^- + 2N_2$ Denitrification using nitrate-nitrogen as electron acceptor	(31.5)
$C_5H_7O_2N + 5.75O_2 \rightarrow 5CO_2 + 0.5N_2 + 3.5H_2O$ With complete nitrification and denitrification	(31.6)

inhibit the nitrifying bacteria, as illustrated by eq 31.4 (with partial nitrification). This occurs where there is only an inconsequential amount of alkalinity in the sludge that is fed to the digester. In this case, there could be a mixture of both ammonia and nitrate produced at the same time, as is evidenced by a number of installations across the country. An example of this could be during the so-called "decant or supernating phase" of digestion, where an accumulation of ammonia in the sludge cannot be removed, because there is not enough alkalinity to drive the reaction, resulting in odors.

If the oxygen in the nitrate can be used as an oxygen source to stabilize biomass, as it is used in the liquid stream processes, then both nitrification and denitrification can be accomplished, as shown in eq 31.5. Oxidizing biomass with nitrates releases ammonia once again (as in eq 31.1) and also produces nitrogen gas plus bicarbonate, which is again a form of alkalinity.

Equation 31.6 shows a balanced stoichiometric equation of nitrification and denitrification. As shown in this equation, oxidized biomass is converted to carbon dioxide, nitrogen gas, and water.

Nitrification and denitrification can be achieved by the following:

1. Turning the air on and off (typically 75% on, 25% off), or
2. Running at low dissolved oxygen and creating conditions to allow simultaneous nitrification and denitrification to occur.

Both of these operating procedures are similar to those used in the liquid stream.

Benefits of the nitrification and denitrification process include the following:

1. Alkalinity recovery and pH balance,
2. Energy savings,
3. Dissolved oxygen and oxygen requirements,
4. Nitrogen removal, and
5. Phosphorus removal.

As shown from studies conducted in Kuwait in a controlled environment at 20 °C and an SRT of 10 days, the length of the anoxic stage was optimized at 8 to 16 h/d (Al-Ghusain et al., 2004). Figure 31.4 shows that total nitrogen removal in the filtrate is optimized at 8 hours of anoxic cycle (Al-Ghusain et al., 2004).

Alkalinity Recovery and pH Balance. If complete nitrification and denitrification occurs, from an alkalinity prospective, the following is true:

Destruction of biomass produces 1 mole of alkalinity	= +1 mole of alkalinity
Nitrification destroys 2 moles of alkalinity	= −2 moles of alkalinity
Denitrification produces 1 mole of alkalinity	= +1 mole of alkalinity
Alkalinity consumption during nitrification and denitrification	= <u>0 moles of alkalinity</u>

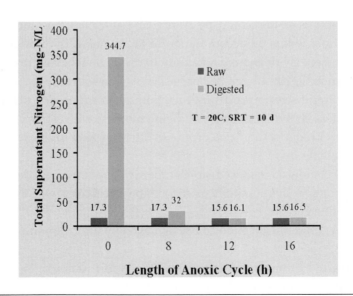

FIGURE 31.4 Evaluating the effect of digestion temperature on the concentration of total nitrogen in the filtrate shows that the highest concentration of total filtrate nitrogen occurred at 30 °C and the lowest at 20 °C (Al-Ghusain et al., 2004).

The end result and benefit of full nitrification and denitrification is that no alkalinity is consumed in the aerobic digestion process.

If sufficient alkalinity is available in the feed solids, an approximately neutral pH will be maintained. If complete nitrification and denitrification occurs by incorporating an anoxic cycle, the residual of ammonia and nitrate is typically in the range 10 to 20 mg/L (Hao and Kim, 1990; Hao et al., 1991; Matzuda et al., 1988; Peddie et al., 1990). This is one reason that pH is a primary indicator of controlling aerobic digestion processes and enables operators to optimize nitrification and denitrification techniques. If sufficient alkalinity in the feed is not available to balance the pH biologically, then an addition of chemicals to the digester will be required. Sodium bicarbonate can be added to the feed sludge, or lime or sodium hydroxide can be added to the digester directly to solve the problem. Figure 31.5 is from a study evaluating the effect of both

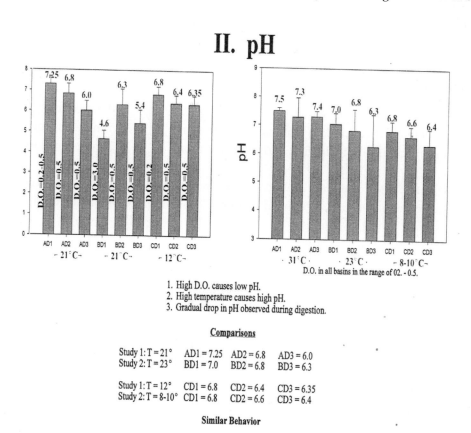

II. pH

1. High D.O. causes low pH.
2. High temperature causes high pH.
3. Gradual drop in pH observed during digestion.

Comparisons

Study 1: T = 21°   AD1 = 7.25   AD2 = 6.8   AD3 = 6.0
Study 2: T = 23°   BD1 = 7.0   BD2 = 6.8   BD3 = 6.3

Study 1: T = 12°   CD1 = 6.8   CD2 = 6.4   CD3 = 6.35
Study 2: T = 8-10°   CD1 = 6.8   CD2 = 6.6   CD3 = 6.4

Similar Behavior

FIGURE 31.5 Effect of pH levels in an aerobic digester (D.O. = dissolved oxygen) (Daigger et al., 1999).

high dissolved oxygen and high temperature on pH fluctuation in a three-stage digester system (Daigger et al., 1999). Observations from the study were as follows:

1. High dissolved oxygen causes low pH. This is most obvious when we compare the pH in AD1 at 7.25 and in BD1 (high dissolved oxygen) at 4.6. Both basins operate at the same temperature (21 °C). This indicates that active nitrification converts the ammonia to acidic nitrate and causes low pH.
2. High temperature causes high pH. Once again, at a higher temperature, there is faster generation of ammonia from the heterotrophic protein degradation.

Energy Savings. The process oxygen requirement is also reduced from 7 moles oxygen/mole biomass destroyed for the fully aerobic process (eq 31.3), to 5.75 moles oxygen/mole biomass destroyed for the anoxic–aerobic digestion process (eq 31.6). If complete nitrification and denitrification occurs from the oxygen source, energy savings can be realized, as shown below:

Destruction of biomass requires	5 moles oxygen
Complete nitrification requires	7.00 moles oxygen
Complete nitrification and denitrification requires	5.75 moles oxygen
Net savings from nitrification and denitrification	<u>1.25 moles oxygen</u>

Approximately 18% of energy savings can be realized based on reduced process energy requirements because of reduced process oxygen requirements during the anoxic phase.

Dissolved Oxygen and Oxygen Requirements. The characteristics of the sludges pumped to the aerobic digesters will determine the oxygen requirements for the process. The oxygen requirements that must be satisfied are those of the microorganisms (cell tissue) in the digestion process and, with mixed sludges, the biochemical oxygen demand (BOD) in the primary sludge.

The viscosity of undigested WAS is much higher than water and changes with sludge concentration and temperature. Undigested sludges at 6% solids have four times the viscosity of digested sludge. Thus, during the critical first stage of digestion, when oxygen demand is highest and the sludge is most viscous, intense mixing is essential to ensure that oxygen transfer occurs.

When prethickened sludges are fed to the aerobic digester, the oxygen uptake rate in the first part of the digester is extremely high, with as much as 70 to 80% of the total oxygen requirement needed in the first 10 days SRT, if the temperature is approximately 17 to 20 °C. The aeration system must be designed so that there is enough built-in flexibility to ensure that the oxygen profile in each basin matches the oxygen demand. If adequate oxygen is not available, anaerobic conditions will occur in parts of the digester, and odors and ammonia released will remain in the system. When calculating air requirements, 0.9 kg oxygen/0.5 kg VSS (2 lb oxygen/lb VSS) should be allowed, based on the expected destroyed volatile solids quantity. Sufficient air should then be provided to ensure that 1 to 2 mg/L oxygen could be maintained during the aerobic phase, if required. This enables the nitrification phase to operate at high dissolved oxygen.

Care should be taken when relating clean water transfer efficiency to prethickened sludge applications. From an operator's point-of-view, it is important to remember that traditional oxygen transfer efficiency factors should be downgraded by at least 25 to 45% when compared with the performance of the same product used in the liquid sidestream. The viscosity correction factor makes an allowance for the additional airflow required to overcome the negative effects of thixotropic sludges. The system should be designed to provide adequate velocity gradient, shearing action, and intense mixing, all required to enable oxygen to be transferred to the waste of extremely viscous materials, such as 4 to 8% solids feed, as required for volatile solids reduction.

Maintaining adequate oxygen levels allows the biological process to take place and prevents objectionable odors. In aerobic digesters, a typical dissolved oxygen level range is from 0.1 to 3.0 mg/L. Inadequate dissolved oxygen levels result in incomplete digestion and, more importantly, odor problems. On the other hand, by using low dissolved oxygen, there is potentially large energy savings resulting from simultaneous nitrification and denitrification.

The effect of low dissolved oxygen in pathogen destruction and volatile solids destruction was evaluated and later reported (Daigger et al., 1999). As concluded from this study and shown in Figure 31.6, low dissolved oxygen did not have any negative effect on either pathogen destruction or volatile solids destruction. A summary of the results are as follows:

1. No apparent drawbacks in pathogen destruction and volatile solids reduction resulting from low dissolved oxygen operation,
2. Low dissolved oxygen gives large energy savings, and
3. High dissolved oxygen causes low pH.

Nitrogen Removal in Biosolids and Filtrate or Supernatant. Although ammonia and nitrogen reduction is not typically a goal of aerobic digestion, design-loading rates for land application of biosolids can be limited by pollutants, such as heavy metals, or

V. D.O.

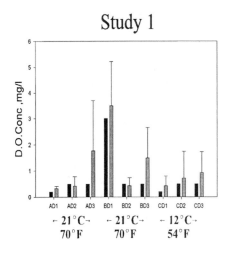
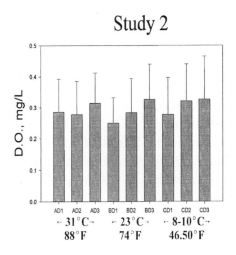

1.   No apparent drawbacks in pathogen destruction and V.S. reduction due to low D.O. operation.
2.   Low D.O. gives large energy savings.
3.   High D.O. caused low pH.

FIGURE 31.6 Effect of low dissolved oxygen in pathogen destruction and volatile solids destruction (D.O. = dissolved oxygen) (Daigger et al., 1999).

by nitrogen. The long-term loading of heavy metals is based on U.S. EPA 503 regulations (U.S. EPA, 1999b). The annual loading rate is typically limited by the nitrogen loading rate. These rates typically are set to match the available nitrogen provided by commercial fertilizers (Chang et al., 1995; Metcalf and Eddy, 2002). Because municipal biosolids represent a slow-release organic fertilizer, a combination of ammonia and organic nitrogen must be included in calculations of the application rate (Metcalf and Eddy, 2002). Conventional aerobic digestion operated in the aerobic–anoxic mode, providing full nitrification and denitrification, is the most ideal of the aerobic digestion processes to meet these requirements.

Another advantage of aerobic–anoxic sludge digestion is the ability to reduce the amount of nutrients, especially nitrogen in the filtrate or supernatant, thus reducing the load to the plant resulting from the recycle of filtrate for further treatment (Al-Ghusain et al., 2004). In a recent study, various forms of nitrogen were analyzed throughout the digestion period. The total nitrogen in the sludge filtrate (nitrite [NO_2], nitrate [NO_3], ammonium [NH_4], and organic nitrogen) was calculated for the reactors.

The data showed that there was no profound effect by the SRT (10 and 20 days), at an elevated temperature of 30 °C and anoxic time of 8 hours, on the concentration of total nitrogen in the reactor. Figure 31.4 shows the buildup of nitrogen species in the aerobic digester (with continuous aeration) in contrast to the much lower levels of nitrogen in the aerobic–anoxic digester at anoxic cycle lengths of 8, 12, and 16 hours, respectively. In the same study, the length of the anoxic cycle was evaluated at 20 °C and an SRT of 10 days, indicating that the use of an 8-hour anoxic cycle provides the optimum trade-off between energy saving and quality of sludge stabilization. Figure 31.7 shows the effect of the anoxic cycle on VSS removal (Al-Ghusain et al., 2004). In the same study (Al-Ghusain et al., 2004), the sludge dewaterability was tested using a filtration test on both aerobically digested sludge and aerobic–anoxic digested sludge. This study showed that aerobic–anoxic digestion reduces the resistance to filtration (i.e., improves sludge dewaterability).

Phosphorus Reduction in Biosolids and Biophosphorus. Biological phosphorus release occurs in the anaerobic zone of the liquid stream, where there is BOD uptake and phosphorus release. The aerobic zone provides the environment where BOD is oxidized and phosphorus is taken back up, increasing the population of high-phosphorus-content organisms, so that when the waste is removed, the waste sludge may have an

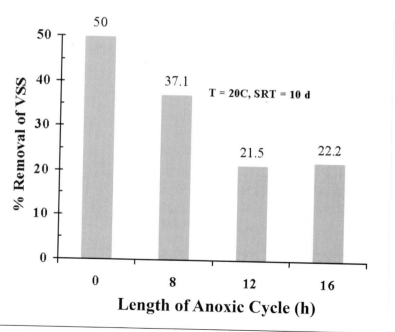

FIGURE 31.7 Effect of anoxic cycle on VSS removal (Al-Ghusain et al., 2004).

overall phosphorus content of 5 to 7%. In the aerobic digester, the organisms destroy themselves through digestion, so the phosphorus that is contained within them is released into solution. As a result, a high concentration of phosphorus may be found in solution. From an overall mass-balance prospective, the phosphorus coming into the digester ends up either in the waste sludge or in the effluent. If the phosphorus is not allowed to move forward with the sludge, then it will be recycled back to the head of the plant. Phosphorus will be released from both anaerobic and aerobic digestion; however, the release is lower from aerobic processes than from anaerobic processes.

It has been proven through experimental work of Lu Kwang Ju (Daigger et al., 2001) and in full-scale installations that cyclic operation of air on and off, or under low dissolved oxygen conditions (which provides simultaneous nitrification and denitrification), minimizes phosphorus release. As shown in Figure 31.8, when sludge was tested from a biophosphorus facility (Ozark WWTP, Kansas), if it was digested under fully aerobic conditions, the phosphorus release was in the range 120 to 150 mg/L. On the other hand, if the same sludge was digested under cyclic operation, it produced phosphorus in the range 70 to 90 mg/L after 500 hours of operation. The rate of phosphorus release and uptake decreases with time. For comparison, under similar cyclic operation

Poly-P Release and Uptake During Sludge Digestion

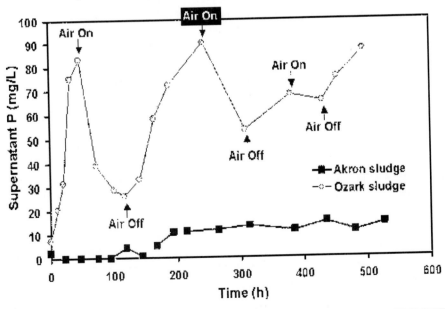

FIGURE 31.8 Polyphosphorus release and uptake of phosphorus during sludge digestion (Daigger et al., 2001).

on sludge from a nonbiophosphorus facility (Akron WWTP, Ohio), the sludge did not show any fluctuation in the phosphorus profile because this sludge did not have any biophosphorus in it; however, it was in the range 10 to 20 mg/L.

Another factor that affects phosphorus, ammonia, and nitrogen is pH. Conclusions from the same study (Daigger et al., 2001) showed that a pH less than 6.0 should be avoided because it encourages dissolution of the inorganic metal phosphates.

Option I: Liquid Disposal—No Restriction of Phosphorus on Land Application. Assuming that a system is designed as a two-stage-in-series, prethickened aerobic digester using a mechanical thickener that can thicken to 6% solids, the expected solids leaving the system would be approximately 3.6%. Figure 31.9 shows a typical prethickened application for liquid disposal with no phosphorus limit restriction on land application per Jim Porteous (Daigger et al., 2000). The necessary steps for option I are as follows:

1. The main design consideration, in this case is that the sludge being fed to the prethickening device remains aerobic.
2. To prevent anaerobic conditions, the sludge should be wasted directly from the liquid side stream to the prethickening device; or
3. If storage is required before prethickening, the detention time should be reduced to the minimum.

Option II: Dewatering, Post-Thickening, and Supernating, with Limit Restriction of Phosphorus on Land Application. This option assumes that a system is designed as a three-stage-in-series, prethickened aerobic digester using a mechanical thickener that can

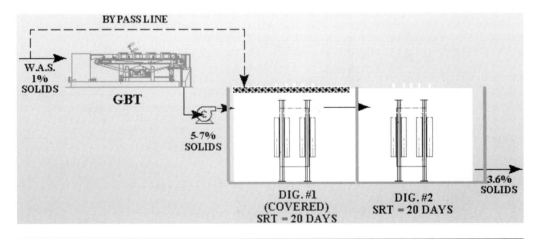

FIGURE 31.9 Option I: prethickened liquid disposal with no phosphorus limit restriction on land application (GBT = gravity belt thickener) (Daigger et al., 2000).

thicken to 6% solids; however, it is also designed to post-thicken. In this case, it is assumed that there is a restriction on land, and, at times, the prethickening device will be bypassed (during the summer). Figure 31.10 shows such a case of a dewatering option, with restriction of phosphorus on land application. The necessary steps for option II are as follows:

1. Phosphorus will be released during digestion and will be returned to the headworks.
2. Nitrification can be controlled in first digester to encourage struvite production, which is to be separated.
3. Add alum or ferric chloride to the supernatant and filtrate. Chemicals are only required for phosphorus removal. There is little else in the supernatant, so chemicals for phosphorus removal are minimized.
4. Phosphorus is now fixed and can be removed with the sludge.

TEMPERATURE CONTROL. If the potential exists to capture the heat generated by volatile solids destruction, the temperature of the reactor can be controlled.

Depending on geographical location and climatic conditions, tanks may or may not require covers. Covers are used in colder climates to help maintain the temperature of the waste being treated. Covers should not be used if they reduce evaporative cooling too much, and the liquid contents become too warm. When the liquid becomes too warm, offensive odors may develop, and the process effluent will have a very poor quality (California State University, 1991).

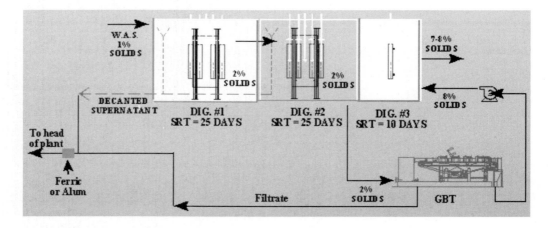

FIGURE 31.10 Option II: dewatering post-thickening, with phosphorus limit restriction on land application (GBT = gravity belt thickener) (Daigger et al., 2000).

The liquid temperature in an aerobic digester significantly affects the rate of volatile solids reduction and pathogen reduction. The rate of volatile solids reduction increases as the temperature increases. As with all biological processes, the higher the temperature, the higher the efficiency. At temperatures less than 10 °C (50 °F), the process is basically ineffective. Figure 31.11 shows highlights from a study focused on pathogen destruction and volatile solids performance of three-stage digester systems operated at temperatures in the range 8 to 31 °C. The following conclusions can be drawn from this study:

1. Above 12 °C, pathogens less than 2 000 000 can be achieved in 7 days; and
2. At temperature <10 °C, a minimum of 25 was required to achieve the same limits.

IV. Temperature

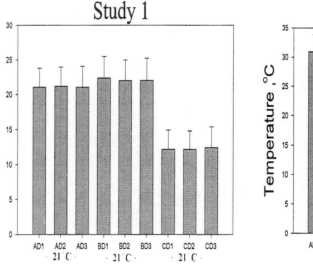
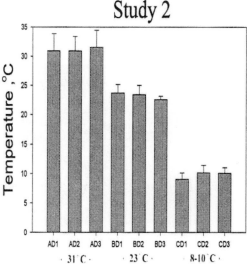

1. Temperature range of studies: 8°C - 31°C.
2. Above 12°C pathogens less than 2,000,000 can be achieved in 7 days.
3. Above 31°C and in 29 days V.S. reduction was 40%.

FIGURE 31.11 Study conclusions of temperature effect in the performance of aerobic digesters (Daigger et al., 1999).

The effect of temperature on volatile solids destruction is shown best on the full-scale installation data from Plum Creek, Colorado per Tim Grotheer (Daigger et al., 1997). This WWTP was a common-wall complete WWTP that was converted to a digestion system, as shown on the flow schematic in Figure 31.12. During this phase of the project, the waste was prethickened and fed to a five-stage-in-series digester system. No covers or insulation were supplied. During phase I of the project, even though the total SRT of the system was 54.5 days, the volatile solids reduction in February was only 16%, when the temperature was as low as 10 °C.

Figure 31.13 shows the volatile solids reduction for the entire system for each month, versus solids concentration and temperature. Figure 31.14 shows the volatile solids reduction for each month between each stage.

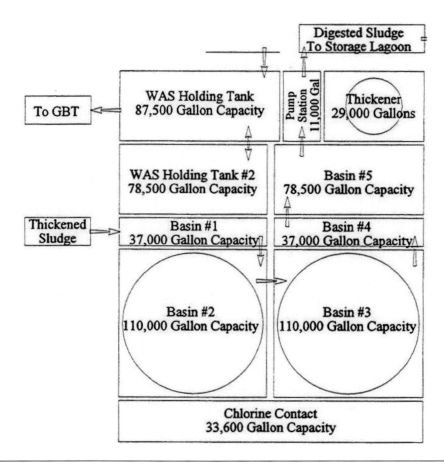

FIGURE 31.12 Flow schematic of sludge treatment process, Plum Creek, Colorado (GBT = gravity belt thickener; gal × 3.785 = L) (Daigger et al., 1997).

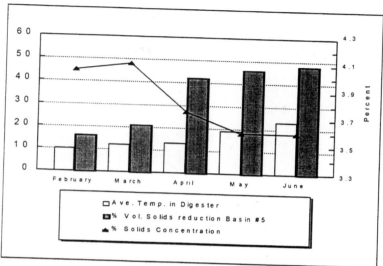

Phase I Digester Study

FIGURE 31.13 Volatile solids reduction of Plum Creek, Colorado, versus solids concentration and temperature (Daigger et al., 1997).

The study by Al-Ghusain (Al-Ghusain et al., 2004) evaluated the effect of temperature on volatile solids destruction at a low SRT of 10 days, with an anoxic cycle of 8 hours, as shown in Figure 31.14. This study agrees with the Plum Creek data, showing VSS destruction of 42.4% at 30 °C, as shown in Figure 31.15.

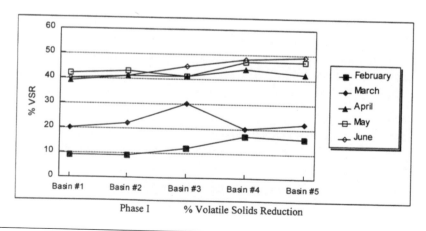

Phase I   % Volatile Solids Reduction

FIGURE 31.14 Volatile solids reduction (VSR) of each in-series stage, at Plum Creek, Colorado (Daigger et al., 1997).

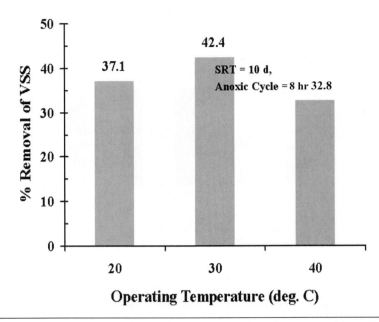

FIGURE 31.15 Effect of temperature on VSS removal with an anoxic cycle of 8 hours (Al-Ghusain et al., 2004).

Another study (Daigger et al., 2001) evaluated the reliability of two systems—one that provides in-loop thickening to achieve 3% solids in the digester compared with a system that prethickens to 3% solids. The focus of this study was to compare the capabilities of each system to maintain 20 °C liquid temperature in the digesters during the winter months, with waste of low temperature. Figure 31.16 highlights the finding from this evaluation. As shown in this figure, for an in-loop system, if the feed is as low as 10 °C, the digester basins and walls need to be insulated to maintain 20 °C in the digester. On the other hand, if the waste sludge is prethickened at the same concentration, covers are more than adequate to provide the same temperature elevation in the basins.

Part of this evaluation was to show that if the potential is there to capture the heat generated by the volatile solids, it is possible to control the temperature of the reactor. This could be very beneficial in cold-weather climates. Because conventional aerobic digestion is classified as a mesophilic process, it is optimized when the temperature is above 20 °C.

FLEXIBILITY OF THE SYSTEM. Flexibility should be built into the aerobic digester design, especially of a staged operation system, where two or more basins are

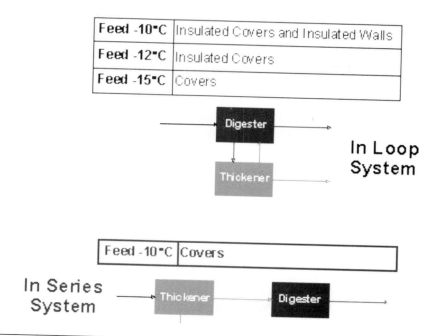

Feed -10°C	Insulated Covers and Insulated Walls
Feed -12°C	Insulated Covers
Feed -15°C	Covers

In Loop System

| Feed -10°C | Covers |

In Series System

FIGURE 31.16 Cover requirements for cold-weather installations with 3% solids feed (Daigger et al., 2001).

incorporated to the design. In addition, if a mechanical prethickening device is used, at a minimum, bypass piping for the digesters should be included to allow for down time and service of the machine. This also allows the operator the flexibility to make process changes, if necessary, resulting from temperature fluctuations, sludge volume, strength of waste to the system, toxicity, and possible extreme weather events, such as heavy rain and snow.

As a minimum, the aerobic digester equipment, blowers, and thickening equipment should have the ability to operate as follows:

1. The option to operate the digesters both in series and in parallel operation.
2. The aeration system and blowers to be designed to allow for incremental aeration for nitrification and denitrification.
3. The aeration system and blowers to be flexible to allow for air to move from basin to basin, as required.
4. The aeration system and blowers to be flexible to allow for air to be provided in one digester while the other digester is off. The blowers should be capable of handling water-level variations.

5. The ability to feed unthickened solids in the summer in two out of three tanks and post-thickening before the third tank. For example, feed 1% solids to the first digester every day, as shown in Figure 31.17, and post-thicken before liquid haul.

6. The ability to have in-loop thickening of the digester content and then post-thicken. This is one way to reduce temperature in the digester during the summer. For example, bypass the prethickening device, and then feed the thickener directly from the first digester. The thickened sludge can either go back to the same digester or to the second digester. If a third digester is present, the sludge from the second digester will be post-thickened and stored in the third basin in a much higher concentration.

7. Bypass the prethickening process a few days a week, and continue prethickening the remaining days of the week. For example, prethicken 3 days per week at 6% suspended solids, and bypass the 0.7% suspended solids 4 days per week to

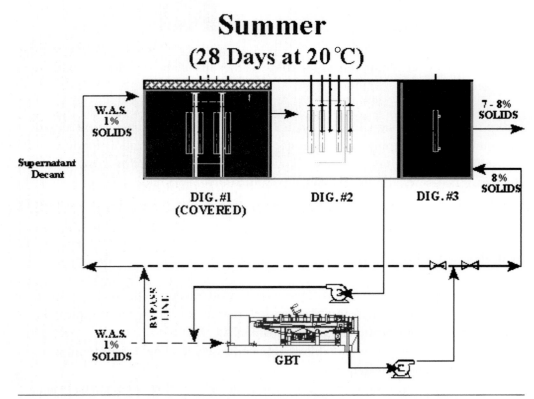

FIGURE 31.17 Reduce first stage to 1% solids in the summer to control temperature and post-thickener before liquid haul (GBT = gravity belt thickener) (Daigger et al., 1999).

the first digester. Figure 31.18 shows a schematic of the operation of a full-scale installation at Paris, Illinois as presented by Richard Yates (Daigger et al., 1997). As can be seen from this example, this approach is also applicable for digesters systems designed to treat a combination of primary and secondary WAS. For plants with high degradable solids, it is typical to have a summer and winter operation to optimize temperature control and process control.

8.   The ability to prethicken and post-thicken. Figure 31.19 shows an example of a possible flow diagram that can provide the necessary piping and machinery to prethicken and post-thicken using the same prethickening device.

EQUIPMENT DESIGN AND SELECTION

TYPES OF REACTORS. Aerobic digesters are either open or closed tanks. Open tanks are more common; however, in the last decade, more designers use covered and deeper tanks. Common-wall construction with a basin wall height of 7.3 m (24 ft) and water level of 6.7 m (22 ft) is typical. If prethickening is included in the design, tanks are typically 6.1 m (20 ft) deep or more; refer to the Diffused Aeration Equipment section for explanation. For round structures, tanks are 6.1 to 9.1 m (20 to 30 ft) deep and 9.1 to 18 m (30 to 60 ft) in diameter.

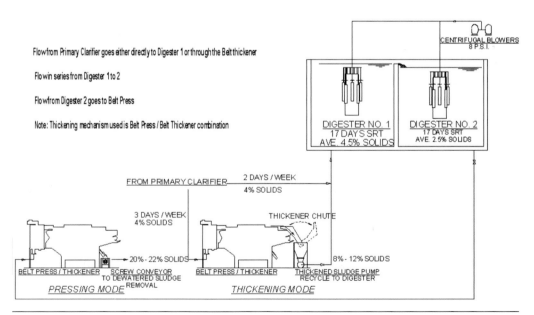

FIGURE 31.18 A schematic of the flow diagram of Paris, Illinois (8 psi = 55 kPa) (Daigger et al., 1997).

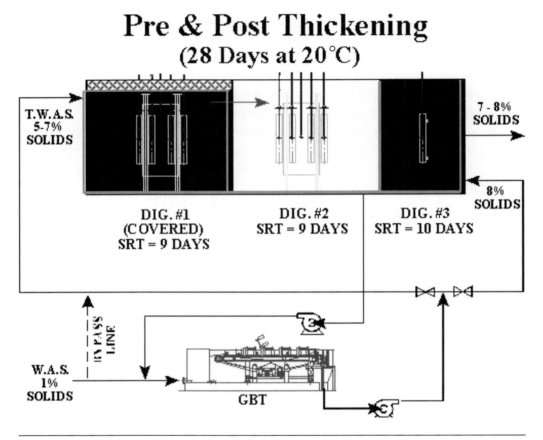

Pre & Post Thickening
(28 Days at 20 °C)

T.W.A.S.
5-7%
SOLIDS

7 - 8%
SOLIDS

8%
SOLIDS

DIG. #1
(COVERED)
SRT = 9 DAYS

DIG. #2
SRT = 9 DAYS

DIG. #3
SRT = 10 DAYS

BYPASS LINE

W.A.S.
1%
SOLIDS

GBT

FIGURE 31.19 A schematic of a pre- and post-thickening system (T.W.A.S. = total waste activated sludge, GBT = gravity belt thickener) (Daigger et al., 1999).

Covers reduce freezing, heat loss, and surface evaporation. As described in the Temperature Control section, covers combined with prethickening can increase the performance of conventional aerobic digestion by increasing both the volatile solids reduction and pathogen destruction. Combined with flexibility, temperature elevations in the thermophilic stage can be avoided, and the process can be controlled all year. In general, if the digester liquid temperature can be maintained at 20 °C all year, after thickening, the covers are not required. If the temperature is expected to be below that, then economics typically favor covers.

Most aerobic digesters are constructed to allow the liquid level to fluctuate very little in the first tank to maximize transfer efficiency and mixing. Some aerobic digesters are designed with sloped floors so that digested sludge is drawn off the bottom;

others have a constant weir overflow design. Aerobic digesters are constructed of re-inforced concrete and steel. In cold climates, steel tanks that are above ground should be insulated. If the steel and concrete tanks are below ground, they are fairly well-insulated by the soil if the groundwater is drained from the exterior of the tank walls.

At least two tanks should be installed to provide process flexibility and equipment flexibility, should any one tank need repairing. Increasing the aerobic digester size to produce an operating safety factor may create operational problems because of a loss of heat and increased energy requirements for mixing, as described in the Rules and Regulations section. The two schematics shown in Figures 31.20 and 31.21 are exam-ples of two sludge treatment processes that are typical for dewatering and liquid dis-posal. Figure 31.20 shows a two-stage dewatering option, and Figure 31.21 shows a three-stage liquid disposal option.

PIPING REQUIREMENTS. Piping requirements should, at a minimum, include provisions for feeding sludge; decanting supernatant, where applicable; withdrawing digested sludge; and supplying air for aeration. If possible, piping arrangements for flushing out sludge lines with WWTP effluent should be provided.

When possible, piping should allow bypass of each digester basin. For more de-tails, refer to the Flexibility of the System section of this chapter.

DIFFUSED AERATION EQUIPMENT. Equipment used for digestion to sup-ply air and mixing includes conventional mechanical aerators, coarse-bubble diffusers, fine-bubble aeration, or jet aerators. The best device not only should be capable of

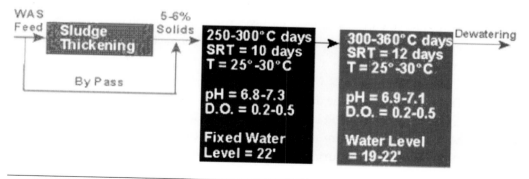

FIGURE **31.20** Two-stage sludge treatment dewatering option (D.O. = dissolved oxygen) (Daigger et al., 1999).

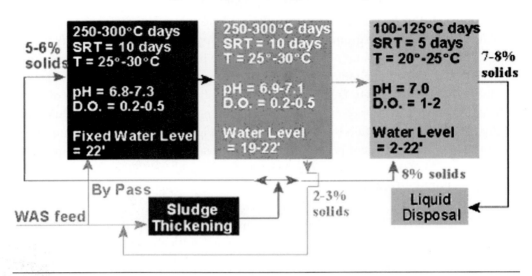

FIGURE 31.21 Three-stage sludge treatment liquid disposal option (D.O. = dissolved oxygen) (Daigger et al., 1999).

developing the required oxygenation and mixing, but should also provide flexibility (Water Environment Federation, 1998). In aerobic digesters, avoid fine-bubble ceramic diffusers because they are especially susceptible to clogging. Even though fine-bubble diffusers can work effectively in aerobic digesters aerated on a continuous basis, as stated by Jim Scisson, there are practical limitations to effectively operating an aerobic digester in an aerobic–anoxic mode (Daigger et al., 1998).

Typically, floor-covering, submerged-header air diffusers are found near the bottom of the digester, toward one side, to induce a spiral or cross-roll mixing pattern. The most widely used air diffusers are the small- and large-bubble types. Submerged header air diffused systems

1. Tend to add heat to the digester,
2 Are not greatly affected by foaming conditions, and
3. May require the entire assembly to be removed for cleaning if it becomes clogged or plugged during the anoxic operation.

An alternative to the floor-covering diffuser systems is a full range of high shear, nonclog aeration equipment designed specifically for high solids concentrations in the range 4 to 8% solids concentration. This aeration system is combined with an adjustable, above-water orifice to allow for profiling the air requirements to meet demand, and nonclog diffusers ensure that plugging will not occur during the anoxic operation. The shear tube and draft tubes that are also incorporated into the design have proven to provide the type of rapid mixing and shearing action necessary to transfer oxygen and achieve volatile reduction with high solids. The limitation of this system is that it works better when the liquid depth is greater than 6.1 m (20 ft). Figure 31.22, per Richard Yates, shows an installation of a system of this kind. Figure 31.23 shows the plan and section views of the technical drawings for the same project.

FIGURE 31.22 Typical installation of National Aeronautics and Space Administration (Washington, D.C.) draft tube at Paris, Illinois. Picture taken after conversion from anaerobic to a prethickened two-stage, in-series, aerobic digester system, treating a combined primary and secondary waste (Daigger et al., 1997).

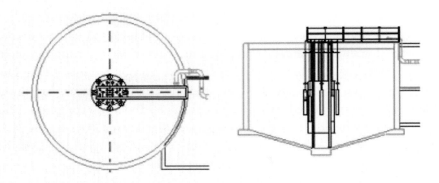

FIGURE 31.23 The plan and section views of a digester basin at Paris, Illinois. The digester basin is 14 m (45 ft) in diameter by 9.4 m (31 ft) deep (Daigger et al., 1997).

In summary, single-drop aeration with shear tubes or draft tubes systems

1. Are specifically designed for higher solids concentrations (4 to 8% suspended solids),
2. Tend to add heat to the digester,
3. Do not require maintenance because they are nonclog systems, and
4. Require deep tanks with greater than 6.1 m (20 ft) of liquid depth.

MECHANICAL SURFACE AERATORS. Both low- and high-velocity mechanical surface aerators, in either free-floating or fixed installations, are efficient in oxygen transfer (amount of oxygen fed by the aerator that has been absorbed by the sludge). Typically, mechanical surface aerators

1. Supply oxygen efficiently to aerobic digesters, if the desired solids concentration in the basin is less than 2.5% solids;
2. Have minimal maintenance requirements;
3. Are greatly affected by foaming conditions; and
4. Are greatly affected during cold weather by ice buildup.

SUBMERGED MECHANICAL AERATORS. Submerged mechanical aerators include a rotating submerged impeller mounted on a drive shaft extending vertically into the aeration basin. Compressed air is supplied beneath the impeller, where it is sheared into bubbles and pumped into the basin (Water Environment Federation, 1998).

Generally, submerged mechanical aerators are unaffected by foaming or icing conditions.

BLOWERS. Centrifugal blowers and rotary positive displacement air blowers are suitable for aerobic digestion. As a minimum, for a two-stage digester system, three blowers are recommended. If rotary positive displacement blowers are selected, then as a minimum, two out of the three blowers should be complete with dual-speed or variable-speed motors to provide maximum flexibility and process optimization. Examples for such a system include a blower designated for the first digester to operate at both maximum and design airflow (dual-speed motor is included in this machine), a blower designated for the second digester to operate at design and minimum airflow, and a standby blower that can be used to supply air to either basin at maximum and design airflow (dual-speed motor is included in this machine also).

Figure 31.24, per Glen Daigger, shows a flow schematic of a typical three-stage, prethickened digester system with a total of 30 days SRT operated at 20 °C. This diagram shows the recommended airflow requirement between the three stages and blower requirements (Daigger et al., 1999).

PUMPS. A number of different types of pumps have been used to transfer sludge to aerobic digesters. Both positive displacement pumps and centrifugal pumps have been used successfully for this purpose. Airlift pumps and nonclog pumps draw off the thickened, digested sludge for transferring between basins, dewatering, or disposal.

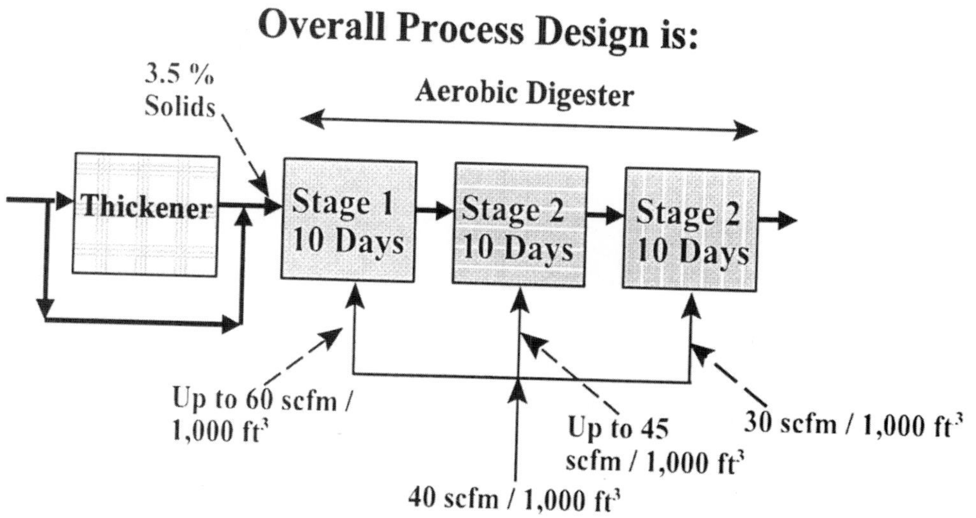

FIGURE 31.24 Airflow distribution and blower requirements for a three-stage, prethickened aerobic digester system (1 scfm/1000 cf = 0.06 m^3/m$^3 \cdot$ h) (Daigger et al., 1999).

An extensive section on pump selection is included in Chapter 8 of this book. Please refer to that chapter for more details.

MIXING AND AERATING EQUIPMENT. Various aeration systems have been used in ATAD systems, including U.S. Filter floor-mounted jet aspirator or jet aeration (U.S. Filter [Siemens], Warrendale, Pennsylvania), Fuchs side-mounted aspirating aeration (Fuchs, Harvey, Illinois), Turborator Technology top-mounted aspirator aeration and pump and venturi system (Burnaby, British Columbia, Canada) (Shamskhorzani, 1998). With all air aspirating systems, the equipment provides both mixing and oxygen transfer (Figures 31.25 and 31.26).

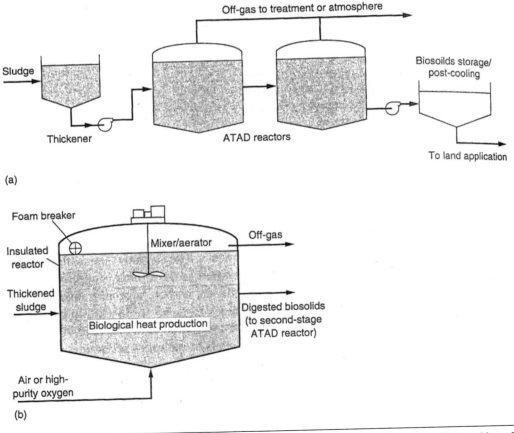

(a)

(b)

FIGURE 31.25 A typical ATAD system schematic and reactor schematic (Metcalf and Eddy, 2002).

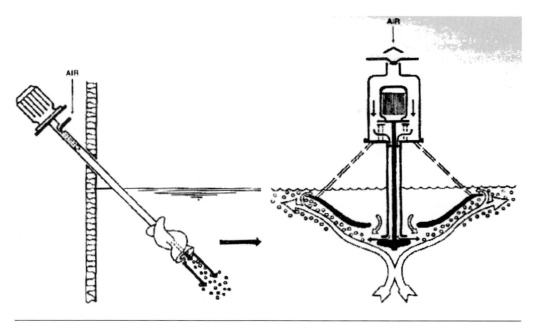

FIGURE 31.26 A schematic of an aerator used in an ATAD system (U.S. EPA, 1990).

CONTROL EQUIPMENT. Aerobic digestion systems often incorporate automatic pH control equipment. The typical automatic pH control system consists of a pH electrode placed within the digester that measures the pH and provides a signal to chemical feed pumps if the desired pH range is not being achieved. The pH control systems can also be used to control the blowers to provide a controlled airflow to the basins, which will provide the operator with the capabilities to optimize the process and operate in a simultaneous nitrification and denitrification mode. Another way that the pH system can be used is to turn the blowers on and off to provide a true aerobic–anoxic operation.

THICKENING EQUIPMENT. The thickening equipment recommended for aerobic digestion is divided into three categories and is described in the Prethickening section of this chapter. Because Chapter 29 is dedicated to this subject in its entirety, please refer to Chapter 29 for more details.

PROCESS PERFORMANCE

RULES AND REGULATIONS. *Standards for the Use or Disposal of Sewage Sludge.* The Standards for the Use or Disposal of Sewage Sludge (40 *CFR* Part 503; U.S.

EPA, 1999b) regulate the use and disposal of biosolids from WWTPs. Limitations are established for items such as contaminants (mainly metals), pathogen content, and vector attraction.

The regulations addressed by 40 *CFR* Part 503 (U.S. EPA, 1999b) were promulgated in 1993 and revised in 1999 by the U.S. Environmental Protection Agency (U.S. EPA) (Washington, D.C.). The modifications and clarifications of the rules are described in the Environmental Regulations and Technology document, "Control of Pathogens and Vector Attraction in Sewage Sludge", as revised by the Pathogen Equivalency Committee (U.S. EPA, 1999a). The regulations cover specifically (1) land applications of biosolids, (2) surface disposal of biosolids, (3) pathogen and vector reduction in treated biosolids, and (4) incineration.

The regulations on land applications and vector reduction will be explored in this chapter to provide a better understanding of the effect of these rules on aerobic digestion systems.

Regulation of Land Application. Land application relates to biosolids reuse and includes all forms of applying bulk or bagged biosolids to land for beneficial uses at agronomic rates (i.e., rates designed to provide the amount of nitrogen needed by crop or vegetation, while minimizing the amount that passes below the root zone). The regulations establish the following levels of quality:

1. Two levels, with respect to heavy metals concentrations;
2. Two levels, with respect to pathogen densities (class A and class B); and
3. Two types of approaches for meeting vector attraction, one being the biosolids processing of choice, and the other the use of a physical barrier.

Definition of Biosolids. *Biosolids* is defined as a wastewater product that is both organic and semisolid and becomes suitable for beneficial use after biological or chemical stabilization.

Class A biosolids are biosolids in which the pathogens (including enteric viruses, pathogenic bacteria, and viable helminth ova) are reduced below required detectable levels (WEF, 1995) Class A biosolids must meet all the specific criteria to ensure that they are safe to be used by the general public and for nurseries, gardens, and golf courses.

The process that most likely will qualify for class A biosolids covered in this chapter is the thermophilic process. Although the regulations for class A are very elaborate, and a number of options are available, the class A biosolids criteria that a thermophilic aerobic digestion system will most likely be designed to meet are as follows:

1.   A fecal coliform density could be less than 1000 MPN/g total dry solids, or
2.   Salmonella sp. density of less than 3 MPN/4 g total dry solids (3 MPN/4 g total solids).
3.   In addition to meeting requirement 1 or 2, the biosolids should be treated by one of the prescribed processes that reduce pathogens beyond detectable levels, described as processes to further reduce pathogens (PFRPs). Thermophilic aerobic digestion that, by definition, processes liquid biosolids agitated either with air or oxygen to maintain aerobic conditions, with a mean cell residence time of 10 days at 55 to 60 °C, qualifies as an acceptable PFRP process.

Class B biosolids are biosolids in which the pathogens are reduced to levels that are unlikely to pose a threat to public health and the environment under specific use conditions (WEF, 1995). Class B biosolids cannot be sold or given away in bags or other containers or applied on lawns or home gardens. They are typically used for applications to agricultural land or disposed of in a landfill.

The process that most likely will qualify for class B biosolids covered in this chapter is the conventional aerobic digestion process. Although the regulations for class B are elaborate, and a number of options are available, the class B biosolids criteria that a mesophilic conventional aerobic digestion system will most likely be designed to meet are the following:

For a single digester basin, the following criteria could be met:

1.   Meet one of the two pathogen reduction requirements, either
     •   60 days detention at 15 °C or 40 days at 20 °C, or
     •   Meet fecal coliform density of less than 2 000 000 MPN/g total dry solids.
2.   Meet one of the two vector attraction reduction requirements, either
     •   Meet at least 38% reduction in volatile solids during biosolids treatment, or
     •   Meet a SOUR at 20 °C (68 °F) of less than 1.5 mg oxygen/g total solids · h.
3.   Sludge can also be accepted as suitably reduced in vector attraction when it shows less than 15% additional volatile solids reduction after 30 days additional batch digestion at 20 °C (68 °F).

For a two stage or more digester system, the following criteria could be met:

1. Meet *both* of the pathogen reduction requirements:
    *   Meet fecal coliform density of less than 2 000 000 MPN/g total dry solids, AND
    *   Provide 42 days detention at 15 °C or 28 days detention at 20 °C. In this case, because approval of the process as a PSRP equivalent alternative is required by the permitting authority, the plant operator should demonstrate experimentally that microbial levels in the product from their sludge digester are satisfactorily reduced and meet one of the Vector Attraction Reduction requirements listed above.
2. Meet the vector attraction reduction requirements, either
    *   Meet at least 38% reduction in volatile solids during biosolids treatment, or
    *   Meet a SOUR at 20 °C (68 °F) of less than 1.5 mg oxygen/g total solids·h.

PARAMETERS USED TO EVALUATE PERFORMANCE OF AEROBIC DIGESTERS.
The following parameters are used to evaluate the performance of aerobic digesters:

1. SOUR,
2. Pathogen reduction,
3. Volatile solids reduction and solids reduction,
4. Solids retention time × temperature product (days °C),
5. Nitrogen removal in biosolids,
6. Phosphorus reduction in biosolids and biophosphorus,
7. Sludge dewatering characteristics, and
8. Supernatant quality of recycled sidestreams.

Standard Oxygen Uptake Rate. The rate of oxygen use by the microorganisms depends on the rate of biological oxidation. The SOUR of 1.5 mg oxygen/g total solids·h at 20 °C (68 °F) was selected by U.S. EPA to indicate that an aerobically digested sludge has been adequately reduced in vector attraction.

The oxygen uptake rate is used to determine the level of biological activity and the resulting solids destruction occurring in the digester. The SOUR is becoming a more common testing procedure for most operators instead of using the traditional volatile solids reduction. However, the volatile solids reduction test still remains the preferred choice for anaerobic digestion, because the SOUR does not apply to anaerobically digested solids.

The SOUR is a quick test and is independent of the initial value in the system or re-duction of SOUR in the upstream process. On the other hand, volatile solids reduction is a percentage of the incoming volatile level to the digester system. During active aer-obic digestion, if primary sludge is added to the first digester, it may exhibit an uptake rate ranging from 10 to 30 mg oxygen/g total solids·h. For staged operation, however, a typical rate of oxygen uptake is from 3 to 10 mg oxygen/g total solids·h in the first stage digester as compared with a range of 10 to 20 mg oxygen/g·h in the active phase of the activated sludge process.

An oxygen uptake rate for a well-digested sludge from the aerobic digestion pro-cess ranges from 0.1 to 1.0 mg oxygen/g total solids·h, well below the required 1.5 mg oxygen/g·h by U.S. EPA standards.

PATHOGEN REDUCTION. As mentioned earlier in this chapter, pathogen re-duction is similar to solids reduction in that little pathogen reduction can be expected at temperatures less than 10 °C (50 °F), while significant reduction may be achieved at temperatures higher than 20 °C (68 °F). Even though it is acceptable by U.S. EPA stan-dards to operate at 15 °C (59 °F), for economic reasons and performance reliability, the author recommends designing and operating the plant at a minimum of 20 °C (68 °F).

VOLATILE SOLIDS REDUCTION AND SOLIDS REDUCTION. Aerobic digestion results in the destruction of VSS. In addition, if a membrane thickening di-gestion option is selected, fixed suspended solids (FSS) can also be reduced. This oc-curs because both the organic and inorganic material in the biodegradable suspended solids are solubilized and digested. However, the components of VSS and FSS are not equal. Consequently, they will not generally be destroyed in the same proportion.

Primary solids and WAS from a system with a short SRT will contain relatively high fractions of biodegradable material, as opposed to WAS from a system with a long SRT, which will contain a low fraction of biodegradable material and a high frac-tion of biomass debris (Grady et al., 1999).

As mentioned earlier in this section, the destruction of biodegradable suspended solids is very much affected by temperature and can be characterized as a first-order reaction. This occurs because the decay of active biomass is a first-order reaction. The decay coefficient of relatively high active biomass is independent of the SRT at which the waste solids were produced. This is because the decay coefficient for the bio-degradable suspended solids is influenced strongly by the decay coefficient for hetero-trophic bacteria, which is relatively constant.

In a study performed to determine the minimum SRT required to meet class B requirements by meeting both pathogens and volatile solids reduction operating at minimum temperatures as concluded by Lu Kwang Ju (Daigger et al., 1999), two systems were evaluated. One system included two basins in series and the other, three basins in series. Table 31.5 shows the volatile solids reductions at different temperatures and different SRTs.

The sludge used in this study had a very low digestible fraction. It was concluded that even though the pathogen destruction was met in all the systems, an SRT of 29 days was required to exceed the required 38% volatile solids reduction. This confirms that volatile solids reduction is dependent on the source of the sludge, and if it has a very low fraction of digestible organic content, it is difficult to meet the minimum requirements by U.S. EPA.

For full-scale data on this subject, please refer to the Temperature Control section.

Solids Retention Time × Temperature Product. A significant factor in the effective operation of aerobic digesters, the SRT is the total mass of biological sludge in the reactor, divided by the mass of solids that are removed from the process on a daily average. Typically, increased SRT results in an increase in the degree of volatile solids reduction.

Based on the discussion above, with respect to temperature, degradable sludge and nondegradable sludge and the effect it can have on volatile solids reduction, the SRT × temperature product (days °C) curve can be used to design digester systems, taking into consideration not only the total days °C that the digester system will have to meet, but also the quality of the source. The original U.S. EPA SRT × temperature curve, developed in the late 1970s (U.S. EPA, 1978, 1979), was updated (Daigger et al., 1999) by incorporating data from two extensive pilot studies by Lu Kwang Ju (Daigger et al., 1999) conducted over 3 years with data from three full-scale installations. Figure 31.27 shows a process design based on 600 days °C, assuming the feed has relatively

TABLE **31.5**   Volatile solids (VS) reduction at minimum operating temperatures and minimum SRT (Daigger et al., 1999).

	VS reduction Data from both Studies in all basins					
°C	**8–10 °C**	**12 °C**	**21 °C**	**21 °C**	**23 °C**	**31 °C**
2 basins in series # of days	19.25	13.75	13.75	13.75	19.25	19.25
VS Removal	27%	31%	31%	31%	28%	31%
3 basins in series # of days	29.25				29.25	29.25
VS Removal	28%				32%	40%

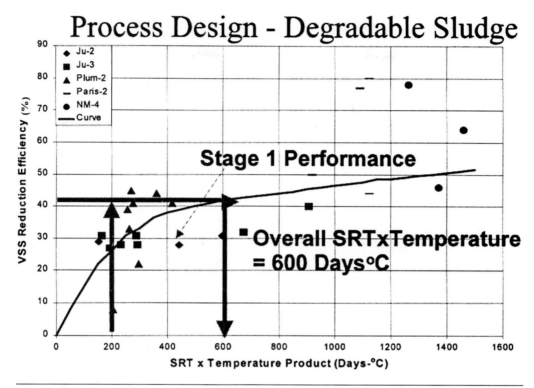

FIGURE 31.27 Selection of SRT × temperature (days °C) product for feed with high degradable solids content (Daigger et al., 1999).

high degradable solids content; and Figure 31.28 shows an alternative operation if the feed sludge has low degradable solids. Both systems can coexist, from a design standpoint, if prethickening is incorporated to the design into allow an increase or decrease of SRT as necessary.

Nitrogen Removal in Biosolids. As mentioned earlier in this chapter, the conventional aerobic digestion process, operated in the aerobic–anoxic mode, provides full nitrification and denitrification and a reduction in total nitrogen. The annual design loading rates for land application of biosolids is typically limited by the nitrogen loading rate. Because nitrification is inhibited in the ATAD process, the most ideal process to meet these requirements is the conventional process operated in the aerobic–anoxic mode.

Phosphorus Reduction in Biosolids and Biophosphorus. Phosphorus is released in both anaerobic and aerobic digestion; however, if the conventional aerobic digestion

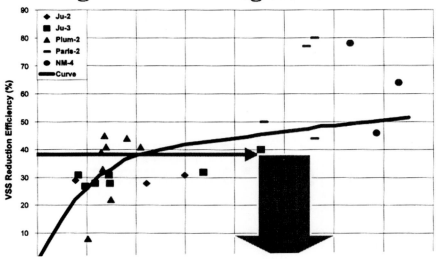

FIGURE 31.28 Selection of SRT × temperature (days °C) product for feed with low degradable solids content (Daigger et al., 1999).

process is operated in the aerobic–anoxic mode or under low dissolved oxygen conditions, the phosphorus release is minimized. As of 2005, there was a land application limit on phosphorus in very limited areas, where the land is saturated with phosphorus. If such rules apply in your area of interest, please follow the guidelines in the Phosphorus Reduction in Biosolids and Biophosphorus section under Aerobic–Anoxic Operation (aerobic–anoxic operation provides phosphorus reduction in biosolids and biophosphorus) and in the Category D—Thickening Treatment Process: Using Membranes for In-Loop Thickening with Aerobic Digestion section of this chapter.

Sludge Dewatering Characteristics. When compared with other type of sludges, the dewatering characteristics of the traditional aerobically digested sludges are typically worse than those of anaerobically treated sludges. In general, it has been observed that sludge type has the greatest effect on the quantity of chemicals required for chemical conditioning.

Factors affecting the selection of the type and dosage of conditioning agents are the properties of solids and type of mixing and dewatering devices to be used. Important solids properties include source, solids concentration, age, pH, and alkalinity. Sources such as primary sludge and WAS (digested and nondigested) are good indicators of the range of probable conditioner doses required. Solids concentrations will affect the dosage and dispersal of the conditioning agent. The pH and alkalinity may affect the performance of the conditioning agents, in particular, the inorganic conditioners.

Sludges that are difficult to dewater and require larger doses of chemicals generally do not yield cake as dry as other sludges and have a poorer quality of filtrate or centrate (Metcalf and Eddy, 2002). Table 31.6 shows typical levels of polymer addition for belt-filter press and solid-bowl centrifuge dewatering. As shown in this table, anaerobically treated primary sludge and aerobically treated WAS, when compared with aerobically treated primary sludge and WAS, required similar quantities of polymer addition.

The data presented in Table 31.6 was accumulated before the adoption of the aerobic–anoxic operation of aerobic digestion processes (aerobic–anoxic operation techniques were adopted in the mid-1990s across the United States). Since then, a number of plants have shown improvement in their sludge characteristics when modifications were made from traditional conventional aerobic digestion to aerobic–anoxic operation. Based on preliminary reports conducted by Stensel and Bailey (2000) in Paris, Illinois, aerobically digested sludge operated in aerobic–anoxic mode produces less solids than aerobically digested sludge operated in full aeration.

Supernatant Quality of Recycled Sidestreams. Careful monitoring of solid–liquid separation in the continuous and batch-feed digesters increases the performance of an aerobic digester. The supernatant liquid or filtrate should be low in soluble BOD, TSS, and nitrogen, for both batch and continuous-flow operations. Table 31.7 lists characteristics of "acceptable" supernatant values from aerobic digestion processes (Metcalf and Eddy, 2002).

TABLE 31.6   Typical levels of polymer addition for belt-filter press and solid-bowl centrifuge sludge dewatering (Metcalf and Eddy, 2002).

Type of sludge	Belt-filter press (g/kg [lb/ton] dry solids)	Solid-bowl centrifuge (g/kg [lb/ton] dry solids)
Primary	1 to 4 (2 to 8)	0.5 to 2.5 (1 to 5)
Primary and waste activated	2 to 8 (4 to 16)	2 to 5 (4 to 10)
Primary and trickling filter	2 to 8 (4 to 16)	—
Waste activated	4 to 10 (8 to 20)	5 to 8 (10 to 16)
Anaerobically digested primary	2 to 5 (4 to 10)	3 to 5 (6 to 10)
Anaerobically digested primary and air waste activated	1.5 to 8.5 (3 to 17)	2 to 5 (4 to 10)
Aerobically digested primary and air waste activated	2 to 8 (4 to 16)	—

TABLE 31.7   Acceptable characteristics of supernatant from aerobic digestion systems (Metcalf and Eddy, 2002).

Parameter	Acceptable range	Acceptable values
pH	5.9 to 7.7	7.0
5-day BOD, mg/L	9 to 1700	500
Filtered 5-day BOD, mg/L	4 to 173	50
Suspended solids, mg/L	46 to 2000	1000
Kjeldahl nitrogen, mg/L	10 to 400	170
Nitrate-nitrogen, mg/L	— 30	
Total phosphorus, mg/L	19 to 241	100
Soluble phosphorus, mg/L	2.5 to 64	25

Although the values are considered "acceptable" values, using the techniques described in this chapter, the process can be controlled and optimized to provide supernatant of higher quality than what is described in Table 31.7. To improve these parameters, it is necessary to operate in aerobic–anoxic operation. Controlling the availability of necessary quantities of carbon source required for denitrification to allow for full nitrification and denitrification will enhance supernatant quality. Maintaining a neutral pH of 7.0 will also enhance nitrification and denitrification. For additional information, refer to Figure 31.4 for the effect of both lengths of anoxic cycles and temperature, with respect to total nitrogen levels in the filtrate.

Table 31.8 shows data from a category C sludge treatment process installation in the city of Stockbridge, Georgia, as described in the Category C—Thickening Treatment Process: Using a Gravity Thickener in Loop with Aerobic Digestion section of this chapter. As described in this section, the thickener and aerobic digester are operating in loop, nitrification occurs in the digester, and denitrification and pH recovery occur in the thickener. Thickener blanket data indicate good settling. Average data for TSS, ammo-

TABLE 31.8   Data from a category C process at the Stockbridge, Georgia WWTP (Stege and Bailey, 2003).

Parameter	Actual data from a category C process treatment	Compared with acceptable typical values from Table 31.7
pH	6.5 to 7.1	7.0
Suspended solids, mg/L	10 to 50	1000
Total Kjeldahl nitrogen, mg/L	2.5 to 4	170
Nitrate-nitrogen, mg/L	—	30
Total phosphorus, mg/L	0.3	100
Thickener blanket	8 to 13.5	10

nium, and phosphorus indicate both nitrification and denitrification and phosphorus removal with TSS removal. The sludge blanket during the collection of these data points varied between 2.4 and 4.1 m (8.0 and 13.5 ft). For comparison, data presented by Elena Bailey (Stege and Bailey, 2003) were used to produce Table 31.8, demonstrating how optimizing the techniques affects supernatant quality.

PROCESS CONTROL

The important factors in controlling the operations of aerobic digestion are similar to those for other aerobic biological processes. The primary process indicators for monitoring aerobic digesters on a daily basis are temperature, pH, dissolved oxygen, odor, and settling characteristics, if applicable. Ammonia, nitrate, nitrite, phosphorus, alkalinity, SRT, and SOUR are secondary indicators that are useful in monitoring long-term performance and for troubleshooting problems associated with the primary indicators.

All the parameters, both primary and secondary indicators, either affect or are used to monitor the operational performance of the process. While control and monitoring of these parameters is important, the degree of control that can be exercised on each parameter varies. Monitoring helps control process performance and serves as a basis for future improvements.

Table 31.9 provides a list of both the primary and secondary monitoring parameter indicators. Analysis frequency should be increased during startup and during those times when large changes are made to operating conditions, such as sludge flowrate, sludge source, change in polymer, large increase or decrease in solids concentration of feed, or large increase or decrease in temperature of feed.

TABLE 31.9 Recommended values for primary and secondary monitoring parameters (Stege and Bailey, 2003).

Monitoring parameter	Frequency	Operating range		
		Minimum	Nominal	Maximum
Temperature, °C	Daily	15	20	37
pH	Daily	6.0	7.0	7.6
Dissolved oxygen, mg/L	Daily	0.1	0.4 to 0.8	2.0
Alkalinity, mg/L as calcium carbonate	Weekly	100	>500	—
Ammonia-nitrogen, mg/L	Weekly	—	<20	40
Nitrate, mg/L	Weekly	—	<20	—
Nitrite, mg/L	As required	—	<10	—
SOUR, mg oxygen/h/g total solids	As required	—	<1.5	—
Phosphorus, mg/L	As required	—	<5	

PROCESS STARTUP
GENERAL GUIDELINES FOR PROCESS STARTUP. The following are general guidelines for process startup:

1. To start an aerobic digester smoothly, do not overload the digester. Proceed carefully by adding feed sludge to the digester on a regular schedule and operating the aeration system equipment continuously when sludge is being fed to the system.

2. An aeration system with adequate capacity ensures dissolved oxygen concentrations of 0.5 mg/L or more. In the startup phase, it will be higher than that, so take care not to cause foaming problems by overaerating.

3. If the initial volume of sludge does not provide sufficient liquid volume to operate the aeration system effectively, additional sludge or plant influent may be fed to obtain the needed volume. The daily sludge feed is then added on a regular schedule, preferably during as long a period as possible.

4. The primary process indicators for monitoring the aerobic digesters on a daily basis are temperature, pH, dissolved oxygen, odor, and settling characteristics, if applicable. Ammonia, nitrate, nitrite, phosphorus, alkalinity, SRT, and SOUR are secondary indicators that are useful in monitoring long-term performance and for troubleshooting problems associated with the primary indicators. All these parameters need to be closely monitored during startup to optimize the system.

5. The startup of aerobic digestion of primary feed sludges takes a substantially longer time and more oxygen than digestion of WAS. The degree of aerobic digestion depends on the concentration of solids in the digester and on the rate of solids feed. For routine monitoring, the total VSS may be used in place of TSS. As organic loading increases, the required SRT and total oxygen requirements increase and have a direct bearing on process efficiency. For a complete-mix, continuous-flow digester, the volumetric loading rate and detention are used to indirectly reflect solids destruction.

6. Regular loading, rather than periodic slug doses or large inputs, improves most biological process efficiencies. Regular feeding reduces both volume and aeration equipment requirements; however, batch-type operations often are used in the aerobic digesters.

7. If digesters are prethickened using a mechanical thickener, such as a gravity belt thickener, drum thickener, or dissolved air flotation unit, supernatant should not be returned to the head of the plant at any time. If, on the other hand, an in-loop digester-thickener process or a membrane thickening system

is used, the supernatant or permeate from these systems will be recycled to the head of the plant. During startup, it is important to make sure that the digester supernatant quality does not impose an excessive solids recycle load at the head of the plant, and the quantities of ammonia, nitrates, and/or phosphorus are within acceptable limits.

PREFERRED PROCEDURE FOR STARTUP. The following two examples are prepared by the author as examples of a step-by-step procedure during startup. Both examples assume that the design incorporates two digesters in series, a gravity belt thickener is used to prethicken to 6% solids 5 days per week, and primary sludge is also added directly in the first digester.

Preferred Choice for Process Startup. The following steps describe the preferred choice for process startup:

1. Step 1. Use clean water to fill both digesters completely.
2. Step 2. Test all the aeration equipment, blowers, and bypass lines to make sure that all the design features required in the Flexibility of the System section of this chapter do, in fact, work.
3. Step 3. Drain 1.5 m (5 ft) from digesters 1 and 2.
4. Step 4. Set up feed and transfer piping to operate in series.
5. Step 5. Prethicken WAS using the gravity belt thickener 3 times per week. For 2 days per week, bypass the gravity belt thickener. If primary sludge is scheduled to be aerobically digested as well, it is recommended to avoid introduction of the primary solids to the digesters the first 10 days of operation. Allow for the process to stabilize by creating a healthy population of nitrifiers before addition of primary sludge.
6. Step 6. Add all the feed to digester 1.
7. Step 7. When digester 1 is full, transfer flow to digester 2.
8. Step 8. Continue operating in series.
9. Step 9. After step 8 is accomplished, start operating the gravity belt thickener on full schedule.

Second Choice for Process Startup. The following steps describe the second choice for process startup:

1. Step 1. Use WAS to fill both digesters completely.
2. Follow steps 2 to 9 as described in the Preferred Choice for Process Startup section above.

OPERATIONAL MONITORING

For the most part, aerobic digestion is a self-regulating process, unless the process is overloaded or the equipment is inoperative. Routine operational surveillance includes analytical testing and periodic checks of electrical and mechanical equipment, such as seals, bearings, timers, and relays.

From an operator's point-of-view, there are two interrelated types of issues that typically could require troubleshooting—equipment and process. Sometimes, equipment malfunction could cause a process upset, and, at other times, because of process conditions, equipment could malfunction. The following troubleshooting guide is categorized based on primary causes of the problem.

EQUIPMENT TROUBLESHOOTING. *Clogging of Air Diffusers.* A buildup of rags or grit on the diffusers causes diffuser clogging when the aeration system is shut off to concentrate the digested sludge before supernatant, decanting, sludge withdrawal, or after denitrification. If the air is off for a long period of time (sometimes 6 to 12 h/d), rags and grit within the digester may lodge on and inside the diffuser mechanisms. When aeration is resumed, the particulates clog the diffusers and eventually will reduce the air discharge, causing an increase in the blower discharge pressure. As mentioned in the Equipment Design and Selection section, fine-bubble diffusers are not recommended for digestion application because they are especially susceptible to clogging problems.

Effective maintenance prevents or cures clogging. While the elimination of diffuser clogging is difficult, equipment modification can reduce its frequency and severity. Jet aerators can be equipped with a self-cleaning system. Alteration of the air header, so that the diffusers may be withdrawn for service without emptying the tank, may also be used. Another alternative would be to use diffusers with above-water orifices, each complete with individual drop pipes. The chance for clogging is reduced because the orifices are located out of the liquid. Otherwise, during the denitrification or anoxic stage, the air is off and solids will accumulate in the pipes.

For systems in which mechanical prethickening is used, the mixing air requirement in the first digester (for in-series digester systems) is the dominating design criterion and should be optimized. The ability to maintain suspended solids in suspension is important, so shear tubes or draft tubes are recommended in these applications.

Figure 31.29 shows a typical system of this type. This project incorporates prethickening with a gravity belt thickener to as much as 8%, has four basins in series, and is covered and insulated because of the high altitude and low winter temperatures. A

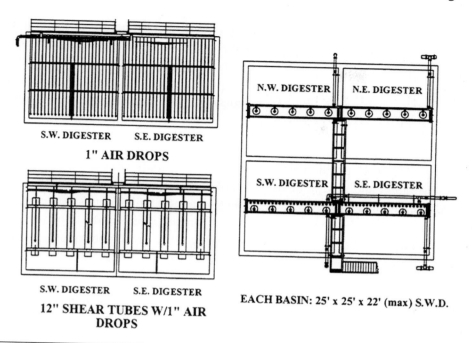

S.W. DIGESTER   S.E. DIGESTER

1" AIR DROPS

S.W. DIGESTER   S.E. DIGESTER

12" SHEAR TUBES W/1" AIR DROPS

N.W. DIGESTER   N.E. DIGESTER

S.W. DIGESTER   S.E. DIGESTER

EACH BASIN: 25' x 25' x 22' (max) S.W.D.

FIGURE 31.29 High solids operation at Los Lunas, New Mexico, in four-stage, prethickened, covered, and insulated digester system (ft × 0.3048 = m; in. × 25.40 = mm) (Daigger et al., 1997).

dual-type aeration system was provided on this project, as shown in the figure, to allow for high- and low-water-level operation (Daigger et al., 1997).

Blowers and Valves. An inefficient aeration system or excessive organic loading to the digester can cause a low dissolved oxygen concentration. Blower, mechanical aerator, or aeration diffuser deterioration results in a loss of aeration efficiency. It is important to routinely check and monitor the air delivery rate and pipeline pressures for potential problems; record the position of all air valves; investigate suspected clogging, leaks, and excessive pressure; and occasionally check the horsepower delivered to mechanical aeration shafts to determine their condition.

Mechanical Aerators. Mechanical aerator efficiency decreases if the liquid level exceeds specified upper and lower operational limits. Consult the manufacturers' engineering data and the operations and maintenance manuals for recommended liquid-level limits.

Freezing. Extended subfreezing weather may lead to ice formation on the liquid surface and on the mechanical aeration equipment. To prevent malfunction and possible breakdown during the winter, examine open digesters for ice formation. Break the ice and remove it before it damages digester appurtenances by wind action or expansion forces. Use warm air to thaw mechanical aerators troubled by ice formation. In extremely cold climates, it may be necessary to construct temporary covers to deter freezing.

Solids Deposition. Deposition may occur when gritty materials enter the digester and when the aeration and mixing devices do not create enough turbulence to resuspend the solids following a supernatant decanting sequence. Solids deposition can be prevented by improving the operation of the grit chamber, if one exists, or by using more powerful aeration and mixing equipment. Adding fillets to tank bottoms helps reduce solids buildup.

PROCESS TROUBLESHOOTING. The operator can evaluate the performance of the aerobic digester by following the guidelines described in the Process Control section; refer to Table 31.9 for the recommended acceptable values. If, during a check, an operator finds any of these parameters to show unusual values, the following troubleshooting guidelines could be used to identify problems with an aerobic digester system.

Increased Organic Loading. Increased organic loading can cause problems when the rate of oxygen required for aerobic digestion reaction is greater than the rate at which oxygen can be transferred to the sludge liquid by the WWTP's aeration system. Typically, the problem can be solved by reducing the organic loading rate and increasing the SRT. This is done by decreasing the influent sludge mass, total volume added, and amount of sludge withdrawn from the digester.

Excessive solids levels in the digester can also cause low dissolved oxygen concentrations. When suspended solids levels reach or exceed 3.5 to 4%, oxygen transfer efficiencies are reduced. This reduction may cause operational problems because of low dissolved oxygen concentrations, depending on the capacity of the aeration system.

Nuisance Odor from Digested Sludge. Inadequate dissolved oxygen is the primary cause of odor from the aerobic digestion process. To correct odor problems, first increase the dissolved oxygen concentration by additional aeration. Second, reduce the organic loading. If both of these fail, the addition of chemicals, such as potassium permanganate or hydrogen peroxide, to help oxidize the odor-causing compounds in the digester, may be needed. This is considered an expensive and temporary solution to the problem. Odor control chemicals can be added to small volumes of sludge on a bench-scale test and proportioned to the digester volume to determine the amount of chemicals needed in the digester.

Excessive Foaming. Foaming can be caused by factors as simple as organic overloading or as complex as filamentous bacterial growth. To reduce foaming, decrease organic loading, install foam-breaking water sprays, reduce excess aeration rates, and use defoaming chemicals. To avoid dilution of the digester contents, operate water sprays sparingly.

Curing filamentous bacterial growth is a difficult task that should be handled at the aeration basins. Some operators have added oxidants, such as chlorine and hydrogen peroxide, both with and without beneficial results. Shocking the system with several hours of anaerobic conditions has also been tried. In most cases, limited sprays and defoaming agents control the foam until natural forces can cause a shift to a nonfilamentous type of bacteria.

Low pH (High Ammonia). The intent of this chapter is to encourage the reader to optimize digestion by using all the techniques available. Assuming that aerobic–anoxic operation is incorporated into the design, then a neutral pH of 7.0 is achieved during the nitrification and denitrification process.

When biomass is fed to the system, an increase in pH will be shown, as a result of the oxidation of biomass and production of ammonia. This event could occur once per day if a gravity belt thickener is used to prethicken; or 8 times per day, which is considered a typical wasting cycle of SBR and MBR systems; or 24 times per day, if a gravity thickener in loop is used in a batch operation. Both healthy nitrifiers and adequate alkalinity are required for nitrification. The pH could increase from initially 7.0 to 7.5 and then decrease to 7.0 again. When additional oxygen is provided to the system to oxidize the produced ammonia, the byproduct is hydrogen, indicating an acidic environment, or a drop in pH below 7.0. If denitrification or addition of chemicals is not added to the process to balance the lost alkalinity, then a drop in pH will continue. If, on the other hand, an anoxic cycle is incorporated into the digester, it is possible to recover the lost alkalinity and increase the pH without the use of chemicals.

Low Dissolved Oxygen. An acceptable range of dissolved oxygen is anywhere from 0.1 to 2, with a target of 0.5. Anything above a 0.1 is considered a positive dissolved oxygen; however, typically, dissolved oxygen probes are neither accurate nor reliable ways to control dissolved oxygen, so a minimum reading of 0.3 is suggested. The dissolved oxygen reading should be evaluated with the ammonia concentration.

Clogging of diffusers, blower malfunction, or organic overload could cause low dissolved oxygen. In this case, check the equipment specifications, air deliveries, pipeline pressures, valving, and organic load. Clean diffusers, repair equipment as necessary, and, if the cause is organic load, reduce the feed to the digester, if possible, or reduce the solids concentration to the digester, reducing the oxygen demand and increasing the residual dissolved oxygen.

DATA COLLECTION AND LABORATORY CONTROL

DATA COLLECTION. Consistent data collection and process control through laboratory analyses are essential in the efficient operation of the treatment system. Analyses for total solids, volatile solids, dissolved oxygen, ammonium, nitrate, phosphorus, pH, temperature, flowrate, airflow rates, solids concentration, fecal coliforms, SOUR, and alkalinity of the raw and digested sludges provide the minimum surveillance for operating the aerobic digester. However, additional analyses would be warranted if operational problems arise.

The author highly suggests an in-depth reading of the Environmental Regulations and Technology document "Control of Pathogens and Vector Attraction in Sewage Sludge", produced by U.S. EPA's Pathogen Equivalency Committee (U.S. EPA, 1999a). The intent of this book is to give better understanding and guidance to the analytical procedures of fecal coliform tests and salmonella, and provide a more in-depth explanation of the basic concepts related to pathogens and vector attraction, as specified under 40 *CFR* Part 503 (U.S. EPA, 1999b).

MAINTENANCE MANAGEMENT PROGRAM. The planned maintenance program for the aerobic digester is similar to the maintenance program for the activated sludge process. Plot key elements include identification of maintenance tasks, scheduling, staffing, and a record system on trend charts. These charts plot various values according to time and show the trends in process changes.

MAINTENANCE TASKS. The key components that require regular attention in most aerobic digestion systems include

1. Aeration or oxygen supply system,
2. Mixing and pumping equipment, and
3. Instrumentation and control equipment.

Aeration and Oxygen Supply System. The principal requirement for the aeration or oxygen supply system is the inspection and service of the diffuser equipment. Once per year, schedule a diffuser mechanism inspection. If it is necessary to drain the tank for inspection and service, summer may be the best time because it is easier to restart.

Maximize the SRT during the spring and early summer before any diffuser inspection. Lower the mixed liquor concentration in the activated sludge tank just before the inspection to allow solids to be stored in the activated sludge process.

During the inspection, discharge the digested sludge to the drying beds or dewatering mechanism and accumulated solids in the aeration tank. Schedule material and manpower so that the tank is returned to service within 2 to 3 days. Restart digestion by seeding from other digesters or by the previously described startup procedure.

Mixing and Pumping Equipment. Annually, inspect the mixing and pumping equipment for worn blades and impellers. Replace faulty components and record critical parts for inventory. Service the seals, packing, and points of lubrication as frequently as recommended by the manufacturer.

Instrumentation and Control Equipment. Use a trained service representative to maintain the instrumentation and control components. Consider using the training courses available from equipment suppliers. Contract maintenance may also be available.

SCHEDULING. Assign specific maintenance tasks to individual staff members. Select staff on the basis of training and experience. An overall WWTP training program provides for upgrading the capabilities of both primary and backup staff members.

RECORDS. The importance of maintaining adequate operation and maintenance records cannot be overemphasized. The purpose of recording data is to track operational information that will identify and enable duplication of optimum operating conditions. The types of records include volume of sludge wasted to the digester, percentage of feed solids, volume of digested solids removed from the digester, and percentage of solids of sludge removed. If sludge is hauled to a site, keep records of hauling cost on a monthly basis.

Additional information includes dissolved oxygen data and pH, ammonia, nitrates, and/or phosphorus, SOUR and/or VSS reduction, pathogen reduction, alkalinity, and salmonella.

Keep a monthly report form. In WWTPs where the aeration system capacity is marginally adequate in providing desirable dissolved oxygen levels in the digester, record dissolved oxygen level data on a trend chart.

If chemicals are added to the digester for pH or odor control, record the type and amount of chemicals added.

If mechanical aerators are used, record power outage. In the case of diffused air systems, airflow records may be of interest. If airflow meters are not available, records of power consumption may be useful. Experimenting with the aeration system often leads to significant savings in power costs.

If the operator has chosen a two-stage digestion process because approval of the process as a PSRP equivalent alternative is required by the permitting authority, the plant operator should demonstrate experimentally that microbial levels in the product from their sludge digester are satisfactorily reduced. A record of pathogens has to be submitted and, once the performance is demonstrated, the process would have to be operated between monitoring episodes at time–temperature conditions at least as severe as those used during their tests.

To save time, use a simplified recording system. Increase completeness by using preprinted forms, computer data entry, or color-coded systems. Design the system as a permanent method to link routine and unscheduled maintenance with inspection procedures, equipment replacement, and accountability. Keep records where they are readily available to all WWTP personnel.

REFERENCES

Al-Ghusain, I.; Hamoda, M. F.; El-Ghany, M. A. (2004) Performance Characteristics of Aerobic/Anoxic Sludge Digestion at Elevated Temperatures. *Environ. Technol.*, **25**, 501–511.

Anderson, B. C.; Mavinic, D. S. (1984) Aerobic Sludge Digestion with pH Control—Preliminary Investigations. *J. Water Pollut. Control Fed.*, **56**, 889.

California State University (1991) *Operation of Wastewater Treatment Plants*, Vol. II; California State University: Sacramento, California.

Chang, A. C.; Page, A. L.; Asano, T. (1995) *Developing of Human Health-Related Chemical Guidelines for Reclaimed Wastewater and Sewage Sludge Applications in Agriculture*; World Health Organization: Geneva, Switzerland.

Daigger, G. T.; Bailey, E. (2000) Improving Aerobic Digestion by Pre-Thickening, Staged Operation, and Aerobic-Anoxic Operation: Four Full-Scale Demonstrations. *Water Environ. Res.*, **72**, 260–270.

Daigger, G. T.; Graef, S. P.; Eike, S. (1997) Operational Modifications of a Conventional Aerobic Digestion to the Aerobic/Anoxic Digestion Process. *Proceedings of the 70th*

Annual Water Environment Federation Technical Exposition and Conference, Chicago, Illinois, Oct. 18–22; Water Environment Federation: Alexandria, Virginia.

Daigger, G. T.; Ju, L. K.; Stensel, D.; Bailey, E.; Porteous, J. (2001) Can 3% SS Digestion Meet New Challenges? *Aerobic Digestion Workshop,* Vol. V; Enviroquip, Inc.: Austin, Texas.

Daigger, G. T.; Novak, J.; Malina, J.; Stover, E.; Scisson, J.; Bailey, E. (1998) Panel of Experts. *Aerobic Digestion Workshop,* Vol. II; Enviroquip, Inc.: Austin, Texas.

Daigger, G. T.; Scisson, J.; Stover, E.; Malina, J.; Bailey, E.; Farrell, J. (1999) Fine Tuning the Controlled Aerobic Digestion Process. *Aerobic Digestion Workshop,* Vol. III; Enviroquip, Inc.: Austin, Texas.

Daigger, G. T.; Stensel, D.; Ju, L. K.; Bailey, E.; Porteous, J. (2000) Experience and Expertise Put to the Test. *Aerobic Digestion Workshop,* Vol. IV; Enviroquip, Inc.: Austin, Texas.

Daigger, G. T.; Yates, R.; Scisson, J.; Grotheer, T.; Hervol, H.; Bailey, E. (1997) The Challenge of Meeting Class B While Digesting Thicker Sludges, *Aerobic Digestion Workshop,* Vol. I; Enviroquip, Inc.: Austin, Texas.

Grady, C. P. L., Jr.; Daigger, G. T.; Lim, H. C. (1999) *Biological Wastewater Treatment,* 2nd ed.; Marcel Dekker: New York.

Farrah, S. R.; Bitton, G.; Zan, S. G. (1986) *Inactivation of Enteric Pathogens During Aerobic Digestion of Wastewater Sludge,* EPA-600/2-86-047; U.S. Environmental Protection Agency, Water Engineering Research Laboratory: Cincinnati, Ohio.

Hao, O. J.; Kim, M. H. (1990) Continuous Pre-Anoxic and Aerobic Digestion of Waste Activated Sludge. *J. Environ. Eng.,* **116,** 863.

Hao, O. J.; Kim, M. H.; Al-Ghusain, I. A. (1991) Alternating Aerobic and Anoxic Digestion of Waste Activated Sludge. *J. Chem. Technol. Biotechnol.,* **52,** 457.

Lee, K. M.; Brunner, C. A.; Farrell, J. B.; Ealp, A. E. (1989) Destruction of Enteric Bacteria and Viruses During Two-Phase Digestion. *J. Water Pollut. Control Fed.,* **61** (6), 1421–1429.

Lue-Hing, C.; Zenz, D. R.; Kuchenrither, R. (1998) *Municipal Sewage Sludge Management: Processing, Utilization, and Disposal;* Technomic Publishing: Lancaster, Pennsylvania.

Matzuda, A.; Ide, T.; Fujii, S. (1988) Behavior of Nitrogen and Phosphorus During Batch Aerobic Digestion of Waste Activated Sludge—Continuous Aeration and Intermittent Aeration by Control of DO. *Water Res.,* **22,** 1495.

Mavinic, D. S.; Koers, D. A. (1979) Performance and Kinetics of Low-Temperature Aerobic Sludge Digestion. *J. Water Pollut. Control Fed.*, **51**, 2088.

Metcalf and Eddy (2002) *Wastewater Engineering: Treatment and Reuse;* McGraw-Hill: New York.

Peddie, C. C.; Mavinic, D. S.; Jenkins, C. J. (1990) Use of ORP for Monitoring and Control of Aerobic Sludge Digestion. *J. Environ. Eng.*, **116**, 461.

Shamskhorzani, R. (1998) Design of Autothermal Thermophylic Aerobic Digestion (ATAD) with Jet Aeration System. *Proceedings of the 3rd European Biosolids and Organic Residual Conference.*

Stege, K; Bailey, E. (2003) Aerobic digestion Operation of Stockbridge Wastewater Treatment plant, GA; Georgia Water Pollution Control Association.

Stensel, H. D.; Bailey, E. (2000) Preliminary Report Evaluation of Paris Sludge Production Due to Aerobic/Anoxic Operation; Enviroquip, Inc.: Austin, Texas.

Stensel, H. D.; Coleman, T. E. (2000) Assessment of Innovative Technologies for Wastewater Treatment: Autothermal Aerobic Digestion (ATAD); Preliminary Report; Project 96-CTS-1; U.S. Environmental Protection Agency.

U.S. Environmental Protection Agency (1990) *Autothermal Thermophilic Aerobic Digestion of Municipal Wastewater Sludge,* EPA-625/10-90-007, Environmental Regulations and Technology; U.S. Environmental Protection Agency: Washington, D.C.

U.S. Environmental Protection Agency (1992) *Control of Pathogens and Vector Attraction in Sewage Sludge,* EPA-625/R-92-013, Environmental Regulations and Technology; U.S. Environmental Protection Agency: Washington, D.C.

U.S. Environmental Protection Agency (1999a) *Control of Pathogens and Vector Attraction in Sewage Sludge,* EPA-625/R-92-013, Environmental Regulations and Technology; U.S. Environmental Protection Agency: Washington, D.C.

U.S. Environmental Protection Agency (1978) *Field Manual for Performance Evaluation and Troubleshooting at Municipal Wastewater Treatment Facilities,* EPA-68/01-4418; U.S. Environmental Protection Agency, Office of Municipal Pollution Control: Washington, D.C.

U.S. Environmental Protection Agency (1979) *Process Design Manual for Sludge Treatment and Disposal,* EPA-625/1-79-011; U.S. Environmental Protection Agency: Washington, D.C.

U.S. Environmental Protection Agency (1999b) Standards for the Use or Disposal of Sewage Sludge. *Code of Federal Regulations,* Part 503, Title 40.

Water Environment Federation (1998) *Design of Wastewater Treatment Plants,* 4th ed., Manual of Practice No. 8, Vol. 3, Chapters 17–24; Water Environment Federation: Alexandria, Virginia.

Water Environment Federation (1995) *Wastewater Residuals Stabilization,* Manual of Practice No. FD-9; Water Environment Federation: Alexandria, Virginia.

Chapter 32

Additional Stabilization Methods

Effective sludge processing is crucial to the reliable operation of a municipal wastewater treatment plant (WWTP). Because WWTPs have become more efficient at producing high-quality effluents, increased quantities of sludge are being generated. The solids treatment processes—thickening, dewatering, stabilization, and disposal—represent a major portion of the cost of wastewater treatment.

Sludge stabilization processes further treat the sludge to reduce odors or nuisances, to reduce the level of pathogens, and to facilitate efficient disposal or reuse of the product. This chapter focuses on the conventional methods of sludge stabilization other than anaerobic and aerobic digestion. Chapters 30 and 31 of this manual discuss anaerobic and aerobic methods. This chapter discusses the following stabilization methods:

- Composting,
- Lime stabilization,
- Thermal treatment,
- Heat drying, and
- Incineration.

All of these stabilization methods (other than incineration) are considered acceptable processes and are widely used for treating sludge to Class A or Class B levels for beneficial reuse and disposal, as referenced in the U.S. Environmental Protection

Agency's 40 *CFR* 503 regulations. The 503 regulations require documentation that one of these processes is used when meeting one of the alternatives listed for pathogen reduction in 503.32(a) for Class A sludge and 503.32(b) for Class B sludge. The resultant product has been referred to in this text as biosolids.

In addition, the 503 regulations reference the U.S. EPA guidance manual *Control of Pathogens and Vector Attraction in Sewage Sludge* (U.S. EPA, 1992) for calculation procedures that are acceptable for mass volatile solids reduction that is required for 503.33—Vector Attraction Reduction.

COMPOSTING

DESCRIPTION OF PROCESSES. Composting is a managed method for the biological decomposition of organic materials. Composting is one of the typical stabilization processes and is most frequently used to stabilize raw sludge; hoewever, applications for further stabilizing digested sludge are also typical. For many years, composting has been used to convert dewatered wastewater sludge into a stable, odorless, humus-like product. The practice of composting has grown largely because of rising costs of, or increasing restrictions on, the alternatives to composting such as landfill, ocean disposal, and incineration. Most solids composted in the U.S. are used for soil conditioning or horticultural application (WEF, 1995).

The four principal objectives of composting are

- Biological conversion of putrescible organics to a stabilized form;
- Pathogen destruction (the heat generated during composting results in disinfection but not sterilization of the end product);
- Mass reduction of the wet sludge quantity through moisture and volatile solids removal (although when a bulking agent is added to facilitate the composting process, overall volume may increase); and
- Production of a usable end product.

As a result of composting, the resulting product becomes a more attractive material for reuse. If the composted solids are sold, the revenues can help offset the composting reduction costs.

Figure 32.1 shows a typical flow diagram for a composting process. Providing a suitable environment in the compost mass requires the initial compost mixture to have a solids content of 40 to 50%. To attain this solids level, raise the carbon to nitrogen ratio (C:N), and provide the desired structural properties of the compost mass (promoting adequate air circulation). A bulking agent is often added. As the organic material in the

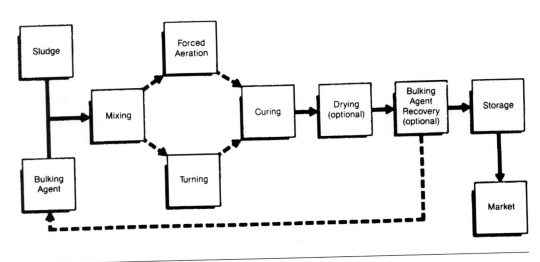

FIGURE 32.1   Compost process flow diagram.

sludge decomposes, the compost heats to temperatures in the pasteurization range of 122 to 158 °F (50 to 70 °C) and destroys most enteric pathogenic organisms.

The quality (that is, the concentrations of certain metals and organic pollutants) of the composted material will determine its suitability for use as a soil amendment. Compost quality needs monitoring on a regular basis to ensure compliance with applicable regulatory standards.

Composting may proceed under either aerobic or anaerobic conditions. Most composting operations seek to maintain aerobic conditions throughout the compost mass. Aerobic conditions accelerate material decomposition and result in the temperatures necessary for pathogen destruction. Anaerobic conditions can produce significant foul odors that are not generated when aerobic conditions are maintained throughout the compost mass.

Composting can be performed by reactor or nonreactor systems. Nonreactor systems are often referred to as open systems. Reactor systems are often referred to as in-vessel, enclosed, or mechanical compost systems. Reactor systems may be classified according to reactor type, solids-flow mechanisms, bed conditions in the reactor, and the method of air supply.

Composting with nonreactor systems is a labor-intensive process involving the addition of a bulking agent, mixing, windrowing or pile building, screening, and other activities. Some of these operations, however, may be automated. With in-vessel systems, most, if not all, operations are automated. The potential for odor generation exists at all composting facilities, especially at poorly designed and operated sites. Also, the potential exists for the generation of significant quantities of dust, which might possibly spread pathogens.

Nonreactor Systems. Nonreactor systems include the windrow system (see Figure 32.2) and the aerated static pile (see Figure 32.3). In the windrow system, aerobic conditions are maintained through convective air flow and periodic turning. The windrow is turned daily except when it is raining. Through several turnings or mixings, the sludge is subjected to the higher interior temperatures for pathogen reduction and stabilization of the organic material. In the static pile system, blowers attached to pipes located in the base of the piles either blow or pull air through the compost pile. After composting for 3 weeks, including 3 days at a minimum temperature of 55 °C (131 °F), the windrow or compost pile is torn down and placed in curing.

Reactor Systems. In a reactor-type composting system, the sludge and the bulking agent are mixed and then aerated in silos (see Figure 32.4), rotating drums (see

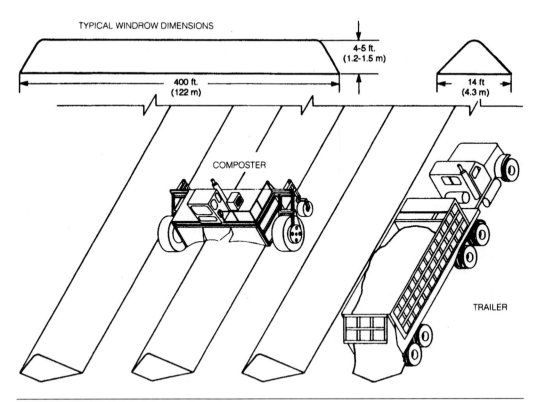

FIGURE 32.2   Windrow composting system.

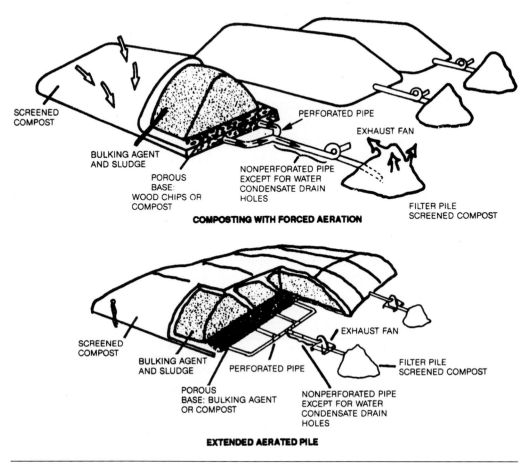

FIGURE 32.3   Two variations of a static pile composting system.

Figure 32.5), or horizontal beds (see Figure 32.6). In these systems, composting takes place in a stationary mode, by plug flow, or in an agitated bed. Regulated aeration controls the oxygen concentration and the temperature regime. Composted material typically is removed after approximately 14 days and placed in curing.

Facilities. The basic facilities for a composting operation include a paved working area; an aeration system; equipment for moving the sludge, bulking agent, and compost; equipment for screening and separating the bulking agent; odor control facilities; areas for bulking agent storage, compost storage, and compost curing; and a system to control runoff (WEF, 1995).

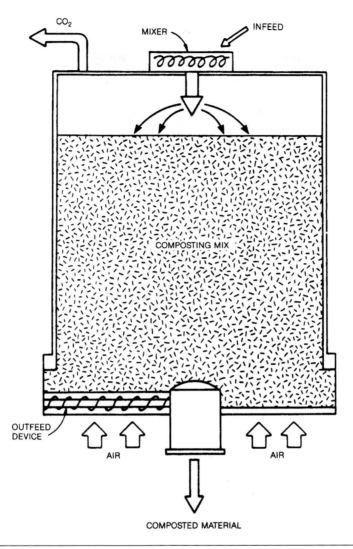

FIGURE 32.4   Silo type in-vessel composting system.

TYPICAL OPERATIONS. Although composting can be accomplished by a variety of methods, the basic process flow remains unchanged (see Figure 32.1). Even though composting systems may differ in appearance, they all include the following basic steps:

- Mixing of the sludge and bulking agent,
- Composting or microbial decomposition of the organic matter,

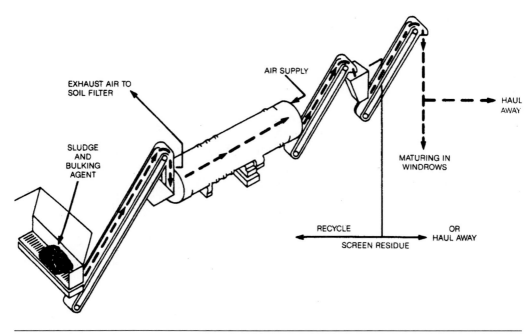

FIGURE 32.5   Rotating drum type in-vessel composting system.

- Recovery of the bulking agent or product recycling,
- Curing,
- Storage, and
- Final disposition of the composted material (which may be distributed or marketed as a humus material).

Homogeneous mixing is an extremely important aspect of composting. Partially dewatered sludge (typically 15 to 25% solids) is mixed with recycled compost or other bulking agents to provide proper porosity, initial moisture content, and nutrients. As sludge is evenly distributed within most compost systems, air flow becomes more uniform, resulting in greater microbial activity and efficiency, odor control, and ease of material handling.

Bulking agent recycle depends on the composting method used and the moisture content of the sludge. Some methods include recycling a portion of the finished compost product.

Curing is an extension of the composting process that results in further organic matter stabilization and volatile solids reduction. The length of the curing period de-

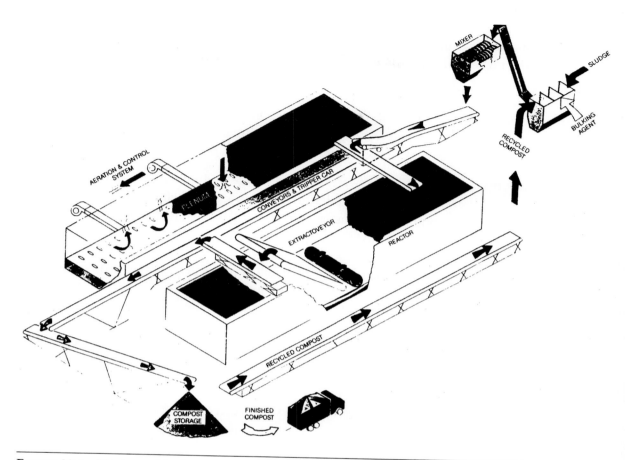

FIGURE 32.6   Horizontal bed type in-vessel composting system.

pends on the product's end use and the process efficiency in pathogen reduction, odor control, and reduction of volatile organic compounds.

Aeration Systems. Aeration systems can be either movable or fixed. Movable aeration equipment includes compost mixers (for windrow systems) or 0.5- to 2-hp blowers connected to perforated pipe over which compost piles are constructed (see Figure 32.3). Permanent systems include blowers that are connected to a pipe or plenum that is imbedded within the compost pile; or, as in a closed vessel system, the pipe or plenum feeds a reactor vessel.

The air supply to each pile or reactor (and in some cases zones within each pile or reactor) should be controlled separately. The air supply may be controlled manually, by time clocks, or by a temperature sensor inside the pile. In some systems the carbon dioxide level in the exhaust air has been used to control the air supply. Also, the system should be able to pull or push air through the pile.

Typical temperatures for active composting range from 55 to 60 °C (131 to 140 °F). These set points may be lowered before compost removal or pile breakdown to increase moisture removal, cool the compost pile, and reduce odors. Aeration requirements for moisture removal may be more than 10 times what is needed to maintain an aerobic environment (Haug, 1980).

Mixing Systems. An effective mixing of sludge and bulking agent is essential for rapid and even composting. A uniform mix reduces the likelihood of anaerobic pockets within the compost mass, thereby decreasing the potential for odor generation.

Two typically used mixing systems are

- A stationary system (that is, a plug mill or rotary drum) using paddles or augers to stir the materials; and
- Moving equipment, such as a wheel loader or composter.

Movable equipment is best for smaller WWTPs and for interim composting operations; however, facilities processing more than 10 dry tons/day should consider a stationary mixing system.

Screening. Screening is important in the production of a uniform fine-grained product and in the recovery of the bulking agent for reuse. Screening compost is difficult because at different moisture contents compost can be sticky (which tends to clog screens) or light (which tends to skip off the screen instead of passing through). Relatively dry compost (less than 48% moisture) produces the best results. To reduce the moisture content for efficient screening, a drying step is sometimes included before screening. However, very dry compost may create a severe dust problem. To counter the problem of clogging, almost all screens use some kind of shaking action or brushing to keep the screen clean.

Trommel, harp, and vibratory screens have been used successfully; however, vibratory screens may be better able to handle wet compost (WEF, 1995). The size of the screen openings depends on the need for bulking agent recovery and the required sizing for product marketing. Typical screen sizes between 6 and 10 mm (0.2 and 0.3 in.) allow for both high bulking agent recovery and desirable product quality. For some commercial applications of compost, screen sizes of up to 25 mm (1 in.) are used. The smaller screen sizes, however, remove objectionable materials that may be present in the raw sludge such as razor blades and plastic tampon applicators.

PROCESS CONTROL. Moisture, temperature, nutrients, bulking agents, and aeration greatly affect stabilization by composting.

Moisture. Moisture affects the rate of biological activity. At less than approximately 40% moisture, activity begins to decrease. At approximately 60% moisture, the air pore space is blocked (WEF, 1995). This affects the aeration efficiency of the system and results in anaerobic zones within the compost bed. Thus, the interaction of moisture and aeration will affect the composting rate and odor production. Moisture content will also affect materials handling and, thus, processing efficiency.

Temperature. Temperature has a dramatic effect on the microbial population (see Table 32.1). Decomposition is fastest in the thermophilic range. Research indicates the optimum temperature to be between 55 and 60 °C (131 and 140 °F); at temperatures more than 60 °C (140 °F) microbial activity decreases (Feinstein et al., 1980). One method for maintaining the temperature within the optimum range is to use forced ventilation or aeration and control the blower rate so as to hold the pile exit temperature to less than 60 °C (140 °F). Maintenance of an adequate temperature of 55 to 60 °C (131 to 140 °F) for at least 3 days is required to achieve pathogen destruction (see Tables 32.1, 32.2, and 32.3). Temperature can also influence the volatilization rate of many organic compounds.

Nutrients. Carbon and nitrogen are the principal nutrients that affect composting. The carbon:nitrogen ratio (C:N) affects the microbial activity and the rate of organic

TABLE 32.1   Populations during aerobic composting (Poincelot, 1975).

	Numbers per gram of wet compost for each stage			Number of groups of microorganisms identified
	Mesophilic, ambient to 40 °C	Thermophilic, 40 to 70 °C	Cooling	
Bacteria				
Mesophilic	10^8	10^6	10^1	16
Thermophilic	10^4	10^9	10^7	1
Actinomycetes				
Thermophilic	10^4	10^8	10^5	14
Fungi				
Mesophilic	10^6	0	10^5	18
Thermophilic	10^3	10^7	10^6	16

TABLE 32.2   Temperature and time required for pathogen destruction in sludge—laboratory data (Stern, 1974).

Microorganisms	Exposure time for destruction at various temperatures, minutes				
	50 °C	55 °C	60 °C	65 °C	70 °C
Cysts of *Entamoeba histolytica*	5				
Eggs of *Ascaris lumbricoides*	60	7			
Brucella abortus		60		3	
Corynebacterium diptheriae		45			4
Salmonella typhi			30		4
Escherichia coli			60		5
Micrococcus pyogenes, var., aureus					20
Mycobacterium tuberculosis					20
Viruses					25

matter decomposition. Microorganisms require carbon for both metabolism and growth, and nitrogen for protein synthesis and cell construction.

The best approach is to maintain the C:N ratio at a level that allows for optimum microbial growth; that is, between 26 and 31 units of carbon/1 unit of nitrogen (Poincelot, 1975). Carbon values greater than 31 tend to slow the process and result in lower temperatures. In most cases, direct measurement of the compost mix to determine the C:N ratio is unnecessary because these ratios typically are obtained with sludge and wood chip mixtures.

Low C:N ratios (less than 20) typically affect the compost product rather than the process. With low C:N ratios, ammonia is released and the nitrogen content of the

TABLE 32.3   Temperature and time required for pathogen destruction in compost—field data (WEF, 1995).

Microorganisms	Exposure time for destruction at various temperatures, hours		
	45 to 55 °C	60 °C	65 °C
Salmonella newport	168	25	
Salmonella	168	48	
Poliovirus type 1		1	
C. albicane		72	
Ascaris lumbricoides		4	1
Mycobacterium tuberculosis			336

compost is reduced (40 *CFR* Part 503). Because one of the major goals of composting is to produce an economical and usable product, it is best to maintain the nitrogen content of the compost as high as possible.

Bulking Agents. When bulking agents are added to the sludge, they control moisture, increase the air voids for proper aeration by giving porosity, provide structural support for the compost mass, and act as an organic amendment for C:N ratio adjustment. Bulking agents can also be used for product modifications. Their addition can change the physical and chemical characteristics of the compost, thereby affecting its range of uses and marketability.

The physical texture of the compost varies depending on the characteristics of the bulking agent and whether it is completely screened out, partially screened out, or left in the final product. The bulking agent can dramatically affect the chemical characteristics of the compost. Sawdust and other slow-degrading, high-carbon materials left in the compost result in an unfavorably high C:N ratio that consumes the nitrogen in the soil and limits the amount available to support plant growth (40 *CFR* Part 503).

Important bulking agent properties include moisture content, particle size, and absorbency. Bulking agent characteristics affect materials handling, processing time, and product quality. Some bulking agents may require shredding or screening. High-carbon or high-cellulose materials that are not recovered and reused may require long curing times. Pilot testing of new bulking agents should be performed to ensure compatibility with process operations.

Aeration. Aeration is important for providing oxygen for the decomposition process, temperature control, and moisture removal. Higher temperatures are achieved under aerobic conditions than under anaerobic conditions. Underaeration may result in low temperatures because of anaerobic conditions; excessive aeration may result in cooling of the compost piles.

Studies suggest that 20 to 50 m^3/h (700 to 1 800 cu ft/hr) per dry ton of sludge results in oxygen levels from 5 to 15% throughout the pile (WEF, 1995). At oxygen levels below 5%, anaerobic conditions can occur, resulting in anaerobic zones in the sludge mass, the generation of odors, and inadequate stabilization. Oxygen levels can be analyzed with a portable oxygen monitor and should be measured simultaneously with each temperature reading. During the early stages of production, higher aeration rates may be necessary to control temperatures and increase microbial activity.

Aeration also provides a means of removing moisture and drying the compost product. Excessive moisture can adversely affect the materials handling and processing operation. If the solids content falls to less than 50%, the static pile process is affected by a decrease in the efficiency of the screening. In turn, this means that less compost product will be produced causing a corresponding increase in the material to be recycled. In addition, excessive moisture can contribute to anaerobic conditions and odor problems during curing and storage. Moisture also increases the costs of distributing the compost product.

Moisture removal using aeration occurs when the air becomes saturated as it moves through the compost pile. As air is drawn through the pile, it is gradually heated by the elevated interior temperatures. This temperature increase is accompanied by an increase in specific humidity that allows the air to increase its water-holding capacity. It is this temperature and specific humidity relationship that allows successful air drying even in regions of high relative humidity.

ODOR CONTROL. In composting operations, odor control measures may be necessary whenever incompletely stabilized sludge or compost is exposed to the air. Processes within the composting operation that have a potential for odor generation include mixing, aeration, tearing down of the composting and curing piles, and screening. Odors during composting are best controlled by maintaining an aerobic environment.

Several methods are used to reduce odors depending on the location of the composting facility in relation to developed areas (especially residential areas). For example, certain facilities—such as mixing and screening—can be enclosed. Also, exhaust from aeration of the compost can be discharged through scrubber piles or air scrubbers.

SAFETY AND HEALTH PROTECTION. Employees working at composting operations should take precautions to reduce exposure to pathogenic organisms. Composting where temperatures reach the thermophilic range can eliminate practically all viral, bacterial, and parasitic pathogens; however, some fungi (Aspergillus fumigatus for example) are thermotolerant and survive the composting process. Large numbers of spores are released into the atmosphere during certain composting operations (compost screening, dumping and mixing of compost, and wood chip dumping). To reduce exposure to airborne pathogens, enclose the cabs of heavy equipment, ventilate all enclosed areas properly, and provide dust masks to employees working in dusty areas.

TROUBLESHOOTING. Table 32.4 presents a troubleshooting guide for forced air static pile (FASP) systems that will help the operator identify problems and develop solutions. If you operate a reactor system, refer to the troubleshooting guide in the operation and maintenance manual provided by the reactor manufacturer.

TABLE 32.4 Troubleshooting guide for composting with the forced air static pile system (U.S. EPA, 1978)

Problem	Probable cause	Solution
Compost pile does not reach 50 to 60 °C (122 to 140 °F) in a few days after construction	Poor mixing of sludge and bulking agent	If oxygen levels are above 15%, reduce operating time of blower
	Compost pile too wet	Pipe hot exhaust air from an adjacent pile into this pile to raise its temperature
	Too much aeration	Reduce aeration rate; if the pile does not come up to temperature within a couple of days after taking these steps, the pile should be torn down, remixed, and reconstructed; if the bulking agent is too wet (more than 45 to 55% moisture) it must be dried, or drier bulking agent can be found deeper within the bulking agent storage pile
Temperature does not remain above 50 to 60 °C (122 to 140 °F) for more than 1 to 2 days	Poor mixing of sludge and bulking agent	Adjust blower cycle to maintain oxygen between 5 and 15%
	Compost pile too wet	Pipe hot exhaust air from an adjacent pile into this pile to try to maintain it at temperature; this should only be done for a few days at a time because of moisture accumulation
	Overaeration	Reduce aeration rate; if temperature does not come back up, the material should be recycled or remixed, and steps should be taken to correct the problems before the next composting cycle
Odors are emitted from a composting pile	Poor air distribution in pile	Check blower for proper operation; check that aeration pipes are not plugged
	Poor mixing of sludge and bulking agent	Typically the odors develop from anaerobic conditions in the pile because of a lack of air; probably the best procedure is to increase the blower "on" cycle or run it continuously until the odors disappear, or turn the windrow more often
	Inadequate or nonuniform air distribution through the pile	Check for water accumulation in the lines and blower; review air distribution system design and layout for causes of high pressure drops
Blower does not operate	Timer failure; power failure; motor failure; fan is "frozen" from corrosion or ice and will not turn	Check each probable cause until trouble is found

LIME STABILIZATION

DESCRIPTION OF PROCESSES. Lime can be used to stabilize raw primary, waste-activated, and anaerobically digested sludges. Lime stabilization can be accomplished either before dewatering (prelime) or following a de-watering step (postlime). Prelime stabilization is more typical; however, postlime stabilization may have significant advantages including reduced lime dosages and the elimination of special constraints on conditioners and equipment used for dewatering (WEF, 1995). The lime-stabilized material may be disposed of in landfills or beneficially used.

The standard process involves adding sufficient lime to liquid sludge to raise the pH of the mixture to 12 or more and to maintain a pH of 12 for a minimum of 2 hours. This typically destroys or inhibits pathogens and the microorganisms involved in the decomposition of the sludge. Therefore, little or no decomposition occurs and few odors are produced by biological activities. The destruction of pathogenic organisms reduces bacterial hazards to a safe level.

In a typical operation, lime slurry and sludge are combined in a mixing tank equipped with either diffused air or mechanical mixing systems. After initial mixing, the treated material is transferred to a contactor vessel where mixing is continued, typically for a period of 30 minutes. If necessary, additional lime can be added in the contactor to maintain a pH of 12 or more for a minimum of 2 hours following mixing. The treated sludge is then dewatered and stored or disposed of immediately.

Differences between various designs may affect the operation at individual WWTPs. For example, in a batch process the mixing, 30-minute contact time, and the required thickening may all occur in one tank (U.S. EPA, 1978). Figure 32.7 presents a typical process flow sheet for lime stabilization.

The postlime stabilization process has several significant advantages over the prelime stabilization process. These advantages include

- The option of using either hydrated lime or quicklime (without slaking);
- The avoidance of high lime-related abrasion, corrosion, and scaling problems with mechanical dewatering equipment when a prelime stabilization process is used; and
- Pathogen destruction through use of heat generated during slaking of quicklime in the lime–sludge mixture.

Some proprietary processes are related to lime stabilization. Sludge is mixed with a setting agent, either cement kiln dust or Portland cement and silicate, thereby producing an alkaline environment in the treated sludge. Typically, the pH of the treated sludge is between 11.5 and 12.5. The chemical reaction between the sludge and the set-

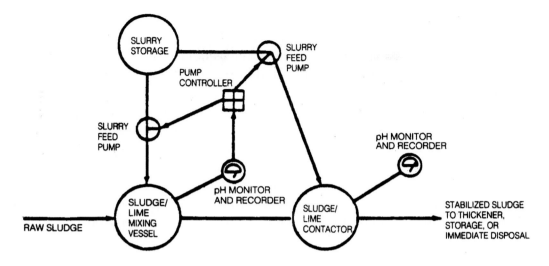

FIGURE 32.7   Lime stabilization process flow diagram.

ting agent also results in elevated temperatures. The combination of elevated temperatures and high pH destroys pathogenic organisms and drives off water. By windrowing and periodic turning of the sludge mass, the solids content of the product can be increased to more than 50% in 2 to 3 weeks. The end product may be suitable for beneficial use.

ADDITIONAL TECHNOLOGY.   One proprietary process uses a mixer designed to gently mix sludge and lime together to produce proper stabilization without turning the material into a pasty-type consistency. The ability to control the consistency of the material is accomplished by adjusting the speed of the augers and the depth setting of a weir located at the discharge. Optimum operation of the mixer is accomplished by controlling the depth of the material within the mixer.

Another proprietary process elevates the solids temperature in accordance with 40 *CFR* Part 503 regulations to reduce pathogens. The process uses lime to increase the pH, to meet the vector attraction requirements. The supplemental heat system is used to reduce the amount of lime necessary to obtain proper temperatures. On a typical sludge containing 20% dry solids, the lime dosage and operating costs for the process are cut in half by the use of supplemental heat. The process includes a heat pulse step that is accomplished within the pasteurization vessel. The pasteurization vessel is a totally enclosed, heated vessel that holds the material at the pasteurization temperature for the time required. The vessel provides a continuous flow of material in and out.

PROCESS CONTROL. Process control is critical to achieve adequate lime stabilization. Proper control techniques are important because the effects of incomplete stabilization are not readily apparent and may not be seen at the WWTP.

Sensors, particularly pH electrodes, need to be properly cleaned, calibrated, and maintained. Special pH electrodes are necessary for routine measurements more than pH 10. Sensitive narrow-range pH paper is acceptable for process monitoring.

The absence of odors at the WWTP plant and acceptable dewatering characteristics are not good indicators of adequate stabilization. Only careful monitoring of the pH will ensure that an adequate pH is being maintained for a period sufficient for adequate stabilization. Microbiological examinations for indicator organisms such as fecal coliforms and fecal streptococci should be performed on a quarterly basis.

Compared to composting, the lime stabilization process is simpler to operate. Some of the principal disadvantages of lime stabilization are that the sludge can become unstable if the pH drops after treatment and a regrowth of biological organisms occurs, that the added lime increases the mass to be disposed, and that lime is expensive.

SAFETY AND HEALTH PROTECTION. Employees should take pre-cautions to reduce exposure to lime when working in a lime stabilization area. The high pH of the lime is very caustic and will damage skin, eyes, and lungs if exposed through contact or inhalation. Use of personal protective equipment should be mandatory for all employees when working in or around the lime stabilization areas, including, at a minimum, gloves, safety goggles, and a ventilator. Proper housekeeping and storage procedures as recommended by the chemical manufacturer should be followed at all times because the lime materials can be very corrosive if not handled properly.

TROUBLESHOOTING. Table 32.5 presents a troubleshooting guide for lime stabilization that will help the operator identify problems and develop solutions.

THERMAL TREATMENT

DESCRIPTION OF PROCESSES. Thermal treatment systems release water that is bound within the cell structure, thereby improving the dewatering and thickening characteristics of the sludge. Thermal treatment systems include the processes of heat treatment and wet air oxidation. Thermal processes are an attractive reduction technique because of the limited volume of the resulting product. However, volume reduction and product quality must be evaluated with regard to fuel costs, equipment expenditures, air pollution control demands, and increased concentration of metals in the end product.

TABLE 32.5 Troubleshooting guide for lime stabilization (U.S. EPA, 1978).

Problem	Probable cause	Solution
Air slaking occurring during storage of quicklime	Adsorption of moisture from atmosphere when humidity is high	Make storage facilities airtight, and do not convey pneumatically
Feed pump discharge line clogged	Chemical deposits	Provide sufficient dilution water
Grit conveyor or slaker inoperable	Foreign material in the conveyor	Check and replace shear pin if necessary; remove foreign material from grit conveyor
Paddle drive on slake is overloaded	Lime paste too thick	Adjust compression on the spring between gear reducer and water control valve to alter the consistency of the paste
	Grit or foreign matter interfering with paddle action	Remove grit or foreign materials, try to obtain lime with a lower grit content, or install grit removal facilities in slaker or slurry line
Lime deposits in lime slurry feeder	Velocity too low	Maintain continuous high velocity by use of a return line to the slurry holding tank
"Downing" or incomplete slaking of quicklime	Too much water is being added	Reduce quantity of water added to quicklime (detention slaker/water to lime ratio = 3.5:1; paste slaker ratio = 2:1)
"Burning" during quicklime slaking	Insufficient water being added, resulting in excessive reaction temperature	Add sufficient water for slaking (see ratios above)
Sludge retains definite offensive odor after addition of lime	Lime dose too low	Check pH in sludge-lime mixing tank and increase lime dose, if needed; check pH for possible malfunction

Heat Treatment Process. In the heat treatment process (see Figure 32.8), sludge is ground to a controlled particle size and pumped at a pressure of more than 2 100 kPa (300 psi) (U.S. EPA, 1986). The sludge is brought to a temperature of approximately 180 °C (350 °F) by heat exchange with treated sludge and direct steam injection. The sludge then "cooks" in the reactor at the desired temperature and pressure. Finally, the hot treated mass cools by heat exchange with the incoming sludge.

The treated material settles from the supernatant before the dewatering step. Gases released at the separation step pass through a catalytic afterburner at 340 to 400 °C (650 to 750 °F) or through a deodorizing device. In some cases, these released gases are returned to the diffused air system in the aeration basins for deodorizing.

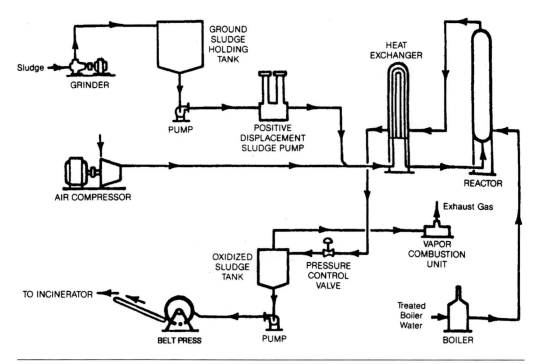

FIGURE 32.8   Heat treatment system flow diagram.

Wet Air Oxidation. The wet air oxidation process resembles heat treatment except that the former operates at higher temperatures (230 to 340 °C [450 to 640 °F]) and higher pressures (8 200 to 11 000 kPa [1 200 to 1 600 psig]) than does heat treatment (U.S. EPA, 1986). The wet air oxidation process is based on the principle that any substance capable of burning can be oxidized in the presence of water vapor at temperatures between 120 and 370 °C (250 and 700 °F). The degree of oxidation achieved varies depending on the temperature, pressure, reaction time, and air supplied to the reactor. The process is used either as a thermal conditioning process or as a relatively complete oxidation system, although the maximum achievable oxidation levels do not approach the degree of organic destruction achieved in a true incineration process. Either method of operation results in a liquid sludge with optimum dewaterability characteristics.

The feed to the wet air oxidation process does not require dewatering as does the feed to the processes involving both heat drying and incineration, discussed later in this chapter. However, the oxidized ash must be separated from the water by vacuum filtration, centrifugation, or some other solids separation technique. The water content of the feed sludge may exceed 95%. The sludge feed mixes with compressed air and passes through a heat exchanger before entering the reactor. The oxidized sludge leaving the reactor supplies the heat in the heat exchanger. If high levels of oxidation are occurring and the sludge has a relatively high fuel value, oxidation in the reactor may be self-sustaining. If this is not the case, heat to the heat exchanger is supplied by injecting steam. Because the temperature in the reactor typically is approximately 260 °C (500 °F) for high levels of oxidation, high pressures (between 6 900 and 20 600 kPa [1 000 and 3 000 psig]) in the reactor may be necessary to prevent water vaporization.

Depending on the installation, either before or after passing the sludge through the heat exchanger, the gases are separated through a pressure control valve and sent to a catalytic gas purifier for odor control. The gases then may be expanded for power recovery or released to the atmosphere. If not done before separation, the oxidized material then is passed through a heat exchanger, released through a pressure control valve, and dewatered. The liquid portion, which is odorous and has a high COD, typically is returned to the WWTP. This side stream may have a significant effect on other treatment processes.

PROCESS CONTROL. Control of the heat treatment and wet air oxidation processes is discussed below.

Heat Treatment. Heat treatment converts suspended solids to dissolved or dispersed solids. These dissolved solids cause a highly polluted liquid from the dewatering process that is recycled through the WWTP for reprocessing.

Increasing the holding time in the thermal reactor increases the breakdown of the sludge cells and degrades the fibrous material. For example, in low oxidation at 180 to 200 °C (350 to 400 °F), the color units of the recycle liquor increase from 2 150 units with a reaction time of 3 minutes to 3 800 units with 15 minutes, and to 5 500 units with 30 minutes (U.S. EPA, 1978). The recycle liquor can be foul smelling, difficult to treat, and disruptive to WWTP processes. Typical recycle liquor characteristics are presented in Table 32.6.

The high concentrations in Table 32.6 illustrate the potential effect that recycle of the liquor can have on the wastewater treatment processes. Accordingly, designers and operators must recognize the significance of the recycle load in the overall WWTP operation.

If the recycled liquors adversely affect other wastewater treatment processes, two potential solutions can be considered:

- Storing the liquors and recycling them back to the WWTP during low-flow or low-BOD (nighttime) conditions, and
- Installing a separate treatment system for the liquors before they are recycled to the WWTP.

TABLE 32.6   Typical recycle liquor characteristics for heat treatment (U.S. EPA, 1978).

Substances in strong liquor[a]	Concentration range, mg/L (except as shown)
TSS	100–20 000
COD	10 000–30 000
BOD	5 000–15 000
NH₃–N	400–1 700
Phophorus	20–150
Color	1 000–6 000 units

[a]TSS = total suspended solids; COD = chemical oxygen demand; BOD = biochemical oxygen demand; and NH₃–N = ammonia-nitrogen.

An equal degree of filterability and settleability can, within limits, be accomplished by various combinations of time and temperature. For instance, high temperature and short reaction time provide results comparable to lower temperature and longer reaction time. Typically, the most economical is the longer reaction time, low-temperature treatment. Overcooking (various combinations of high temperatures and long reaction times) actually breaks down the fibrous material itself (as compared to simply releasing the cell water) and produces a material that is more difficult to dewater.

Cooling of heat-treated sludges before atmospheric exposure can reduce, but will not eliminate, odor problems. Increasing the solids content of the sludge feed to the heat treatment process decreases operating costs but increases the concentration of dissolved COD, nitrogen, and phosphorus in the recycle liquor and may reduce the dewaterability of the treated material.

Wet Air Oxidation. The wet air oxidation process may be operated intermittently or continuously. Heat for start-up is provided by steam that is injected into the heat exchanger upstream of the reactor.

Operations will vary from one facility to another depending on the degree of oxidation that is selected. Any degree of organic destruction may be accomplished and is dependent on the temperature, pressure, reaction time, and air supply. Increasing the values of these parameters typically results in increased oxidation of organics.

It has also been noted that the percentage of solids in this feed will affect operations and fuel requirements. A given installation may be able to reduce costs and increase the maximum capacity of a wet air oxidation WWTP by increasing the percentage of solids in the feed, provided there is sufficient air capacity.

The four important physical variables that control the performance of wet oxidation units are

- Temperature,
- Air supply,
- Pressure, and
- Feed solids concentration.

Controls typically are provided for regulating reactor temperature, pressure, and the air supply.

The reactor pressure, temperature, and quantity of air used determine the extent and rate of oxidation. Much higher degrees of oxidation are possible at higher pressures and temperatures. The reactor temperature and pressure affect the quality of the recycle water (liquor) and the dewaterability of the oxidized sludge. Reaction temperature is kept as low as possible, consistent with adequate conditioning. Higher temperatures cause more complete breakdown of the sludge particles, releasing more cell water, and releasing more BOD into the solution (U.S. EPA, 1978). Higher temperatures produce a treated sludge that dewaters readily but has a poorer-quality recycle liquor. This recycle reduces the capacity of the wet end process.

As with conventional incinerators, discussed later in this chapter, an external supply of oxygen (air) is required to attain nearly complete oxidation. The air requirement for the wet oxidation process is determined by the heat value of the sludge being oxidized and by the degree of oxidation desired. Thermal efficiency and fuel requirements are functions of air input; therefore, it is important that the air flow not be higher than needed. Because the input air becomes saturated with steam from contact with the liquid in the reactor, it is also important to control the air flow to prevent excessive loss of water from the reactor.

ADDITIONAL TECHNOLOGY. An additional proprietary technology uses an aerated thermophilic pretreatment phase in conjunction with conventional mesophilic anaerobic digestion to produce a Class "A" biosolid. This system is suited particularly to upgrading the solids handling capability of WWTPs with existing anaerobic digesters. This system uses the biogas production from the digesters to heat the sludge. The net fuel energy requirement for the combined system typically is no greater than the digester system without this type of process added. This process may enhance the settlability and dewaterability of the biosolids, allowing a greater solids concentration to be maintained in the digestors, and an increase in the solids concentration of cake from the dewatering equipment, thus reducing hauling costs and possibly enhancing beneficial use prospects.

SAFETY AND HEALTH PROTECTION. Typically, these types of processes are contained in enclosed vessels with minimal exposure potential for employees. However, the high temperatures at which these facilities operate could be hazardous, and basic precautions should be taken by employees working in or around the equipment, valves, boilers, reactors and heat exchanger units.

TROUBLESHOOTING. Table 32.7 presents a troubleshooting guide for thermal treatment systems that will help the operator to identify problems and develop solutions.

HEAT DRYING

DESCRIPTION OF PROCESSES.
The process of heat drying sludge removes moisture from the sludge to practical limits and thus reduces the total volume, retains the fertilizing properties of wet sludge, destroys pathogenic organisms, and produces a nonodorous product.

Heat drying raises the temperature of the incoming sludge to remove moisture and reduce total volume. The temperature remains low enough to retain the nutrient properties of the sludge. The end product contains soil nutrients but no pathogenic organisms. Auxiliary fuel or waste heat is required for heat drying.

The three principal processes of the heat drying method of stabilization are low-temperature heat drying, flash drying, and rotary kiln drying.

Low-Temperature Heat Drying. Low-temperature heat drying involves raising the temperature of the incoming sludge to 100 °C (212 °F) to evaporate moisture. Either a cylindrical drum or a multiple hearth is used for this process.

Flash Drying. Flash drying produces an instantaneous removal of moisture by introducing the wet sludge into a hot gas stream. The process has the flexibility to operate as a heat dryer for the production of fertilizer or as an incinerator by returning the dry solids to the furnace as a fuel. Figure 32.9 represents a typical flash dryer system.

Before introduction to the flash dryer, the sludge undergoes thickening and dewatering. The product from the flash dryer may be sold as fertilizer; the incinerator ash is amenable to landfilling.

The incoming dewatered sludge is blended with a portion of the previously heat-treated sludge in a mixer that provides the consistency necessary for pneumatic conveying. Then, hot gases from the furnace (650 to 700 °C [1 200 to 1 300 °F]) are mixed with the blended sludge in a drying tower before introduction to the cage mill. Agitation in the cage mill dries the sludge to approximately 2 to 10% moisture and reduces the temperature to approximately 150 °C (300 °F) before conveyance to cyclone separa-

TABLE 32.7   Troubleshooting guide for thermal treatment (U.S. EPA, 1978).

Problem	Probable cause	Solution
Odors	Odors being released in decant tanks, thickeners, vacuum pump exhaust, or in dewatering	Cover units, collect air, and deodorize it before release by use of incineration, adsorption, or scrubbing
		Cover open tank surface with small floating plastic balls to reduce evaporation and odor loss
	Odors being released when recycle liquors enter treatment wastewater tanks	Preaerate liquors in covered tank and deodorize offgases
Scaling of heat exchangers	Calcium sulfate deposits	Provide acid wash in accordance with manufacturer's instructions
	Operating temperatures too high—causing baking of solids, if there is insufficient air	Operate reactor at temperatures below 199 °C (390 °F) for heat conditioning of sludge and increase air
		Use hydraulically driven cleaning bullet to clean inner tubes
Steam use is high	Sludge concentration to heat treatment unit is low	Operate thickener to maintain 6% solids if possible; 3% minimum
Solids dewater poorly	Anaerobic digestion before heat treatment	Discontinue anaerobic digestion of sludge to be heat treated
	Temperatures not maintained high enough	Temperature should be at least 177 °C (350 °F)
High system pressure	Blockage in reactor	Remove blockage
		Check pressures and temperatures to note any discrepancies from normal
	Pressure controller set too high	Reduce set point on pressure controller
	Block valve closed	Check system for proper valving
	Improper temperature control setting	Adjust setting
	Heat exchanger scaling	Acid wash heat exchangers
Feed pumps not pumping adequate flow	Improper control setting	Adjust control setting
	Leakage or plugging in product check valves	Repair or replace check valves
	Air trapped in pump cylinders	Bleed off air
System pressure is dropping	Pressure controller set too low	Set pressure controller at proper level
	Pressure control valve trim is eroded	Replace valve

TABLE 32.7   Troubleshooting guide for thermal treatment (U.S. EPA, 1978) (*continued*).

Problem	Probable cause	Solution
Oxidation temperature is rising	Inlet temperature too high	Reduce temperature by diluting incoming sludge with water
	Sludge feed rate is too slow	Increase sludge feed rate
	Improper control setting	Appropriately adjust control setting
	Pump stopped or slowed	Start pump or increase rate
	Volatile matter such as gas or oil being pumped through the system	Switch from sludge to water and stop the process air compressor
	Pneumatic steam valve not functioning properly	Repair malfunctioning valve
Oxidation temperature is falling	Heat exchanger fouled	
	Reactor inlet temperature is too low because of low-density sludge	Reduce dilution of incoming sludge
	High flow rate being pumped through system	Reduce flow rate at high-pressure pump
	Pneumatic steam valve not functioning properly	Repair malfunctioning valve
	No signal air to the temperature control valve	Check instrument air supply
	Boiler not functioning properly	Consult boiler manufacturer's instruction manual for corrective action
Filter cake difficult to feed into incinerator	Filter cake too dry for pumping	Reduce temperature (and pressure) of the treatment system or adjust dewatering process
Low system pressure	High-pressure pump, process air compressor, or boiler stopped	Restart pump or air compressor
	Intake filter clogged	Clean or replace filter
	Pressure controller set too low	Increase set point on pressure controller
	Any of the blowdown valves may be partially opened	Check compressor valving
	Leaking interstage trap	Check trap for proper operation
	Slipping drive belts	Adjust belt tension
High air temperature	Inadequate water flow	Adjust water flow
	Cake too dry	Adjust dewatering
	Leaking cylinder valves	Repair, clean, or replace
	Intercooler or jackets plugged	Clean intercooler or replace

TABLE 32.7   Troubleshooting guide for thermal treatment (U.S. EPA, 1978) (*continued*).

Problem	Probable cause	Solution
	No flow from the force-feed lubricators	Add oil
		Repair lubricator
		Tighten loose belt, or replace if worn
Air compressor safety valve relieving	Pressure controller set too high	Reduce set point on controller
	No signal air pressure to pressure control valves	Check instrument air supply
	One or more block valves in the system are closed	Check system for proper valving
	Plugged pressure control valve	Switch to standby pressure control valve and clean plugged valve

tion of the solids from the gases. At this point the solids may be sent either to storage or to a furnace for incineration.

The gases from the cyclone separators are conveyed by the vapor fan to the deodorization preheater in the furnace where the temperature is raised to approximately 650 to 760 °C (1 200 to 1 400 °F). The deodorized gases release a portion of the heat to the incoming gases and release more heat in the combustion air preheater. The temperature is reduced to approximately 260 °C (500 °F) before the gas is scrubbed for particulate removal and discharged.

If the dried solids are not used in the furnace as a fuel, auxiliary fuel such as gas, oil, or coal is then necessary. The combustion air fan introduces preheated air into the furnace. If sludge incineration is practiced, ash may be sluiced from the bottom of the furnace.

Rotary Kiln Dryers. The kiln is a cylindrical steel shell mounted with its axis at a slight slope from the horizontal (see Figure 32.10). When used for sludge drying, refractory lining typically is unnecessary.

The feed to the unit requires previous dewatering and is introduced into the breech or upper end. A portion of the dried sludge is mixed with the feed cake, thereby reducing its moisture and dispersing the cake. The unit is directly fired at the upper end by a long-flame burner mounted in the hood. Gases are removed at the lower end, providing countercurrent movement of the sludge and the gases. Axial ledges that are provided along the interior wall of the dryer pick up the material, then spill it off in the form of a thin sheet of falling particles as the dryer rotates. This motion is intended to provide contact between the sludge and gases to promote rapid drying.

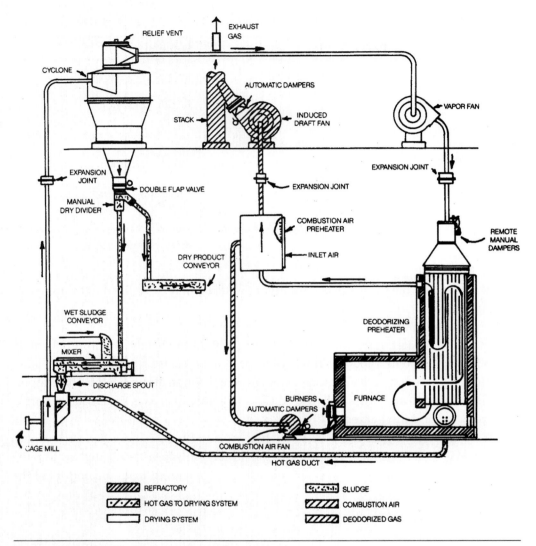

FIGURE 32.9   Flash dryer process schematic.

The dried material may consist of varied particle sizes that require grinding before use as a fertilizer. The internal temperature of a rotary kiln dryer typically is maintained at approximately 370 °C (700 °F). Deodorization of the exhaust gases by after burning at approximately 650 to 760 °C (1 200 to 1 400 °F) is necessary if odors are to be avoided. Also, scrubbers must be used to remove particulates from the exhaust gases.

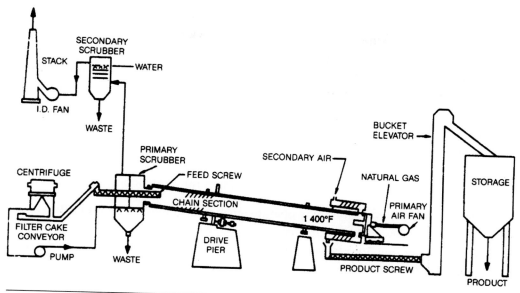

FIGURE 32.10 Rotary kiln dryer schematic.

PROCESS CONTROL. Efficient and consistent operation of heat drying equipment depends on frequent monitoring, both sensory and analytical. By periodically inspecting the drying equipment during the shift, any irregularities in the operating temperatures can be detected, such as pressures and flow rates. After gaining some operating experience, it is possible to recognize any strange sounds or changes in pitch that may indicate a problem.

The four major variables that affect the operation of the heat drying equipment are

- Percent solids in the wet sludge feed,
- Ratio of dried sludge:wet sludge,
- Quantity of hot combustion gases used for drying, and
- System temperatures.

Controlling the dewatering process that precedes the heat drying equipment results in a product of the desired moisture content range of between 2 and 10%. Efficient operation depends on a consistent solids concentration in the sludge feed. The percent solids in the feed is also critical to the economy of operation of heat dryers. The higher the moisture content of the incoming sludge, the more fuel is burned to evaporate the added moisture.

As presented in Figures 32.9 and 32.10, the incoming sludge is mixed with previously dried sludge to create the proper consistency for pneumatic conveying equipment.

A significant change in the incoming percent solids concentration changes the ratio required for proper operation. This ratio varies depending on the sludge used and the incoming percent solids and is determined through trial and error.

For efficient operation, the quantity of hot combustion gases used for drying is just enough to dry the cake to the desired percent solids. This depends on the ratio of the dried to wet sludge and the sludge flow rate, and is determined through operational experience.

Refer to the manufacturer's manuals for their suggestions on system operating temperature maintenance. Operating temperatures that are too low will not properly dry the sludge mixture, while high operating temperatures are inefficient and costly.

Flash Dryers. Smaller facilities may operate on an intermittent basis, typically 8 hours per day for 5 or fewer days per week. Intermittent operation typically requires a warmup period of one hour before drying. Extra fuel is consumed during this period as compared with continuous operation.

Controlling the dewatering process can result in a product at the desired moisture content of approximately 2 to 10%. Control consists of varying the amount of dried sludge mixed with the wet cake and varying the quantity of hot combustion gases for drying. To ensure optimum operation, closely monitor and control the temperature conditions throughout the system.

The exhaust gases are afterburned to eliminate odorous materials and scrubbed to remove particulates. Temperature controls on the afterburner indicate typical operation at approximately 650 to 760 °C (1 200 to 1 400 °F). Analysis of particulates both before and after the scrubber or monitoring of the pressure differential across the scrubber will indicate deviations in scrubber operations.

Rotary Kiln Dryers. The following typical operating conditions of the flash dryer also apply to the rotary dryer.

- The facility may operate continuously or intermittently.
- The product typically compares with that from flash drying except that grinding is necessary.
- A portion of the dried product is mixed with the dewatered cake before it enters the dryer.
- The stack gases are controlled. In contrast to the flash dryer, the rotary kiln dryer is direct fired; the temperature around the cake is controlled at approximately 370 °C (700 °F); and the dryer rotates at approximately 4 to 8 rpm to ensure mixing as opposed to the rapid mixing provided in the cage mill of a flash dryer.

ADDITIONAL TECHNOLOGY. Additional technologies being used for drying sludge are a multipass sludge dryer and a paddle dryer/processor. The multipass dryer uses a combination of both direct and indirect heat transfer to continuously process and thermally dewater sludge cake. The resulting weight and volume reduction significantly reduces hauling and disposal costs. As the sludge is conveyed through the dryer on stainless steel belts, it is not tumbled or suspended, thereby holding particulate emissions to a minimum. Recirculation of the hot exhaust air and water vapor within the dryer before discharge also reduces odors.

The paddle dryer/processor uses hollow wedge-shaped blades mounted on a hollow shaft with the heat transfer medium flowing through them. This method provides a high ratio of heat transfer surface area to process volume, using heat transfer media ranging in temperature from $-40\,°$ to 650 °F.

SAFETY AND HEALTH PROTECTION. The equipment used for these processes is similar to other equipment in the solids-handling area, including pumps, valves, centrifuges, conveyors, fans, and exhaust systems. Use of personal protective equipment, good housekeeping and storage procedures, proper labeling of containers to include detailed information on handling and first aid, posting of warning signs to alert employees to hazardous conditions and special precautions, posting of emergency procedures and instructions at critical operations, and training should all be a part of the employee's operating procedures for these facilities.

These processes use high temperatures. The equipment, valves, and fans should be protected from casual contact by the use of guardrails, covers, and signs to prevent possible burn injuries.

INCINERATION

DESCRIPTION OF PROCESSES. Incineration involves heat from combustibles in the sludge and from auxiliary fuel. Incineration removes all moisture from the sludge, thereby reducing the total volume. It destroys all organic matter by complete combustion; destroys pathogenic organisms; and produces a nonodorous product (WPCF, 1988).

Air emissions from incinerators should be monitored on a specific frequency to determine that no toxic or objectionable substances escape destruction. The U.S. EPA regulations for sludge disposal (40 *CFR* Part 503) limit the pollutant limits in the feed sludge, as well as in any air emissions from burning sludge for disposal. Specific pollutants limited by the 40 *CFR* Part 503 regulations include lead, cadmium, chromium, and nickel in the feed sludge, and beryllium, mercury, and hydrocarbons in the air emissions from burned sludge. The firing of sludge in an incinerator cannot violate the requirements of the National Emissions Standards for these pollutants, outlined in 40 *CFR* Part 61, Subparts C and E.

Multiple-hearth furnaces and fluidized bed incinerators are the most typical types used for sludge combustion in the U.S.

Multiple-Hearth Incinerators. As in the case of rotary kiln and flash dryers, the sludge feed to the multiple-hearth incinerator unit is dewatered and fed in as a cake. The furnace consists of a cylindrical steel shell surrounding a number of solid refractory hearths, and a central rotation shaft to which rabble arms are attached. Upper hearths typically have four rabble arms and more teeth than lower hearths (U.S. EPA, 1986).

Because the operating temperature in the fuel-burning hearths may reach 1 100 °C (2 000 °F), the central shaft and rabble arms are cooled with compressed air fed from the bottom (see Figure 32.11). A portion of the cooling air is drawn from the top of the furnace and returned to the bottom of the furnace to act as hot combustion air.

The multiple-hearth incinerator basically consists of three zones—the top hearths that comprise the drying zone; the middle hearths that comprise the combustion zone; and the bottom hearths that comprise the cooling zone. If auxiliary fuel is required, gas or oil burners are provided in the middle or combustion hearths. The countercurrent flow of combustion air to the sludge and the constant mixing of the sludge by the rabble arms provide maximum contact between the sludge particles and the hot gases.

Scrubbing the exhaust gases from the top hearth removes particulates. Deodorization may or may not be required, depending on the governing codes and standards and the operation of the incinerator. Deodorization may be accomplished in the incinerator itself, provided the cake does not dry too quickly on the lower-temperature hearths near the top of the incinerator.

Fluidized-Bed Incinerators. Fluid bed incineration sets graded silica sand particles in fluid motion in an enclosed space by passing combustion air through the bed zone so that all particles are in homogeneous boiling motion (U.S. EPA, 1986) (see Figure 32.12).

The reactor is a closed cylindrical vessel with refractory walls. Fluidizing and combustion air enters the unit and is dispersed upward through the orifice plate and grid that support the sand. Sufficient air is used to keep the sand in suspension but not enough to carry it out of the reactor.

The dewatered sludge cake is injected into the fluidized sand bed. The boiling action of the bed results in a rapid dispersion of the sludge solids and optimum contact with the combustion air. The sand bed retains the organic particles until they are reduced to ash. The rapid boiling action comminutes the ash to prevent clinker buildup.

Auxiliary fuel is injected into the bed as required to maintain the combustion temperature. Gas, fuel oil, or other fuel sources are used as auxiliary fuels.

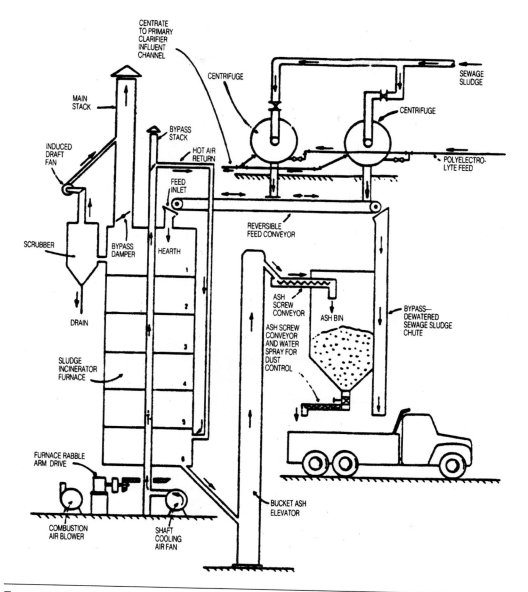

FIGURE 32.11   Multiple-hearth incinerator schematic.

The combustion and fluidizing air quantities supplied are determined by the requirements to

- Fluidize the bed to the proper density without blowing sand and incomplete combustion products out of the bed;

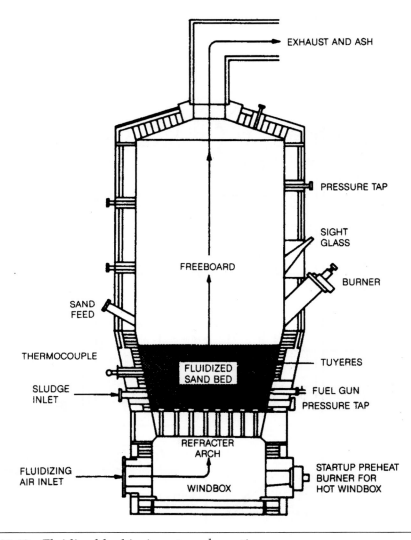

FIGURE 32.12   Fluidized-bed incinerator schematic.

- Provide minimum oxygen requirements for complete combustion of all volatiles in the sludge; and
- Provide temperature control to ensure complete deodorization and protection of the refractory, heat exchanger, and flue piping.

Because of scouring in the bed, periodic replenishment of the sand is necessary. A port is provided on the side of the reactor into which sand is added as needed.

PROCESS CONTROL. For effective and efficient process control, the following components of the incineration process need to be considered.

Dewatering. Optimization of the dewatering process before drying or incineration to conserve fuel is an important consideration. As the percentage of the volatile solids (VS) in the cake increases for a given sludge, the cost of auxiliary fuel per ton of solids decreases (U.S. EPA, 1986). If the cake VS are high enough, autogenous combustion will result and no additional fuel is necessary. Figure 32.13 presents the relationship between auxiliary fuel required and feed sludge solids concentration.

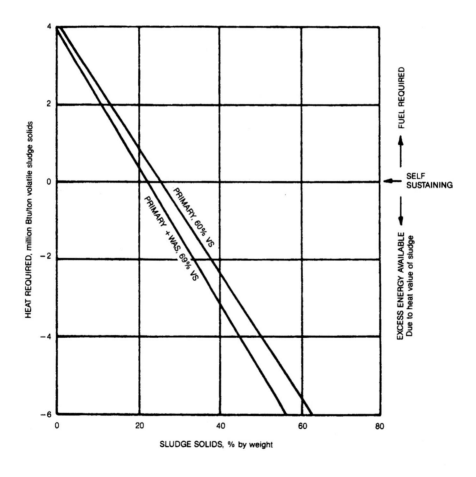

—Assumed heat value of sludge: 10 000 Btu/lb of volatile solids
—Curve assumes that afterburner is not used.

FIGURE 32.13 Auxiliary heat required for incineration.

Heat Recovery. Heat may be recovered by recycling process gases to preheat combustion air. For a given percentage of cake solids, only enough heat is available in the sludge to raise the temperature of the moisture and combustion products to some maximum temperature. With heat recovery, however, the temperature may be raised to a higher temperature, which may provide deodorization.

Air Requirements. The air requirements for combustion include the stoichiometric amount of excess air that is necessary to ensure complete combustion. The amount of excess air will vary with the type of incineration, the type of sludge, and the disposition of stack gases. Because excess air requires fuel to heat it, excess air is provided in the minimum amount that will ensure complete combustion and deodorization. Too little air will result in incomplete combustion and deodorization.

Sludge Fuel Value. The fuel value in the sludge is important in reducing auxiliary fuel consumption. The combustible elements in sludge are carbon, sulfur, and hydrogen. They primarily exist in organic sludges as grease, carbohydrates, and protein. These materials combine with oxygen to release heat and form carbon dioxide, sulfur dioxide, and water. Typical heat values for various sludges and their constituents are presented in Tables 32.8 and 32.9. The percentage of volatiles in the sludge is reduced by previous digestion, which can greatly reduce the heat value of the sludge. As the percentage of volatiles increases, the auxiliary fuel consumption decreases. The volatile content of a sludge is maximized by removing inorganics such as grit; by avoiding the use of inorganic chemicals such as ferric chloride and lime in dewatering processes; and by avoiding biological processes such as digestion before incineration.

Multiple-Hearth Incinerators. As in the drying processes, multiple hearths may be operated continuously or intermittently. When not in operation, however, auxiliary fuel is fed to the multiple hearth to maintain temperatures to protect refractory materials (U.S. EPA, 1986).

TABLE 32.8   Effects of prior processes on fuel value (U.S. EPA, 1986).

Type of sludge	Heating value, Btu/lb (dry solids)[a]
Raw primary	10 000–12 500
Activated	8 500–10 000
Anaerobically digested primary	5 500
Raw (chemically precipitated) primary	7 000
Biological filter	8 500–10 000

[a]Btu/lb $\times$ 2.326 = kJ/kg.

TABLE 32.9   Representative heating values of some solids (U.S. EPA, 1986).

Material (dry solids)	Combustibles, %	Heating value, Btu/lb[a]
Grease and scum	88	16 700
Raw wastewater solids	74	10 300
Fine screenings	86	9 000
Ground garbage	85	8 200
Digested sludge	60	5 300
Chemically precipitated solids	57	7 500
High organic grit	30	4 000

[a]Btu/lb × 2.326 = kJ/kg.

While the temperatures vary in different facilities, the temperature on any given hearth is kept constant to ensure uniform operations. Temperatures typically are interlocked either with the sludge feed and air feed or the auxiliary fuel feed and air feed systems. If the temperature drops, more sludge or fuel and air are added, and vice versa.

In typical operation, a multiple-hearth furnace provides three distinct combustion zones:

- Two or more upper hearths on which most of the free moisture is evaporated,
- Two or more intermediate hearths on which sludge volatiles burn at temperatures exceeding 820 °C (1 500 °F), and
- A bottom hearth that serves as an ash cooling zone by giving up heat to the cooler incoming air.

During evaporation of moisture in the first zone, the sludge temperature is not raised to more than approximately 60 °C (140 °F). At this temperature no significant quantity of volatile matter is driven off and no obnoxious odors are produced. Distillation of volatiles from sludge containing 75% moisture does not occur until 80 to 90% of the water has been driven off. By this time, the sludge has traveled far enough into the incinerator to encounter gases hot enough to burn the volatiles that cause odors. Typically, when fuel is required to maintain combustion in a multiple-hearth furnace, a gas outlet temperature of more than 480 °C (900 °F) indicates that too much fuel is being burned.

An analysis of stack conditions is used to determine typical operations. Oxygen, carbon dioxide, and carbon monoxide are monitored automatically in the stack and compared with preset values. An increase of the carbon monoxide level indicates that incomplete combustion is occurring. If, during the same period, the oxygen level compares well with the preset value, then either the mixing of sludge and air combustion is inefficient or the temperature has been reduced by the addition of cake that is wetter than normal. When introducing air above the hot (combustion) zone, false high readings can occur. In any event, an analysis of stack gases may be used to determine normal operation and indicate problems. Appropriate steps can then be taken to correct the problem and return to normal operation (see Table 32.10).

Fluidized-Bed Incinerators. The basic principles that apply to the operation of multiple-hearth incinerators also apply to the operation of the fluidized bed incinerators (U.S. EPA, 1986). The fluidized bed does have the advantage of operating intermittently without auxiliary fuel wastage because the sand bed acts as an enormous heat reservoir that holds the heat in the furnace longer. Also, temperature control is simpler in the fluidized bed.

The quantity of fluidizing air injected into the reactor is an important variable. An excessive quantity of air would blow sand and incomplete products of combustion into the flue gases, and would then result in needless fuel consumption. Insufficient air results in unburned combustibles in the exhaust gases. Fluidized-bed systems typically are operated with 20 to 40% excess air. In practice, this rate is controlled by measuring the oxygen in the reactor exhaust gases and adjusting the air rate to maintain 4 to 6% oxygen.

Because the theoretical amount of air is never enough for complete fuel combustion, excess air is added. The extra air is expressed as a percentage of the theoretical air requirement. For example, if a fuel requires 1 000 standard cubic feet of air per minute (SCFM) based on theoretical air requirements, and the actual air rate is 1 200 SCFM, the percent excess air is

$$\frac{(\text{Actual air rate} - \text{Theoretical}) \times 100}{\text{Theoretical air rate}} = \frac{(1\ 200 - 1\ 000) \times 100}{1\ 000} = 20\% \text{ Excess air} \qquad (32.1)$$

During start-up, auxiliary fuel is used to raise the temperature of the sand bed to approximately 650 °C (1 200 °F). As soon as sludge feed to the furnace begins, the auxiliary fuel rate is adjusted downward to achieve the maximum heat release from the sludge to avoid wasting fuel. This is done by gradually reducing the fuel feed rate to the minimum needed to maintain bed temperatures in the range of 680 to 700 °C (1 250 to 1 300 °F).

During typical operations, the fluidized bed without an air preheater requires more fuel than a multiple-hearth unit; however, if deodorization is required on the multiple-hearth unit, both will use approximately the same amount of fuel.

SAFETY AND HEALTH PROTECTION. Typically, these types of processes are contained in enclosed vessels with minimal exposure potential for employees. How-

TABLE 32.10   Troubleshooting guide for incineration (U.S. EPA, 1978).

Problem	Probable cause	Solution
Multiple-hearth incineration		
Furnace temperature too high	Excessive fuel feed rate	Decrease fuel feed rate
	Greasy solids	If fuel is off and temperature is rising, this may be the cause; raise air feed rate or reduce sludge feed rate
	Thermocouple burned out	If temperature indicator is off scale, this is the likely cause; replace thermocouple
Furnace temperature too low	Moisture content of sludge has increased	Increase fuel feed rate until dewatering system operation is improved
	Fuel system malfunction	Check fuel system; establish proper fuel feed rate
	Excessive air feed rate	If oxygen content of stack gas is high, this is likely the cause; reduce air feed rate or increase feed rate
Oxygen content of stack gas is too high	Sludge feed rate too low	Remove any blockages and establish proper feed rate
	Air feed rate too high	Decrease air feed rate
	Air feed excessive above burn zone	Check doors and peepholes above burn zone; close as necessary
Oxygen content of stack gas is too low	Volatile or grease content of sludge has increased	Increase air feed rate or decrease sludge feed rate
	Air feed rate too low	Check for malfunction of air supply and increase air feed rate, if necessary
Furnace refractories have deteriorated	Furnace has been started up and shut down too quickly	Replace refractories and observe proper heating and cooling procedures in the future
Unusually high cooling effect from one hearth to another	Air leak	Check hearth doors, discharge pipe, center shaft seal, air butterfly valves in inactive burners, and stop leak
Short hearth life	Uneven firing	Check all burners in hearth; fire hearths equally on both sides
Center shaft drive shear pin fails	Rabble arm is dragging on hearth or foreign object is caught beneath arm	Correct cause of problem and replace shear pin
Furnace scrubber temperature too high	Low water flow to scrubber	Establish adequate scrubber water flow
Stack gas temperatures too low (260 to 320 °C [500 to 600 °F]); odors noted	Inadequate fuel feed rate or excessive sludge feed rate	Increase fuel or decrease sludge feed rates

TABLE 32.10   Troubleshooting guide for incineration (U.S. EPA, 1978) (*continued*).

Problem	Probable cause	Solution
Stack gas temperatures too high (650 to 870 °C [1 200 to 1 600 °F])	Excess heat value in sludge or excessive sludge feed rate	Add more excess air or decrease fuel rate
Furnace burners slagging up	Burner design	Consult manufacturer and replace burners with newer designs that minimize slagging
	Air-fuel mixture is off	Consult manufacturer
Rabble arms are drooping	Excessive hearth temperatures or loss of cooling air	Maintain temperatures in proper range and maintain backup systems for cooling air in working condition; discontinue scum injection into hearth
Fluidized bed incineration		
Bed temperature is falling	Inadequate fuel supply	Increase fuel feed rate or repair any fuel system malfunctions
	Excessive rate of sludge feed	Decrease sludge feed rate
	Excessive sludge moisture	Improve dewatering system operation
	Excessive air flow	Reduce air rate if oxygen content of exhaust gas exceeds 6%
Low (<4%) oxygen in exhaust gas	Low air flow	Increase air blower rate
	Fuel rate too high	Decrease fuel rate
Excessive (>6%) oxygen in exhaust gas	Sludge feed rate too low	Increase sludge feed rate and adjust fuel rate to maintain steady bed temperature
Erratic bed depth readings on control panel	Bed pressure taps plugged with solids	Tap a metal rod into the pressure tap pipe when reactor is not in operation
		Apply compressed air to pressure tap while the reactor is in operation after reviewing manufacturer's safety instructions
Preheat burner fails and alarm sounds	Pilot flame not receiving fuel	Open appropriate valves and establish fuel supply
	Pilot flame not receiving spark	Remove spark plug and check for spark; check transformer, replace defective part
	Pressure regulators defective	Disassemble and thoroughly clean regulators
	Pilot flame ignites but flame scanner malfunctions	Clean sight glass on scanner, replace defective scanner
Bed temperatures too high	Fuel feed rate too high through bed guns	Decrease fuel flow rate through bed guns

TABLE 32.10   Troubleshooting guide for incineration (U.S. EPA, 1978) (*continued*).

Problem	Probable cause	Solution
	Bed guns have been turned off but temperature still too high because of greasy solids or increased heat value of sludge	Raise air flow rate or decrease sludge feed rate
Bed temperature reads off scale	Thermocouple burned out or controller malfunction	Check the entire control system; repair as necessary
High scrubber temperature	No water flowing in scrubber	Open valves
	Spray nozzles plugged	Clean nozzles and strainers
	Water not recirculating	Return pump to service or remove scrubber blockage
Reactor sludge feed pump fails	Bed temperature interlocks may have shut down the pump	Check bed temperature
	Pump is blocked	Dilute feed sludge with water if sludge is too concentrated
Poor bed fluidization	During shutdowns, sand has leaked through support plate	Once per month, clean windbox

ever, the high temperatures at which these facilities operate could be hazardous, and basic precautions should be taken by employees working in or around the equipment, valves, fans, blowers, reactors, and heat exchanger units. The equipment, valves, fans, and blowers should be protected from casual contact by the use of guardrails, covers, and signs to prevent possible burn injuries.

TROUBLESHOOTING. Table 32.10 presents a troubleshooting guide for incineration that will help the operator identify problems and develop solutions.

COMPARISON OF SLUDGE STABILIZATION PROCESSES

Table 32.11 lists the advantages and disadvantages of the five sludge stabilization methods discussed in this chapter.

TABLE 32.11   Comparison of sludge stabilization processes (WEF, 1995).

Process	Advantages	Disadvantages
Composting	High-quality, potentially saleable product suitable for agricultural use Can be combined with other processes Low initial cost (static pile and windrow)	Requires 40 to 60% solids Requires bulking agent or carbon source Requires either forced air or turning (labor) Potential for pathogen spread through dust High operational cost; can be power and labor intensive May require significant land area Potential for odors
Lime stabilization	Low capital cost Easy operation Good as interim or emergency stabilization method	Increases volume of sludge for disposal Land application may not be appropriate where soils are alkaline Chemical intensive Overall cost site specific pH drop after treatment can lead to odors and biological regrowth
Thermal treatment	More readily dewaterable sludge Effective disinfection of the sludge	Ruptures cell walls of biological organisms releasing water and bound organic material Treatment of this high-strength organic waste may require special consideration
Heat drying or incineration	Major reduction in sludge volume for disposal Total destruction of pathogens	Complex mechanical and control equipment Requires auxiliary fuel source Requires time to raise the operating temperature Requires relatively uniform dewatered sludge feed May generate odors

REFERENCES

Feinstein, M.S., et al. (1980) Sludge Composting and Utilization Rational Approach to Process Control. Final Report, Project No. C-340-678-01-1, Dep. Environ. Sci., Rutgers Univ., New Brunswick, N.J.

Haug, R.T. (1980) *Compost Engineering, Principles and Practices.* Ann Arbor Science Publishing, Ann Arbor, Mich.

Poincelot, R.P. (1975) The Biochemistry and Methodology of Composting. Conn. Agr. Exp. Sta., H.H.

Poincelot, R.P. (1977) The Biochemistry of Composting. *Proc. Natl. Conf. Composting Munic.* Residues and Sludge. Information Transfer, Inc., Rockville, Md., 163.

Stern, G. (1974) Pasteurization of Liquid Digested Sludge. *Proc. Natl. Conf. Munic. Sludge Manage.*, Information Transfer, Inc., Rockville, Md., 163.

U.S. Code of Federal Regulations (1993) Title 40, 40 *CFR* Part 503.

U.S. Environmental Protection Agency (1978) *Operations Manual: Sludge Handling and Conditioning.* EPA-430/9-78-002, Office Water Program Operations, Washington, D.C.

U.S. Environmental Protection Agency (1986) *Heat Treatment/Low Pressure Oxidation Systems: Design and Operational Considerations.* Office Municipal Pollut. Control and Office of Water, Washington, D.C.

U.S. Environmental Protection Agency (1992) *Control of Pathogens and Vector Attraction in Sewage Sludge.* EPA-625/R-92-013, Office of Research and Development, Office of Science, Planning, and Regulatory Evaluation, Washington, D.C.

Water Environment Federation (1995) *Wastewater Residuals Stabilization.* Manual of Practice No. FD-9, Alexandria, Va.

Water Pollution Control Federation (1988) *Incineration.* Manual of Practice No. OM-11, Alexandria, Va.

Chapter 33

Dewatering

INTRODUCTION

Each of the dewatering methods described in this chapter has advantages and disadvantages. Economics should drive the selection process. It is often helpful to list the specifics of the situation in terms of needs and wants. Needs are things that must be the design. Wants are things that would be good to have, but are negotiable. For example, if space is a severe limitation, then a disposal method is needed that does not require much space; a dewatering method that requires very little operator attention might be a good choice. The rational buyer should speculate on what changes might be made in the ultimate disposal method in the next 5 to 10 years. If, for example, development is rapidly heading for the plant, dewatering methods that require a lot of land and might cause odor are not a long-term solution. Likewise, for a small plant that will dewater 1 or 2 days a week and haul the biosolids to a nearby land application, high-performance dewatering equipment is probably not cost-effective. Ideally, the selection of major equipment is a cost-effective life-cycle evaluation—the present worth sum of the cost of equipment, maintenance, operation, and disposal. It is tempting to make an exhaustive list of costs, but, typically, the top three (capital, operating, and disposal costs) drive the decision. The Water Environment Federation® (WEF) (Alexandria, Virginia) publication *Engineered Equipment Procurement Options to Ensure Project Quality* (WEF, 1995) can help guide equipment selection.

The process and equipment descriptions given here are necessarily general and are not intended to describe any particular design or manufacturer. Changes to the recommended maintenance procedures and operation of equipment should be made with the advice of the manufacturer. It is important to read the manuals that came with the equipment. If the manuals are missing or incomplete, replacements are available from the manufacturer.

WHAT MAKES BIOSOLIDS DIFFICULT TO DEWATER?

Unfortunately, management's eye is generally on the wet end of the plant because the purpose of the plant is to receive wastewater and produce clean water. The dewatering operations are often treated as an unfortunate consequence of producing a satisfactory plant effluent. There is a tendency to feel that, if only the people in the dry end would do their job, there would not be any problem. Unfortunately, for this point-of-view, the quality of the biosolids sent to dewatering and the equipment available largely determine the cost to dewater. To minimize the overall cost of treatment, the wet end needs to be operated with an eye toward the consequences in the dry end.

DIFFICULTY IN DEWATERING BIOSOLIDS. Biosolids vary greatly regarding the ease with which they can be dewatered. In a paper that speculated about how the various ways that water could be held in the biosolids could affect its dewaterability, Vesilind (1994) realized that sludge is very complex, but sought an analogy or model that would provide insight to cause and effect. Vesilind pointed out that there are several ways that water could be held by biosolids, and this logic can be used to understand the way biosolids are generated and show which treatment alternatives result in more difficult-to-dewater biosolids. It should be noted that more recent research suggests that sludge is much more complicated than Vesilind's model; however, for the purposes of understanding why some sludges are more difficult to dewater than others, Vesilind's model is convenient to use.

Free Water. Free water is not connected in any way with biosolids. This water is the easiest of all to remove. Typically, the free water is removed in clarifiers and therefore does not reach dewatering. The exception is a plant where no attention is paid to a process, such as a clarifier skimmer pumping clear water to dewatering.

Bound Water. Bound water is water held in the particle either by chemical bonds or within a membrane. An example of this might be aluminum hydroxide, which precipitates as a hexihydrate. Each molecule of aluminum has 6 molecules of water attached to it. The City of Windsor (Ontario, Canada) uses ferric and/or aluminum hydroxide to remove phosphorus from wastewater. When the plant switched from ferric hydroxide to alum, the dewatered cake solids dropped 3 percentage points. Some time later, when the plant switched back from alum to ferric hydroxide, the cake dryness increased 3 percentage points (Macray, 2000).

Likewise, bacteria consists of a membrane encasing mostly fluid. Mechanical dewatering cannot break the membrane to release the fluid. Primary solids can be dewatered to 35 to 40% solids, whereas the amount of activated biosolids in the same plant rarely exceeds 25 to 18% solids. From this, it can be concluded that anything that increases the percentage of bacteria in the biosolids will result in wetter biosolids. Likewise, the practice of eliminating the primary clarifiers—while it does save capital costs and simplify the flow diagram—greatly increases the difficulty of dewatering the resultant biosolids. In effect, easy-to-dewater primary solids are broken down by the bacteria and thus increase the volume of bacteria going to dewatering. By anecdotal observation, eliminating the primary clarifier results in 5 percentage points wetter cake, with all other conditions being equal.

Minneapolis Wastewater Treatment Plant (Minnesota) has published data to show that the polymer dose increases and cake dryness decreases as the ratio of primary to secondary biosolids goes from 50:50 to 30:70 (MWWTP, 1999).

Vicinal Water. Vesilind (1994) postulated that water in the vicinity of the surface of a solid particle is attached to the particle by electrostatic and Van der Waals forces. As a result of these forces, this water is very difficult or impossible to remove by mechanical means. From this, it can be concluded that the greater the surface area per kilogram (pound) of solids, the more vicinal water will remain in the dewatered cake. Under a microscope, it can be observed that waste activated biosolids have a lot of surface area—remarkably more than primary biosolids. This profusion of surface area is another reason why secondary biosolids are harder to dewater than primary biosolids.

If a stone is broken into pieces, the weight of stone is unchanged, but the surface area clearly increases. In general, larger particles have less surface area per kilogram (pound) than smaller particles and thus have less vicinal water associated with them. Therefore, it is not good practice to make little particles from big particles, which is what happens when the primary clarifier is absent or not operating properly.

Interstitial Water. Interstitial water is water trapped in the spaces between the biosolids particles. There are several mechanisms to explain the reduction in the amount of interstitial water. Anyone filling a canister with flour or cereal has observed that shaking the canister causes the contents to move closer together and displace some of the interstitial air from between the cereal flakes. In this example, cereal does not change shape, and the packing density increases somewhat to reach a terminal packing density. If the cereal was flexible, then applying pressure to it would press the flakes closer together, causing them to conform to one another, and press the interstitial air out. In wastewater treatment, as the biosolids form a blanket in the clarifiers, the weight of one particle upon another squeezes the biosolids particles together, and squeezes water out. As the biosolids blanket becomes deeper, the cumulative weight of the particles on the lowest layer becomes greater, progressively more water is removed, and the biosolids blanket becomes thicker. Following the biosolids to dewatering, filters or centrifuges apply a great deal more pressure to press the solids further together, expressing even more water from the biosolids. It is interesting that in dewatering biosolids in dry solids centrifuges, various techniques are used to make the cake layer in the centrifuge as deep as possible. This increases the pressure of the solids, one upon the other. It is also true that to obtain dryer cakes requires an increase in the polymer dosage. Simply adding more polymer without increasing the cake depth does not make much difference in cake dryness.

Capillary Water. Capillary water is water held in crevices within the biosolids and is contained by capillary attraction. Generally, mechanical means cannot remove this water, although perhaps if the biosolids particles are not too rigid, squeezing the particles may crush the capillaries and extrude the water, much like squeezing toothpaste from a tube. Certainly, adding sufficient pressure to nonrigid particles could express capillary water out of the biosolids.

Some water is easier to remove than other water, and considering this allows one to profitably speculate on ways to make the biosolids easier to dewater. Unfortunately, efforts to quantify the amount of water held in these various states have not been immediately useful, but the concept itself is quite helpful. For example, in observing that dewatering biosolids from 98% water to a cake containing 72% water, the dewatering process has failed to remove the great majority of the water in the biosolids. By expressing cake solids as a percent, it is implicitly suggested that 100% dryness is possible. Vesilind (1994) has pointed out that it is not possible to mechanically remove

100% of the water, or 50% of the water, or perhaps even as little as 33% of the water in the biosolids. As with nearly everything, as the efficiency improves, further improvements are progressively harder to achieve.

IMPROVING THE QUALITY OF BIOSOLIDS. There are a number of practices that have been shown to improve dewatering, including the following:

- Increasing the capture of suspended solids in the primary clarifiers. Primary biosolids are easier to dewater than secondary biosolids; thus, reducing the amount of solids undergoing aeration will reduce the amount of biological biosolids produced. If the plant does not have primary clarifiers, one could consider adding them in the next plant redesign.
- Avoiding holding or storing biosolids in the plant. Dewaterability changes if biosolids become septic. Gravity thickeners are often a problem, as are tanks that store biosolids for days. The exception is anaerobic digesters, where longer residence times typically reduce the polymer dosage.
- Not allowing uncontrolled bacteria growths. If the temperature in anaerobic digestion is allowed to drift, it causes major changes in the bacteria population, resulting in difficult-to-dewater biosolids. In aerobic digestion, excessive aeration may reduce the volume of biosolids, but often the biosolids that remain are more difficult to dewater.
- Recent work suggests that, while short anaerobic digestion times may be sufficient for volatile reduction, longer digestion times result in lower polymer doses.
- Changing to two-stage digestion, starting with thermophilic anaerobic digestion, followed by mesophilic digestion. Research shows reduced biosolids volume and reduced polymer dosage.
- Avoiding the use of alum in the plant. Ferric salts result in dryer cakes. Mixing alum water treatment biosolids with wastewater biosolids should be avoided. The alum water treatment biosolids contain a high proportion of inorganics, often a high proportion of alum, and sometimes substantial amounts of grit.

TRACKING DOWN THE CAUSE OF A CHANGE. No dewatering operation responds well to change. If the feed biosolids varies, then the operators must determine how it has varied and respond to the change. Some changes can be predicted in advance. If one digester is just coming online, then biosolids from that digester will behave differently in dewatering. Other changes are more difficult. Many plants observe that the biosolids properties change from summer to winter. Process problems upstream are also a problem. It is important to look for changes in the following:

- Feed total solids, total volatile solids, pH, and temperature;
- Alkalinity;
- Conductivity;
- Polymer type, delivery, reaction, and concentration;
- Plant load;
- Customer loadings;
- Biosolids ratio;
- Weather;
- Plant bypass;
- Mixed liquor;
- Sludge volume index; and
- Biosolids age.

It is also important to look for changes in digestion properties, including the following:

- Residence time,
- Volatile acids,
- Alkalinity,
- pH,
- Temperature variation , and
- Feed variation.

THINKING AHEAD: INORGANIC CHEMICAL ADDITION TO ACHIEVE CLASS A BIOSOLIDS. The need to achieve a Class A biosolids through lime addition is more common in recent years, and the previous information provides a good basis to consider how to add lime and other alkaline agents, such as fly ash and kiln dust. There are two general ways to admix the lime.

The wet method is the simplest, where the pulverized lime (calcium oxide) is mixed with water, and the lime slurry is pumped into the biosolids feed line ahead of dewatering. This is a neat, trouble-free, low-maintenance system.

Alternately, the pulverized lime is conveyed by screw conveyor or air to a paddle mixer, where it is blended directly into the dewatered biosolids. This releases large amounts of ammonia, dust, and sulfides. Operators cannot work in this atmosphere, which mandates an odor control system.

From both an operating and an equipment point-of-view, the wet method is a much less expensive option than mixing powdered lime directly into the dewatered cake.

Liming Before Dewatering.

1. When the lime (calcium oxide) reacts with biosolids, it absorbs one molecule of water to form calcium hydroxide. This water is taken from the feed and is water that otherwise would leave the dewatering system in the filtrate, but now moves to the cake and must be hauled away as biosolids.
2. The lime has surface area and therefore will carry surface moisture with it through dewatering.
3. Relatively few polymers react well with high-pH biosolids and, when they do, the polymer dosages are higher. This alone will complicate polymer selection and increase the polymer costs.
4. The addition of lime before dewatering will increase the amount of lime needed, because it must also increase the pH of the filtrate/centrate that recycles back into the plant.

Liming After Dewatering.

1. When the lime (calcium oxide) reacts with biosolids, it absorbs one molecule of water to form calcium hydroxide. This water comes from the dewatered biosolids and is water that would have had to be hauled away.
2. Biosolids to dewatering are unchanged, so there is no problem with polymer selection and use.
3. Only the pH of the dewatered biosolids is raised; no lime is wasted on the filtrate/centrate

In filtration devices, the addition of lime reduces the specific resistance to filtration, which, in turn, would improve performance, so there is an additional benefit not present for centrifuges. In water and wastewater centrifuges, filtration is not a separation mechanism; therefore, there is no such mitigation. In both cases, the amount of material to be hauled is greater with preliming—in part, because the lime must be hauled away, but also because all of the surface area of the biosolids is covered with vicinal or surface water. Adding lime increases the amount of surface area and therefore adds to the surface water that moves to the biosolids.

Cost should be the major factor driving this decision. It is expensive to buy special equipment to place chemicals in the sludge, which increases the volume and weight of the sludge, only to throw the sludge away. The obvious up-front costs need to be balanced by the disposal costs. Regardless of how the lime is added, the product is of value to most agriculture reuse. In addition, limed sludge can be stored for a long time without odor problems. This is a great benefit when the recycle of biosolids to farmland is not possible during the winter or when crops are in the field.

SAFETY. All of the alkaline agents used are irritants to the skin, eyes, and lungs. Calcium oxide is more aggressive than calcium hydroxide. It is important to follow the precautions on the Material Safety Data Sheet.

OBTAINING HELP. It is often difficult to find the cause of problems because there are so many problems. Plants are operated in a certain way, and it takes time to change from what has worked in the past. Also, most operators have relatively little experience at other plants and so have little to compare with their own operation. Thus, from time to time, help is needed. All help comes with a bias. The manufacturer's solution will tend to involve new equipment and/or upgrades. The engineer's solution will tend to involve design work, and so on. This is why visits to other plants are invaluable. It is helpful to join the state chapter of water and wastewater operators and visit the host plant, but perhaps more important is get to know the operators of local plants. Local plant operators have probably encountered many, if not all, of the problems one may have, and they are an invaluable resource of knowledge and experience; the only thing they are likely selling is their self image of competence. Best of all, the only cost is reciprocation.

KNOWLEDGEABLE EXPERTS. The equipment manufacturer has staff who know wastewater treatment and dewatering. They can often provide considerable help over the phone, at no charge. Likewise, there are outside experts available who are specialists in various equipment and processes. The annual Water Environment Federation Technical Exposition and Conference (WEFTEC) and the WEF Residuals and Biosolids Conference have extensive technical programs. Presenters are another source for names of experts. Again, phone calls are typically free; however, not all problems are solved over the phone. The cost to bring an expert to the plant is typically modest in proportion to the cost of living with the problem. Some experts are better than others. One can get some idea of the depth of a person's knowledge by talking the problem over with them. Do not be afraid to ask for references, then call them and talk to them. Most people are reluctant to be frank about anything negative, especially to a stranger. However, they will rarely fail to answer a direct question candidly, so ask such questions. If it seems, from the preceding, that the person is experienced and helpful, pay that person to come out to the plant—and not someone on the person's staff. If there must be a switch, interview the person planning to come out and, if there is no great confidence in that person, demand someone else. It is hard to reject someone, but better to do that than to pay someone who cannot fix the problem.

SUPPLIERS. The people one interacts with regularly can often be of very good help. For example, the polymer salesman sees countless plants and can often point to problem areas within a plant. Often, representatives of firms that do not have a particular plant's business are more willing to point out obvious problems to impress that plant's operators with their knowledge; wishing to diminish the competitor's business volume is also a motive. Equipment representatives often have in-depth experience with a plant's specific equipment and processes and can be a good source of information and suggestions regarding who in the industry can help. If nothing else, they can put one in touch with other plants that have had similar problems.

WATER ENVIRONMENT FEDERATION. One resource often overlooked is WEF. One can search their database for articles and papers on almost every topic. There are numerous publications, such as this book, that will help. Especially useful are the discussion groups at the WEF website, where people can post problems and questions and discuss them with other members of WEF.

U.S. ENVIRONMENTAL PROTECTION AGENCY. The U.S. Environmental Protection Agency (U.S. EPA) (Washington, D.C.) has a series of fact sheets that are useful summaries on various topics (http://www.epa.gov/owm/mtb/biosolids/index.htm). Website organization changes, so it may be necessary to search the U.S. EPA website (http://www.epa.gov).

OPERATING PRINCIPLES FOR DEWATERING

Current management information systems are typically more focused on receiving information than on disseminating it. Each Supervisory Control and Data Acquisition (SCADA) screen focuses on a small section of the process, and there is not enough thought given to what to display and where to display it.

MANAGEMENT OF INFORMATION. Most plants have not taken advantage of modern information-transfer capabilities. Operating information, laboratory analyses, and economic information typically moves up the chain-of-command, when it is at least as important that it move horizontally and downward. The goals of the plant include compliance with regulations and, at the least, cost. Dewatering operators cannot operate at the least cost when they do not know what the costs are. Most SCADA systems are a series of descriptive screens that convey a small amount of sometimes useful information to the operator amidst a clutter of worthless information. The screens look nice, and because once they are designed every such system can use the same screen, they become a generic standard. Unfortunately, they are designed by people who have little experience actually running a dewatering operation. Operating personnel should consider what information is available, both upstream and downstream of their plant, consider what would be useful to them, and find a way to display this information on their screen. If this means that the pretty illustration on the screen needs to go, it will not be missed. It is important to consider the importance of wet-end information to the downstream dewatering operation.

No dewatering operation works well if the dewatering characteristics of the biosolids change without warning. To some extent, variations are unavoidable; after all, the production rate of primary biosolids varies during the day, and therefore the resulting ratio of primary to secondary biosolids blend varies also. Nevertheless, wet-end operators should make an effort to minimize the magnitude of such changes and tell the dry end of the plant when a change is coming. For example, if it is decided that the mixed liquor suspended solids are too high, an unthinking supervisor may make a dramatic change in the wasting rate and bring the mixed liquor down to the target in 2 hours. The resulting large and sudden change from a typical 50:50 blend to perhaps a 30:70 primary to secondary blend will cause a major upset in dewatering. In a few hours, the operators will get the dewatering process under control, and when the wasting rate suddenly returns to normal and the biosolids are once more a 50:50 ratio, there will be yet another upset. This sort of operation demoralizes the dewatering operators—in part, because it is unnecessary and, in part, because the biosolids seem to be unpredictable. Better management of the information would have given the operator changing the wasting rate a "pop-up" telling him that such a large change would upset the dry-end operation, and suggesting that the wasting rate be increased moderately and the mixed liquor be brought down to the target over 24 hours. If there is no overwhelming reason for a sudden change, presumably the operator will be persuaded to opt for a more gradual change in wasting rate. Whatever the decision, when a change was made in the wasting rate, ideally a pop-up on the dewatering operator's screen would show the coming change in biosolids quality; in 45 minutes, the ratio of primary to secondary biosolids will change from 50:50 to 47:53. Some small adjustment could be made to the dewatering devices, and the process would not have gone out of control. If the wasting rate had been increased moderately and the mixed liquor brought down to the target over 24 hours, the operation would have gone much more smoothly. This is an example in which the SCADA system needs to communicate horizontally.

Of course, the operator may be able to call up the wet-end screens, do a trend line on the secondary wasting, and discover that the wasting rate changed; however, realistically, no one has the idle time to do this, especially just at the moment the rate was changed. Also, no one should have to, when the need for the information is clear, and the SCADA system can handle it automatically, if only it was programmed properly.

It may well be worthwhile for a plant to redo the SCADA screens to reflect what is important for the operators to know, and to relegate the less important data to a secondary screen.

CHANGING VALUES. The wastewater plant is a dynamic system, and often it is the changes that are of interest, and not so much the actual value of some parameter at the moment. No one can use all of the data collected all of the time, so some sort of reduction of information needs to be done. Much of what is seen is useful only if the viewer possesses an additional piece of information that is not displayed. For example, knowing that the conveyor torque is 15 Nm is not helpful unless it is also known that the torque trip point is 17 Nm, indicating that the device is about to go down on high torque. The torque reading would have been much more user-friendly if the display read 88% load. Similarly, one's response if the torque is rising is quite different than the response if the torque is falling. The absolute number is often not as useful as information that the reading is steadily changing. A trend line is forensic analysis; after the crash, one looks to see what led up to it. The SCADA system could automatically look for trends and display a summary on the main screen of only those trends that are unusual. This would help spot trouble before it arrived.

Undoubtedly, everyone benefits from prompt feedback. The samples sent to the laboratory are analyzed and entered into the SCADA system, but rarely do they pop up on the screen in dewatering, where the information can be used. Likewise, the dewatering operators often grab samples and analyze them, and then write the results down on a piece of scrap paper. If they entered this data into the computer, then the computer could calculate the percent recovery, polymer dose, and even dewatering cost. The computer could generate a trend line of dewatering cost by shift. It would store the data, which would help other operators on subsequent shifts and easily be retrieved even in 6 to 12 months. Instead, operators rely on logbooks, data sheets, and the like, which disappear into storage every few months, never to be seen again. It would seem, in the age of searchable databases, that operators should write their logs into the computer database.

In summary, plant controls need to be tailored by the operator, for the operator. One-size-fits-all generic displays are not good enough. In the future, it would be very beneficial if operators could click, drag, and paste on the screens and customize the screens to suit themselves.

INORGANIC CHEMICAL ADDITION

FERRIC CHLORIDE AND ALUM. Ferric and aluminum salts are often added in the wet end of the plant because both react with the phosphate in the water to form an insoluble precipitate. The use of these chemicals in the wet end will have a noticeable effect on dewatering; the use of ferric chloride will improve the dewatering process, whereas aluminum will have either no change or result in more difficult-to-dewater biosolids. The effect on dewatering will be proportional to the amount added; therefore, the dewatering operation needs to be advised of any changes in the dosing rate. Likewise, when comparing the cost of ferric chloride and alum, the effect on dewatering also needs to be considered. When ferric chloride is diluted with water, it hydrolyzes, forming positively charged soluble iron complexes that neutralize the negatively charged biosolids particles, causing them to aggregate. Unlike polymers, alum and ferric chloride do not require aging time, and special mixing devices are not used. Ferric chloride also reacts with the bicarbonate alkalinity in the biosolids to form hydroxides that also act as flocculants.

STRUVITE CONTROL. Some anaerobic digesters have a problem with struvite, which is a crystal comprised of calcium ammonium phosphate. Struvite can precipitate out in the digester anywhere carbon dioxide can flash off or where there is a lot of turbulence. The crystals sometimes are a serious problem in the centrate/filtrate lines, where they can block a pipeline in 1 week or so. In theory, struvite can be controlled by altering the amounts of the component ingredients or, to bind up phosphate, one of the key ingredients of struvite. Ferric chloride is commonly added because it ultimately reacts with phosphates and coincidentally lowers the pH of the sludge. Ferric chloride does improve dewatering; the bad news is that it is expensive and adds to the weight of material hauled away.

Another approach to struvite control is to acidify the feed to dewatering, lowering the pH to the point where struvite will not form. Any acid will do (sulfuric, hydrochloric, and nitric). Sometimes spent acid can be found at a good price—even free—but it is important to check it to see if it contains anything such as heavy metals that might be a problem to the plant. In general, adding acid to the biosolids gives little benefit to dewatering. As an acid, ferric chloride is more expensive than the more common acids, so the value of the other benefits must be weighed against its higher cost. Ferric chloride generally improves dewatering, whether added in the primaries or just ahead of the dewatering device. At current prices, it rarely pays to add it solely to improve the dewaterability of the biosolids; the justification is the sum of its benefits.

ODOR CONTROL CHEMICALS. Odor control chemicals are sometimes needed. There are several types—oxidants, such as ozone, hydrogen peroxide, chlorine dioxide, permanganate, hypochlorite, and chlorine (Lambert and McGrath, 2000). These chemicals provide oxygen, which, in turn, reacts with hydrogen sulfide and other odor constituents to convert them to compounds with a higher odor threshold. The various oxidants have different levels of effectiveness, and some may be better than others in dealing with a specific odor problem, so it is worthwhile to experiment. There is some evidence (Rudolph, 1994) that potassium permanganate has a side benefit of reducing the polymer dosage, but reduction of the polymer cost is not the reason it is used. The oxidant feed point should be located upstream of the polymer addition point, to provide a contact time of 1 minute or more before polymer addition. Odor neutralizers are oils and surfactants that bind the odor-causing chemicals to themselves and thus reduce the unpleasant odor. These are commonly sprayed into the air in areas where odors emerge. Some of these chemicals have their own pronounced odor, so they need to be tried out. Nutrients, such as ferric salts, precipitate sulfides, and nitric acid adds oxygen to the sludge, which allows organisms to oxidize fatty acids. Bacterial dosing and enzymes are also reported in the literature.

BUYING CHEMICALS. The chemicals mentioned above are bulk commodities. When buying small amounts, it makes sense to buy through a distributor. In return, for applying a markup, a distributor supplies chemicals in the form and volume needed, may provide technical support, and is cooperative with the purchasing system. If these services are not of value to the operator, it may be possible to save money by buying directly from the manufacturer. In any event, it is worth doing an internet search and finding out the bulk price of the chemicals being purchased and compare it to what is being paid. If the chemical cost can be cut by 25% by accepting 2 large shipments a year, rather than 12 small ones, then it may be worthwhile to increase the storage capacity.

ORGANIC FLOCCULANTS

BACKGROUND. Organic flocculants are widely used in many industries and processes involving the separation of solids from liquids. These liquid–solid separation applications may involve processes related to the recovery of finished products, clarification or purification of liquids, and volume reduction of waste materials. Several major applications are identified in Figure 33.1, which also indicates the relative electronic charge and molecular weight of organic flocculants typically used in the respective application.

The use of organic polyelectrolytes, frequently called *organic flocculants* or *polymers*, in municipal wastewater plants began during the 1960s and increased rapidly as wastewater facilities expanded to include secondary treatment. The use of these products is widespread in the estimated 16 000 publicly owned wastewater treatment works, which process approximately 130 tril. L (34 tril. gal) of household and industrial wastewater each day in the United States, while generating some 6.2 mil. Mg (6.9 mil. dry tons) of wastewater residuals, sludge, or biosolids each year. Figure 33.2 shows the various application points for organic flocculants within the typical municipal wastewater treatment operation.

Initially, ferric chloride and lime were commonly used to condition wastewater residuals before thickening and dewatering. However, in today's sophisticated municipal dewatering operations, these products are rarely used, with the exception of any remaining vacuum filtration and recessed-plate filtration processes.

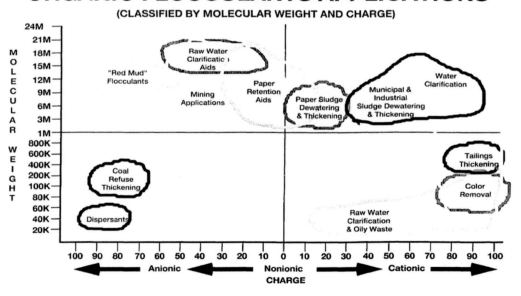

FIGURE 33.1   Typical points of addition of organic flocculants in wastewater treatment.

TYPICAL WASTEWATER TREATMENT FLOW DIAGRAM

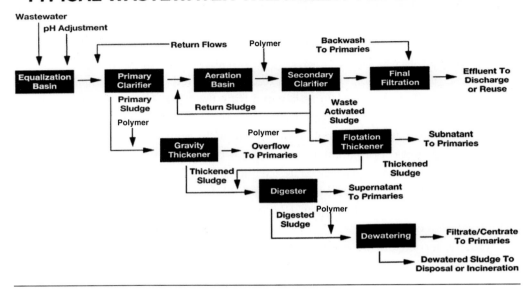

FIGURE 33.2   Typical wastewater treatment flow diagram.

While organic polyelectrolytes are commonly used in applications involving liquid–solid separation, the processes of wastewater sludge thickening and dewatering are completely dependent on their use. These municipal processes consumed approximately $188 million worth of complex, proprietary, cationic polyacrylamide (PAM) flocculants in North America in 2002, specifically for sludge conditioning.

Several factors have contributed to the widespread use of PAM flocculants as conditioners for sludge dewatering, including the following:

- Development of more sophisticated dewatering equipment, such as belt filter presses (BFPs) and centrifuges;
- Technological advances in polymer chemistries to provide more convenient, user-friendly, and high-performance products;
- Development of convenient and effective polymer make-down units; and
- Increased emphasis on the costs of solids processing and disposal.

Because of the complex and proprietary nature of these PAM flocculants and their varying levels of performance or effectiveness with different sludges, each manufacturer offers a full range of products based on product characteristics and forms. The remainder of this discussion will focus on cationic PAMs, as they represent virtually 95 to 100% of the organic flocculants for conditioning wastewater sludge residuals before dewatering.

POLYMER CHARACTERISTICS. The product characteristics of these complex and proprietary PAM flocculants may vary according to the following:

- Electronic charge (anionic, nonionic, or cationic);
- Charge density;
- Molecular weight (standard viscosity); and
- Molecular structure.

As shown in Figure 33.3, the "families" or types of organic flocculants are depicted according to molecular weight and electronic charge and charge density. It is useful to review Figure 33.3 in conjunction with Figure 33.1 to identify the family or type of organic flocculants, which are frequently used in each respective application.

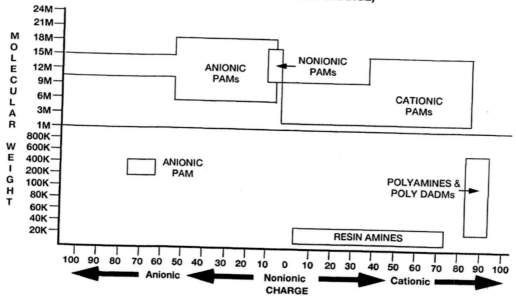

FIGURE 33.3 Types of organic flocculants according to electronic charge density and molecular weight.

Electronic Charge (Type). A variety of organic flocculants are manufactured with three distinct types of available electronic charge—anionic (negative), cationic (positive), and nonionic (no charge)—each suitable for specific liquid–solid separation applications. For conditioning wastewater residuals before dewatering, virtually all of the organic flocculants used are cationic PAMs, carrying a positive charge. There is almost an unlimited number of cationic PAM flocculants that may be produced. A variety of cationic monomers may be co-polymerized with acrylamide in varied molar ratios relative to acrylamide to produce flocculants with a wide range of charge density.

Cationic PAMs are formed either by co-polymerizing the acrylamide monomer with a positively charged monomer to create the cationic polymer or by modifying the acrylamide monomer itself following polymerization. This latter method of cationizing the acrylamide monomer is performed by reacting formaldehyde and dimethylamine with acrylamide, which yields aminomethylated PAM, the traditional Mannich solution PAM named after the inventor of the chemical reaction.

A variety of cationic monomers may be incorporated to the PAM backbone at various molar ratios, thereby giving rise to various degrees of cationic charge or charge density. Typical molar ratios of acrylamide monomer to the cationic monomer are 80:20, 60:40, and 45:55. The weight ratios differ substantially from the molar ratios as a result of the higher molecular weights of cationic monomers to that of the uncharged acrylamide. The names and chemical structures of these cationic monomers may be found in the Water Environment Research Foundation (WERF) (Alexandria, Virginia) publication *Guidance Manual for Polymer Selection in Wastewater Treatment Plants* (WERF, 1993).

Charge Density. Manufacturers of organic flocculants refer to charge density in relative terms. While there is no standardized representation for charge density, the information presented in Table 33.1 may be used as a typical convention for relative charge density.

The charge density of cationic flocculants used for dewatering municipal sludges generally fall into the broad range 10 to 80%. The majority of polymers, which perform effectively to condition most municipal sludges comprised of blends of digested pri-

TABLE 33.1   Charge density.

Relative charge density	Charge density (mole %)
Very high	>70 to 100
High	>40 to 70
Medium	>10 to 40
Low	<10

mary and waste activated sludge (WAS) for dewatering in centrifuges or on BFPs, fall into the category of medium-to-high charge—typically 20 to 60%.

While there is no substitute for laboratory testing and full-scale machine evaluations to determine effective products, the following comments may be considered as general guidance:

- Higher concentrations of primary sludge will require lower-charge products;
- Higher concentrations of biological solids or WAS will require higher charge;
- Sludge with smaller particle size will typically require higher polymer dose and possibly higher charge; and
- Older, septic sludge will require higher charge and higher polymer dose.

Molecular Weight. Similar to that for charge density, most manufacturers of organic flocculants refer to molecular weight in relative terms. While there is no standardized representation for molecular weight, the information presented in Table 33.2 may be used as a typical convention for relative molecular weight.

Variations in the chain length of the polymer may result in products with an extremely wide range of molecular weight, which is typically indicated by standard viscosity data. Further, these polymers may be chemically modified for improved performance by cross-linking the polymer chain or incorporating branching to the polymer backbone.

The molecular weight data can only be estimated. As an indicator of molecular weight, many manufacturers provide standard viscosity data. Unfortunately, standard viscosity measurements are not comparable across different product chemistries. As a side note, viscosity is not a measure or an indicator of polymer content.

In general, lower-molecular-weight products tend to be more soluble and less viscous. The majority of polymers, which perform effectively to condition most municipal sludges comprised of blends of digested primary sludge and WAS for dewatering in centrifuges or on BFPs, fall into the category of medium-to-high molecular weight, ranging from an estimated 0.8 to 6.0 million units and represented by standard viscosities ranging from 0.0025 to 0.0035 Pa (2.5 to 3.5 cP).

TABLE 33.2   Molecular weight.

Relative molecular weight	Molecular weight
Very high	>6M to 18M
High	>1M to 6M
Medium	>200K to 1M
Low	<200K

POLYMER FORMS AND STORAGE AND HANDLING. There are three physical forms of cationic PAMs, which are typically used in conditioning municipal sludge for thickening and dewatering applications.

- Dry PAMs (DPAM),
- Emulsion PAMs (EPAM), and
- Solution PAMs (SPAM), sometimes called *liquid PAMs*.

There is a fourth form, the gel log, which is rarely used in wastewater treatment.

Factors that may affect the determination of which form is most suitable for an operation include the following:

- Cost-effectiveness,
- Storage and handling of neat product,
- Polymer make-down and aging equipment and solution feeding capability, and
- Personnel safety considerations.

Dry Polyacrylamide. Dry PAM may be provided in granular (pellet), flake, or bead (pearl) forms. These may affect storage and handling considerations, as each has different moisture adsorption (caking), flowability, and wetting for polymer make-down characteristics. Dry PAMs generally remain effective for more than 1 year, provided that they remain free of moisture. Most manufacturers suggest a maximum shelf life of 1 year.

Dry PAMs are not 100% active polymer, as is frequently reported. Dry PAMs generally contain 88 to 96% active polymer, with the balance being residual moisture and inert salts. Dry PAMs, therefore, are advantageous for having the lowest shipping cost per unit weight of active ingredient. In many cases, additional salts are added to enhance product life or flowability. Accordingly, a simple solids determination probably does not accurately reflect the active polymer content.

The effective make-down of dry polymers is critical for efficient product use. Each polymer particle must be wetted to ensure complete dissolution into the polymer solution and to avoid product agglomeration and gels. Any gels or "fish eyes" that do not properly dissolve into the polymer solution are not useable and therefore are wasted.

Storage and handling of DPAMs must be carefully considered. Dehumidified areas are recommended for product storage and make-down, as excessive humidity or moisture may cause caking and handling problems in the storage and conveying equipment. Conversely, dust must also be controlled, and personnel may be required to wear protective respiratory equipment. Some DPAM products have a tendency to cause more dust than others.

Spills of DPAM must be contained and promptly swept or vacuumed away for proper disposal. Moisture and water cause the polymer to become a hazardous, gelatinous, and slippery goo and should be strictly avoided.

Consult the polymer manufacturer for storage and handling, humidity and moisture control, dust control, product make-down and solution concentration, and cleanup recommendations.

Emulsion Polyacrylamide. Emulsion PAMs (emulsion or dispersion) appear as a milky liquid, although some emulsions or dispersions, with much smaller particle size, may appear somewhat clearer or cloudy. The emulsion product provides for a higher concentration of active polymer in a convenient, easy-to-handle, liquid form. The EPAMs typically contain 25 to 60% of active polymer.

Most emulsion products have a recommended shelf life of 6 months when stored under normal conditions. Samples stored in a refrigerator have considerably longer shelf lives. Emulsions consist of two distinct phases—oil and water; there is the potential for phase separation or layering. Users should consider slow mixing or recirculation of the polymer storage tank to maintain uniformity of the neat polymer. Similarly, drum or semibulk containers of emulsion polymers should be checked for uniformity and mixed, if necessary, before use. The manufacturers have mixing recommendations for maintaining neat polymer homogeneity and can supply mixers for drums and the semibulk containers.

In an emulsion, the organic polymer is actually contained in a mineral oil phase of droplets, which are stabilized in an aqueous media by emulsifiers or surfactants. These additives also assist in the inversion of the polymer during make-down. However, there is an additional cost associated with the oils and surfactants used to make the convenient product package.

Emulsion PAMs are generally shipped and stored in bulk tanks or semibulk, returnable containers (totebins). The EPAMs may also be available in 208-L (55-gal) steel or fiber drums. Handling takes place within pipes and hoses, which are a closed system. The make-down of emulsion polymers is generally convenient and reliable, using any of several readily available, packaged systems offered by numerous manufacturers. The bulk viscosity of EPAMs may vary considerably, thereby affecting pumping capability and calibration. This means that the equipment must be recalibrated when polymers are changed.

When lines, hoses, or other equipment containing EPAM must be disassembled, they should be flushed with lightweight machine oil, such as SAE 10 to 30, to remove residual polymer. If water is used to flush the neat emulsion lines, the polymer will gel into a sticky mass.

Similarly, spills should be quickly contained and adsorbed in any one of a number of clay-based adsorbents. A number of commercial cleaners are available for cleaning spills and belt washing. Any polymer supplier can recommend specific types. Avoid using water during the cleanup of spills outside of a contained and properly drained area, as it will result in a slippery, gelatinous mass.

Consult the polymer manufacturer for specific recommendations regarding product uniformity and mixing, storage conditions and product stability, neat product viscosity and pumping, product make-down and solution concentration, and cleanup.

Solution Polyacrylamide. As the name implies, SPAM is a true solution of polymer in water, appearing as a clear to cloudy liquid, highly viscous syrup that is almost gel-like. These products are also referred to as *liquid PAMs* or *Mannichs*, named after the inventor of the chemical process. Solution PAMs have a very high molecular weight and range from approximately 3 to 8% active polymer content. The neat SPAM product may also contain inert salts or diluents, so that a simple solids determination may not accurately reflect active polymer content.

These products are very dilute, so they are typically sold in bulk tanker loads. The user will need storage tanks to hold the large volume inventory. The storage time should be limited to 1 month or less, even though some manufacturers suggest a shelf life of 3 months or more. The SPAM products are more susceptible to biological and chemical degradation than the other forms of polymers. The SPAM products may continue to react in storage, increasing viscosity even to the extent of gelling. At the least, this causes pumping and product make-down difficulties; at worst, the tank has to be cleaned out and the product discarded.

Bulk storage and handling are convenient and are generally conducted in closed systems, although the pump that is used for emulsion probably will not be large enough to deliver the volume of SPAM that is needed. Because the polymer exists in an aqueous solution, traditional make-down, such as that required for the dry and emulsion forms, is not necessary. Simple dilution of the neat product with water to a recommended concentration for use, followed by adequate mixing, should be adequate.

The active polymer concentration is very low—typically 5% or less—which results in very high transportation costs. The SPAM products are generally not cost-effective beyond a 400-km (250-mi) radius of the manufacturing plant.

Spills of SPAM should be readily contained and adsorbed in appropriate media for disposal. A number of commercial cleaners are available for cleaning spills and belt washing. Avoid water during cleanup of spills outside of a contained and properly drained area, as it will result in a larger amount of slippery, gelatinous mass.

Consult the polymer manufacturer for specific recommendations regarding neat polymer storage conditions and product stability, neat product viscosity and pumping, product make-down and solution concentration, and cleanup.

GENERAL SAFETY. Review the manufacturer's product literature and Material Safety Data Sheet for each polymer for detailed handling and safety precautions. In general, polymers used in sludge dewatering are not particularly toxic, but it is prudent to avoid contact. Safety problems are primarily related to slip hazards and dusting. Of course, personnel should take appropriate precautions to avoid getting dust into their eyes, ears, nose, and mouth. Any dust that attaches to a moist body part will become sticky and slippery and will be difficult to wash off. Accordingly, use protective equipment and disposable clothing when working in dusty polymer conditions.

Dust from dry polymers is extremely hazardous when wetted. Simply walking in a minor amount of polymer dust with wet shoes can cause a fall. Similarly, dust tracked or carried by drafts into an unexpected area can await the unwary for months. Polymer dust should be carefully contained and thoroughly swept or vacuumed and discarded before it can result in a hazardous, slippery condition.

Similar slippery conditions may result from leakage or spillage of liquid solution or emulsion polymer or from tracking liquid polymer into an unexpected area. When a liquid polymer dries, it is relatively harmless. However, once rewetted, it quickly becomes a slippery hazard. Cleanup of liquid polymer is difficult, and it must be thorough. Adsorbents may be used, and steam cleaning can be effective. The addition of rock salt or bleach (sodium hypochlorite) may help to break the polymer chain before washing with water. A number of commercial cleaners are available for cleaning spills and belt washing. Any residual polymer will result in the same hazardous, slippery condition when contacted again with water.

The best precaution is to avoid polymer leaks, spills, and dusting. Prevention is much easier than containment and cleanup.

POLYMER SPECIFICATIONS AND QUALITY CONTROL. Along with the product identification and type and form of product, the following standard product specifications should be obtained to determine storage conditions, pumping requirements, and potential hazards:

- Total solids (percent),
- Specific gravity,
- Bulk viscosity,

- Flash point,
- Freezing point, and
- Shelf life.

The product specifications should also include the related quality control limitations or variability of the manufacturer, each of which is critical to the usage and performance of the product in the dewatering process.

- Active polymer content (percent),
- Charge density (mole percent), and
- Standard viscosity (hertz [cycles per second]).

Active Polymer Content. Active polymer content (percent) should be obtained by the end-user with total solids. In addition, the end-user should receive, from the manufacturer, the quality control capability, in the form of the minimum and maximum active polymer (percent), to be contained in any individual shipment.

Active polymer content is the specific measure of the real product that is responsible for dewatering performance. This measurement of active polymer eliminates inert solids, which are included in total solids, but do not contribute to dewatering performance. The procedure for determining active polymer involves solvent extraction and is quite involved. Unfortunately, there is no industry-wide standardized procedure. It is prudent to ask for a copy of the manufacturer's procedure and consider spot-checking the deliveries.

Charge Density. Similarly, cationic charge density should be obtained by the end-user for each product. The charge density should be a quantitative assessment expressed as the mole percent of cationic charge, rather than a relative reference, such as low, medium, or high. Also, the end-user should receive from the manufacturer the quality control capability, in the form of the minimum and maximum cationic charge density (mole percent), to be contained in any individual shipment. Charge density is an indicator of the ratio of positive charges relative to the polymer backbone. For any given application or sludge type, if the charge density falls below a certain level, it will be necessary to use more polymer to achieve the same performance. Therefore, it is important to identify the minimum charge density that will be provided by the manufacturer for a selected optimum product, which has been identified through full-scale field evaluations, and the actual charge density of the products tested to obtain the bid information.

Standard Viscosity. Standard viscosity is also important to the end-user with the manufacturer's relative classification of molecular weight. Standard viscosity may be used as an alternative, quantitative comparison for molecular weight.

Standard viscosity is different than bulk viscosity or the viscosity of the neat product. Bulk viscosity is determined on the neat product and is useful for determining pumping and flow and related equipment requirements. Standard viscosity is determined at various make-down concentrations of the neat polymer product and is useful as an indicator of molecular weight. For standard viscosity, there are no industry-wide standardized procedures, so ask the manufacturer for copies of the analytical procedures they use.

PRODUCT MAKE-DOWN. A wide variety of very effective, packaged polymer make-down systems for handling emulsion or dry polymers are available from any number of independent equipment manufacturers and from the polymer producers. Unfortunately, most facilities are set up to effectively handle one form of polymer and are not readily able to convert to another product form without a lot of re-engineering. Flexibility in a polymer make-down system and related polymer handling equipment is very desirable. It is impossible to predict what form of polymer will be the most cost-effective in the future. A number of make-down systems are available that provide universal make-down capability for dry, emulsion, and solution polymer.

To be truly effective, a universal make-down system must include the appropriate ancillary equipment, such as metering pumps and storage tanks, capable of handling varied polymer concentrations and the related volumes and flowrates. As such, universal product make-down systems are not always practical and are a lot more expensive. Installation of equipment to handle the dry and emulsion products is a good compromise between price and value. Sometimes the form of product currently being used dictates the make-down system used, and the built-in capabilities of ancillary equipment preclude the easy conversion from one product form to another.

Accordingly, thorough consideration should be given to a variety of issues, including the following:

- Neat product storage and handling,
- Personnel and safety,
- Make-down capability and space limitations,
- Convenience,
- Dewatering performance and economic criteria,
- Changes in the polymer market and competitive landscape,

- Potential for sludge changes, and
- Anticipation of future developments.

All of these factors should be considered before any final decisions are made relative to the form of polymer and type of make-down unit to be used.

At a minimum, every system should include the make-down device and a separate agitated storage (day) tank for feeding the diluted polymer solution to the dewatering operation. The day tank should be sized to hold the equivalent volume of polymer to satisfy the maximum demand for at least 1 hour, thereby providing effective aging time with gentle mixing of 30 minutes minimum. Tanks should be fitted with clear and calibrated sight tubes for monitoring polymer solution draw-down.

Low-head centrifugal pumps are acceptable for one-time transfer of the polymer solution from the make-down unit to the day tank. Rotary gear pumps and progressive cavity pumps are better for metering the dilute polymer solution to the dewatering process. Inline flow meters are important for each polymer solution feed pump, and each metering pump should be outfitted with an appropriately sized calibration cylinder for confirmation of the feed rate. Sufficient water pressure (minimum of 280 kPa [40 psig]) and flow are required to overcome system head pressure and to operate an atmospheric eductor for aspirating dry polymer.

Dry Polyacrylamide. For dry polymers, care must be taken to ensure that the neat polymer remains dry and free from humidity. The polymer make-down system or dispersing unit must meter the dry polymer at a controlled rate and effectively wet the individual dry polymer particles, to produce a uniform dilute polymer solution without polymer agglomeration, gels, or fish eyes.

Generally, cationic dry polymers are made-down to a concentration ranging from 0.1 to 0.5%, with a target of 0.2% of polymer as supplied by the vendor. The dilute solution should be mixed with a low-speed mixer (400 to 500 r/min) for at least 30 minutes for proper aging and uncoiling of the polymer structure before feeding to the dewatering process. Higher concentrations of polymer and colder makeup water will probably need longer mixing time for proper aging of the polymer.

Emulsion Polyacrylamide. Recall that emulsion polymers may separate or layer over time. Products in storage or containers should be checked for uniformity and mixed, if required, before make-down.

Emulsion polymers must be activated before they can be effectively used. The first step is called *inversion* and occurs relatively quickly. During inversion, the make-down system imparts significant energy, in the form of turbulence or shear, into the emulsion and water mixture to invert the polymer from the oil droplet phase into the aqueous phase. The second step, termed *aging*, occurs over a period of up to 30 minutes, during which quiescent mixing allows the polymer structure to thoroughly uncoil to become completely effective.

There are some emulsion products that may be fully effective without aging, and some equipment manufacturers claim that their make-down units are so effective that aging is not required. Nonetheless, adequate aging capability is strongly recommended in every system to provide the flexibility to use many products.

Cationic emulsion polymers are typically made-down to a concentration of 0.5 to 2.0% of polymer as supplied by the vendor, which typically amounts to approximately 0.2% on an active basis. As with the dry products, the storage (day) tank should be sized for dilute polymer solution to hold the polymer solution for a minimum of 1 hour before refilling.

Polymer Degradation. Intense mixing, pumping, and shearing of the dilute polymer solution will reduce the molecular weight and increase the dosage. Temperatures higher than 50 °C (120 °F) are also to be avoided. The quality of water used for polymer make-down is important. Suspended and dissolved solids found in nonpotable water result in increased polymer demand. It is also important to check the pH of the water to ensure that it is close to 7.0.

Cationic flocculants perform effectively over the typical range of pH (6.5 to 7.5) found in municipal sludge dewatering; however, as pH increases above this level, polymer effectiveness generally decreases. In the case of Mannich SPAM, a pH of 7.5 and higher may result in decomposition of the polymer itself, thereby releasing odor-causing amines. Other substances that may interfere with the performance or cause higher-than-normal doses of cationic PAMs include sulfides, hydrosulfides, phenolics, and chlorides. For a more detailed discussion of make-down systems, including materials of constructions, schematics, and pump-and-valve specifications, the reader may consult *Design of Municipal Wastewater Treatment Plants* (WEF, 1998), the polymer suppliers, and the respective manufacturers of specific make-down systems.

SOLIDS CONDITIONING. It cannot be emphasized enough that polymer conditioning evaluations should be conducted on the actual sludge or sludges that are found in the full-scale operations. A BFP or centrifuge will not perform effectively unless the biological sludge to be dewatered has been effectively conditioned with organic flocculants.

The polymer must contact each sludge particle to capture those particles in a floc. Interfacing of polymer with sludge particles is affected by three mixing factors—the sequence of chemical addition, intensity of mixing, and duration of mixing. The major objective is to get a viscous polymer solution thoroughly dispersed with what may also be a viscous sludge to affect polymer interaction with sludge particles.

This sludge-conditioning process is a two-stage activity. The first, charge neutralization or stabilization, occurs very quickly and eliminates the negative electrostatic charge on the sludge particle. The second stage, flocculation or bridging, occurs within a few seconds thereafter and creates a sludge floc or agglomeration of sludge particles, with many individual sludge particles held together by coiled strands of organic polymer.

Consequently, short-duration, high-intensity mixing is required for dispersing the polymer solution effectively into the sludge, with longer-duration (few seconds), low-shear mixing to flocculate the particles into larger agglomerates. Conversely, as soon as flocs of polymer–sludge particles are formed, excessive shear or mixing should be avoided.

There are a number of mechanisms that assist this sludge–polymer interfacing. The first is simply to add dilution water to the polymer solution to reduce its viscosity and provide more efficient mixing and contact of the polymer with the sludge. Dilution of made-down polymer solution is not an alternative to the proper design of appropriately sized polymer tanks and feed pumps. Rota meters are generally suitable for controlling the flow of dilution water. Dilution water pressure and volume must remain constant to prevent the polymer dosage from fluctuating and to overcome pressure drops in the system. There should be adequate inline mixing of the polymer solution and dilution water before the diluted polymer solution contacts the sludge stream.

One of the more common recommendations that dewatering equipment manufacturers make is to have three or more alternative feed points for adding the polymer to the sludge stream. In the case of centrifuges, the last feed point is typically at the sludge injection tube already inside of the feed end of the centrifuge. Other feed points may be as far back as 9.1 to 15 m (30 to 50 ft) in the sludge line, thereby providing for more mixing at elbows and additional time for sludge–polymer reaction.

Belt filter presses use a mechanical device to accomplish the same effect. They recommend a polymer injection ring, followed by a manually adjusted butterfly mix valve. This is a convenient, trouble-free system that is frequently found in sludge feed systems to BFPs, but, on centrifuges, it is rarely used. Why this is the case is the cause of speculation, but it is certainly not because it has not been tried.

Static mixers are commonly used to mix the polymer with sludge. Dynamic, motorized mixers that provide variable mixing energy sludge–polymer streams are also commercially available; however, their limited use suggests that they may not be cost-effective.

PERFORMANCE OPTIMIZATION (DEWATERING). The economics of dewatering performance is the bottom line. In sludge dewatering, there are typically four economic factors that comprise overall dewatering performance. These are

- Polymer cost (polymer dose times unit price),
- Cost of sludge disposal or downstream processing (dependent on cake solids percent),
- Cost of recycle (function of capture percent), and
- Throughput (feed rate).

It is very important that operating personnel have an understanding of these cost factors and have the ability to calculate these costs with real-time operating data. If operations personnel do not have real-time analytical data for feed solids (percent), cake solids (percent), capture (percent), and polymer dosage, they cannot maintain optimum performance. Laboratory results for cake solids, sludge feed solids concentration, and suspended solids in recycle, which may be available 24 hours later, do not provide the information necessary to make immediate operational adjustments.

Similarly, it is important that sample points for collecting feed sludge, recycle (filtrate/centrate), and dewatered sludge cake are convenient for operators. It should not be necessary to walk down two flights of stairs into the basement to collect a sample of sludge feed or to collect dewatered cake on a different level than recycle (centrate). Ideally, the operator can see the centrate/filtrate running into a drain while walking along a well-traveled route.

Cost of Polymer. Simply stated, polymer cost is polymer dose in weight as supplied by the vendor per weight of solid processed times the unit price of the polymer (price per weight delivered). Polymer dose is typically determined from formal polymer evaluations under standardized conditions. This is discussed in further detail in the section entitled "Outline for Conducting Effective Polymer Evaluations and Product Selection."

Cost of Solids Disposal or Downstream Processing. The cost of sludge disposal or downstream processing is a function of the cake solids (percent) achieved in dewatering times the unit cost of disposal, such as incineration or transportation and landfilling. In the case of further solids processing for reuse, such as composting or lime stabilization, it would include these processing costs plus the cost for transportation and land application. More broadly, the basic disposal cost is the sum of the polymer cost plus the disposal cost. If the plant SCADA system does not calculate and display these costs, then the operators have no way of determining if they are operating at the minimum cost per Mg (ton).

Cost of Recycle (Capture). There is a quantifiable cost of solids recycled to the front end of the treatment operation. Unfortunately, most facilities do not identify this cost. A simple gross method would be to assign to recycled solids the same cost assessed to industrial sewer customers who have excess suspended solids. Alternately, the plant budget could be divided by the amount of suspended solids and biochemical oxygen demand (BOD) processed. The result is a fair representation of the cost to treat the recycle solids. Most plants treat recycle simply on a pass or fail basis and arbitrarily assign a capture requirement of ±95%. They use this assigned value as an operating standard and make operational adjustments accordingly.

Throughput (Feed Rate). Throughput or feed rate is often overlooked as an operational parameter related to overall economics in running the dewatering unit. Most of the reference books treat this parameter as an equipment design factor. Dewatering equipment is specified with sludge throughput parameters based on hydraulic loading (volume) or on solids loading mass of dry solids per hour or per meter per hour for BFPs. For effective process control, operators should know the throughput on a mass basis, generally in dry tons of solids per hour. Accordingly, operators must have access to accurate real-time data for hydraulic loading (flow) and solids concentration (%) in the feed sludge. Throughput varies significantly with the type of sludge being dewatered. Specific data related to throughput rates for various sludges are referenced in standard wastewater engineering manuals and WEF publications. In general, primary sludges yield the highest throughput, with digested WAS yielding the lowest. Blends of raw or digested primary sludge with WAS or digested WAS generally yield maximum throughputs between the two extremes. Unfortunately, the equipment capability is limited by the following:

- The intrinsic design of the equipment,
- The way the equipment is set up and operated,
- The type and quality of the sludge, and
- The strength of the reaction between the polymer and the sludge.

From an operational perspective, it is intuitive that the best economics are achieved when the equipment is used to its maximum throughput, while still achieving the desired performance for cake solids and solids capture. Similarly, it is intuitive that this approach would provide for optimum economics relative to power consumption, labor, and maintenance costs. Unfortunately, intuition is not a good substitute for carefully obtained process data, evaluated with good cost numbers.

Nevertheless, many plants established their dewatering throughput on the basis of the quantity of sludge required for downstream processing rates or to satisfy hauling conditions at some time in the distant past. Another factor often used to determine the throughput rate is the hours necessary to fulfill the personnel schedule. On occasion, multiple units are all being operated, at less than maximum throughput, even when the total volume could be readily handled by fewer machines running at a higher throughput, thereby saving power and equipment wear.

There are a number of factors to be considered when determining the optimum throughput at a particular facility. Operations personnel should spend some serious time and effort determining the optimum performance. Once the optimum performance is established, the operators need training so that they know how to keep operations at the optimum throughput rate and achieve desired performance, while minimizing the costs associated with electricity, equipment wear, labor, and maintenance.

AUTOMATION. There has been much technological development on a variety of automated process sensors, controllers, and related software for managing dewatering operations for optimum performance and polymer usage. These systems are coming into their own in sludge dewatering. The best success occurs when the instrument, installation parameters, and application are considered together. Automation is not a sensor in a box. Operator attention will always be necessary for routine maintenance, such as probe cleaning and recalibration, but, on the whole, not a lot of time is required. In exchange, automation promises to smooth out the variation in unit operations throughout the plant.

The WERF and STOWA, a European wastewater association, are contributing to the research effort, and wastewater plants themselves, often in conjunction with automation engineering consultants, are applying an increasing number of instruments in the field. There are several publications that may provide more detailed information about the state of automation for dewatering operations (e.g., Gillette and Scott, 2001; Pramanik et al., 2002; WERF, 1995, 2001).

POLYMER SELECTION. Selecting the most cost-effective polymer for use in dewatering wastewater residuals is one of the most demanding, frustrating, and time-consuming experiences for many operations personnel and process engineers in the field of municipal wastewater treatment. It seems that every experienced operator and engineer has an individual horror story related to polymer evaluations and they often extend bad contracts rather than face the polymer trials. This section will provide some suggestions for conducting more effective evaluations and eliminating some of the frustrating experiences.

Traditionally, municipalities have used a two-step approach, incorporating product evaluations and competitive bidding to the product selection process. This selection process is intended to differentiate the cost performance of each product under standardized conditions. Unfortunately, the conclusions of these evaluations often result in inappropriate product selection or unrealistic performance expectations. As much as 30% (representing $56 million per year) of the polymers used in municipal sludge conditioning applications results from overuse or improper product selection.

To make matters worse, it is not uncommon for traditional product evaluations to be conducted over an extended period of time, often 4 to 6 weeks or longer. The internal cost for all of the in-house personnel involved in a complete series of evaluations is very significant—estimated to be $6,000 to 10,000. The argument that the plant gets free polymer for the trials is misguided; the cost is folded into the price for the selected polymer.

These evaluations typically contain extensive deficiencies and, more often than not, conclude with the selection of a particular product, influenced primarily by ancillary factors rather than the product's actual performance. These factors are

- Trial conditions and
- Trial management.

The purpose of the two-step selection process is to differentiate the cost performance of each product under standardized conditions. The emphasis is that standardized conditions are absolutely critical to effective product evaluations. Standardized trial conditions mean that the sludge consistency must be maintained throughout the course of the entire product evaluation period. As such, it is virtually impossible to conduct effective trials over a 4- to 6-week period or longer. Formal product evaluations conducted under the old notion that "the sludge is the sludge" are totally inappropriate for effective trials. The sludge processed on Monday, February 17, will most likely vary significantly from that processed on Friday, March 7—a short 3 weeks apart. Nevertheless, it is quite common for formal product evaluations to be conducted by municipalities over extended periods of time, with one product being tested during 1 week and another product tested some 3 weeks or even 5 weeks later. Some municipalities test the polymers on the sludge at one plant, and use the polymer selected at several plants. It is difficult to justify such practices.

Trial conditions, including sludge consistency and equipment modifications, are manageable by the operations personnel. While there are many additional details associated with standardized trial conditions, all are quite manageable, and the most significant factor relates to shortened trial duration. Competitive products can be evaluated in less than 8 hours; five products can be effectively tested in less than 1 week.

With respect to trial management, there are a lot of factors that influence performance results other than the actual product itself. These ancillary factors must be eliminated. Traditionally, municipalities involve vendors, who have a vested interest in the results of the performance evaluations, to participate in the formal evaluations of their respective products. At a minimum, this results in trial techniques that differentiate individual companies or support personnel, rather than the products themselves. And, in worst cases, participation of vendors may result in distorted samples or analyses and polymer concentration and dosage.

Another trial management issue is the trial design or protocol. In particular, performance based on averages over time are most subject to influence by trial personnel. Averages of results do not demonstrate optimum or maximum product performance, but are simply a reflection of the vendor personnel or municipal staff making adjustments and managing the trial. An even worse situation involves product dosages, which are determined by totalizer readings over a specific increment of time. These types of trial design are particularly subject to manipulation by trial participants, thereby resulting in distorted performance results, which are not reproducible under normal operating conditions, and frequently improper product selection.

The purpose of the evaluation is not to identify the company or individual that can conduct or manage the best trial, but solely to differentiate product performance. Vendor participation in formal product evaluations represents a natural conflict of interest. The vendor wants to achieve the best performance to differentiate his or her product, while the municipality wants standardized conditions to allow each product to demonstrate its own respective performance. The WERF *Guidance Manual for Polymer Selection in Wastewater Treatment Plants* (WERF, 1993) acknowledges this natural conflict of objectives when it states that the "primary role of a sales representative is advising customers with the obvious goal of increasing product sales."

Outline for Conducting Effective Polymer Evaluations and Product Selection. There are many details that must be considered to conduct effective polymer evaluations for municipal sludge dewatering. A few principal suggestions to lead one to more efficient and effective polymer evaluations are as follows:

1. Identify proper performance criteria.

    Define economic values.

2. Minimize the time period to complete all product evaluations.

    Allow sufficient time for vendors to conduct informal laboratory testing and possibly informal full-scale testing to recommend their best product.

    Reduce the entire formal, full-scale trial period from 4 to 6 weeks to 1 week.

3. Standardize the sludge conditions.

    Do not operate under the premise that "sludge is sludge." The sludge on Monday, February 17, will not be comparable to that on Friday, March 7.

    Ensure that all variations affecting sludge characteristics are thoroughly understood and controlled throughout the evaluation period.

4. Eliminate the participation of all unnecessary personnel from the formal trial activity.

    Conduct all formal product evaluations solely with in-house staff, supplemented with independent consultants, as necessary.

    Minimize the number of staff personnel involved in the product trials and provide thorough training and written procedures—especially for the polymer make-up calibrations—to ensure consistency.

    Prevent vendor personnel from coming on-site during the formal product trials.

5. Standardize sampling and analytical activities and generate real-time performance data.

    Real-time analysis and data for feed solids, cake solids, and capture is imperative to making appropriate operational adjustments to identify the product's optimum performance and operating range. If the plant does not have a microwave oven, buy or borrow one.

6. Analyze performance data in a manner that assesses actual product performance rather than trial conditions.

    Determine optimum performance based on a dose–response curve, such as cake dryness versus polymer dose.

    Do not average individual data points to determine performance.

    Do not determine polymer dosage based on total polymer usage over the trial period for the respective product. Do not use totalizer readings.

By following these suggestions for more effective polymer trials, operations personnel and process engineers will be better equipped to select the most cost-effective polymer based on overall product performance and related economics, established with criteria determined by the municipality. Also, the product performance will be reproducible under normal day-to-day operating conditions.

CASE HISTORIES. Following are several case histories that demonstrate the types of conditions that often lead to ineffective product evaluations or, even worse, to improper product selection.

Trial Conditions (Sludge and Equipment Management). Some dewatering facilities receive large bulk (barge) transfers of sludge from other wastewater facilities. At other plants, sludge transfers may be made on a periodic basis from digesters or secondary treatment trains to a wet well or sludge holding basin before dewatering. These types of transfers should be monitored and controlled to maintain the consistency of sludge during formal polymer evaluations. More frequent transfers of smaller volumes are less likely to skew the results.

Barge Transfers of Sludge. A barge transfer of wastewater sludge containing a significant quantity of water treatment residuals, such as alum or ferric chloride sludge, produces a significant influence on the charge demand and polymer dosage, relative to that determined from 100% organic or WAS. This is the type of condition that must be managed during the course of a formal polymer evaluation.

Managing In-Plant Sludge Transfers. The periodic transfer of WAS from individual treatment trains to a sludge wet well is frequently conducted in massive slugs to accommodate pumping schedules and personnel convenience. At a particular wastewater treatment facility, the wet-well capacity is 379 000 L (100 000 gal) and the transfer pump was routinely run for 1 hour at the beginning of each shift to transfer 265 000 L (70 000 gal) of sludge from a single treatment train. This slug transfer of sludge repeated itself on each successive shift for each of three treatment trains, resulting in significant differences in sludge consistency each time a new transfer was made. A more effective sludge management protocol during formal product trials was to operate the transfer pump for 20 minutes on each of three treatment trains during each shift, while maintaining the sludge level in the wet well at a high level. This approach transferred the same volume and ensured a uniform mixture and consistency of sludge in the wet well, which was being fed to the dewatering unit.

Managing Sludge Transfers from Digesters. Transfers of sludge from digesters to the wet well represent significant potential for sludge inconsistencies. At one facility, these transfers were made blindly on the basis of volume, without regard for the sludge consistency, solids content, and volatiles. In a particular extreme situation, the transfer of sludge from a digester resulted in a reduction in feed solids to the dewatering unit, from an average of 2.5% to less than 0.5%.

Upon further investigation, sludge mixing in the digester was inadequate, and sample ports and decant lines along the wall of the digester were inoperable. This type of situation is not nearly as uncommon as might be imagined, and it represents an intolerable condition for effective product evaluations. Furthermore, this type of condition also represents extremely inefficient day-to-day operating results and should be promptly corrected.

Equipment and Design Limitations. At one facility, the agitated day tank of the make-down system was adequately sized. However, the upper and lower level controls were set such that only one-third of the tank was used, causing polymer to be batched more frequently, while providing for reduced aging time. This was not a limitation with the house product; however, it did provide limitations for alternative products, which required lower make-down concentrations. In addition, at the same facility, the capacity of the polymer feed pumps was limited such that they were not able to handle lower polymer concentrations. Thus, the system design and equipment limitations severely limited the ability of the facility to effectively evaluate or handle alternative products.

Equipment Limitations. During the competitive bidding and product selection process, polymer vendors frequently suggest that a facility evaluate different product forms than the facility is designed to accommodate. For convenience, polymer vendors may offer to provide their own equipment to facilitate the use of such products during formal trials. Unfortunately, this approach adds another dimension and lack of control over standardized procedures for the product evaluation activity. Such evaluations are valuable, but formal polymer trials are not the time to evaluate alternative product forms that the facility is not functionally capable of handling. Most facilities have an open-door policy and are happy to accommodate vendors' suggestions and informal product tests, but such testing needs to be done at a time when the plant operators can focus on the hardware testing.

Trial Management. Trial design or protocol frequently includes the use of averages or volumes from totalizer readings to determine dosage or performance. These types of designs are most subject to manipulation by trial participants. Trials evaluated in this fashion do not effectively determine the performance of the specific product, but instead represent the results of the personnel most adept at managing the trial.

A dose–response curve is more effective for analyzing performance than is the average of data points. A dose–response curve provides an indicator of the optimum product performance and is much more accurate.

Real-Time Analyses and Data. To effectively determine the optimum product performance, the operator must make adjustments based on real-time performance data. Laboratory data, which may be provided 24 hours later, is useless for determining whether the polymer rate should be increased or decreased. Virtually all vendors provide their own solids analyzers, if none are available at the facility, when conducting formal product evaluations.

Manipulation of Trial Design. Following is an example of a trial that was based on the averages of data points, with the polymer dosage determined from totalizer readings. As is typical, vendors participated in the evaluations of their respective products and were permitted to manage the polymer feed adjustments throughout the trials. This particular trial demonstrates the manipulation of trial design, which resulted in the selection of this product for a long-term award. Unfortunately, the product was never able to reproduce comparable trial performance, which resulted in extensive cost overruns, relative to the costs predicted from the trial throughout the term of the contract.

As is readily evident from the table of data shown in Table 33.3, the specific data points collected at the 10:45 sampling (highlighted) show the lowest feed solids and an extremely low polymer dose—less than 50% of the average of all test dosages. Yet, this same reading shows a polymer feed pump setting at the same level as all of the other

TABLE 33.3   Data from polymer evaluation demonstrating manipulation of trial design.

Time	Feed solids (%)	Sludge flow (gpm[a])	Poly flow (gpm)	Poly pump (%)	Poly dose (lb/dry ton[b])	Cake solids (%)	Capture (%)
8:30	1.95	392	13	50	20.61	28.77	94.72
9:15	1.87	484	10[c]	50	13.16[c]	26.44[c]	94.01
10:00	1.88	366	17	50	29.46	27.93	94.02
10:45	1.85	364	6[c]	50	9.85[c]	28.79[c]	93.39
11:30	1.87	562	27	50	31.06	26.71	94.01
12:15	1.93	615	18	50	17.63	27.83	94.67
Average	1.89	464	15	50	20.29[c]	27.75	94.14
Average dosage without the 10:45 run					24.69		
					22%[c]		

[a]gpm × 5.451 = m³/d.
[b]ton × 907.2 = kg.
[c]Readings based on polymer totalizer.

readings. Obviously, the polymer feed pump settings could not remain the same throughout the trial, while yielding such wide extremes in polymer flows and dosages, as shown in the data. Ironically, the cake solids corresponding to this lowest polymer dose were shown to be the highest of the entire evaluation, which is extremely inconsistent with typical polymer performance.

It should also be noted that the dewatering unit would not even operate at the two lowest polymer dosages (5 and 6.6 g/kg [9.9 and 13.2 lb/dry ton]), as shown in this trial. This is an example of manipulation of the trial design based on averages and the use of totalizer readings to determine polymer dosage.

Figure 33.4 represents a dose–response curve of the same set of trial data from the preceding table. This dose–response analysis demonstrates the inconsistency of polymer performance in this trial. This is a scatter chart of data points connected by a smooth, solid, thin line. Note that the specific data points indicate a very erratic dewatering performance, as indicated by cake solids, for this particular polymer. The hatched line represents a logarithmic regression analysis (trend line) of the same data points. It is important to note that this trend line is inverted, suggesting that higher dose will yield

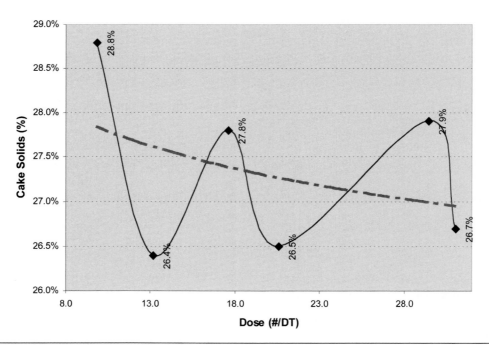

FIGURE 33.4 Dose–response curve of manipulated trial data.

lower cake solids. This trend line is contrary to the normal bell-shaped curve depicting polymer performance, whereby higher doses generally yield increasing cake solids. This depiction of the trial data demonstrates that the results represented by the trial data are reflective of something other than the product performance itself.

Figure 33.5 shows two graphs that represent the dose–response relationship of typical polymer performance. As in Figure 33.4, each of these figures represents a

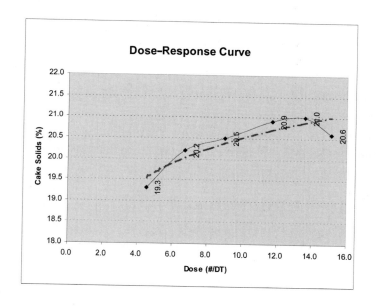

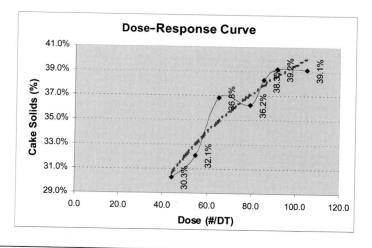

FIGURE 33.5   Typical dose–response curves.

scatter chart of data points connected by a solid, thin line. Note that these data points indicate an increasing performance (cake solids) with higher dose. Similarly, the hatched line represents a logarithmic regression analysis (trend line) of the same data points and shows the traditional bell-shaped appearance. Similarly, the hatched line in each graph represents a logarithmic regression analysis (trend line) of the same data points. The trend lines in each of these graphs shows the typical bell-shaped curve representative of increasing polymer performance (as indicated by cake solids) corresponding to increases in polymer dose up to a maximum level. After that maximum level, performance generally levels off and sometimes declines as polymer dose continues to increase.

Trial Management (Sludge Manipulation). One particular facility was designed to feed sludge directly from digesters or a lagoon via a manually controlled valve at a splitter box. The lagoon served as a buffer to absorb fluctuations in sludge generation and sludge dewatering. During formal product evaluations, the trials were conducted with the mindset that "sludge is sludge." All of the products were to be evaluated on sludge from the lagoon, as there was a significant excess of sludge in storage. None of the prospective vendors were aware of this protocol and conducted their evaluations on the sludge as provided by the facility. Only the incumbent supplier was aware of the sludge control valve and was able to use freshly digested sludge during the evaluation of his product.

The results of the final performance evaluation of all products demonstrated this difference, showing the polymer dosages of the new prospective suppliers to be almost double that of the incumbent product when all dosages were compared on an active polymer basis. Upon further investigation, the polymer dose of the incumbent product used to process lagoon sludge before and after the formal trial was found to be comparable with that of the prospective new suppliers as determined during the trial.

Trial Management (Sampling and Analysis). On a thickening trial at a major facility, sludge from a gravity thickener was sampled on an hourly basis to determine polymer performance over an extended 24-hour period. These samples were collected by laboratory personnel, in conjunction with the vendor's staff of the respective product being evaluated. Separate from the trial activities, operators also pulled sludge samples on each shift for routine operating records, while correspondingly recording sludge blanket depths and comments related to sludge on the continuous circular charts for the underflow sludge pumps.

The overall performance results showed one particular product to have exceedingly high solids in the thickened sludge. In fact, the solids were so high as to raise

speculation. Upon further investigation, it was found that several of the sludge samples from the trial, which were collected during the night for this particular product,
contained exceptionally high solids. The solids content reported for these samples far
exceeded the maximum design capability of the underflow pumps. Ironically, corresponding to the time of these samples, the operator logs and pump charts indicated
that the sludge blanket depths were so low that there were virtually no solids in the
basin, and the underflow pumps had, in fact, been turned off.

Trial Management (Sampling). At a trial at another major facility, centrate/filtrate recycle water samples were routinely collected by operators from the dewatering device.
The operators instinctively knew that it was not desirable to have excessive solids in
the recycled stream. Accordingly, one particular operator would collect his sample of
recycle, pause for a moment to allow the solids to float to the surface, then tip the sample container to pour off the solids on the surface before capping the sample container
and sending it to the laboratory for analysis.

CONCLUSION. These case histories and suggestions in the previous section demonstrate the capability to conduct effective polymer evaluations and make better product selections. By following these recommendations, operations personnel and process
engineers are better equipped to select the most cost-effective polymer based on overall product performance and related economics established with criteria determined
by the municipality. Also, the product performance will be reproducible under normal
day-to-day operating conditions.

The opportunities are readily available to each individual facility to improve dewatering performance and reduce polymer costs.

AIR-DRYING BIOSOLIDS

Air-drying refers to dewatering methods that remove moisture by natural expiration,
evaporation, or induced drainage. Biosolids to be air-dried are first stabilized by aerobic or anaerobic digestion. Air-drying processes are less complex, easier to operate, and
may be more energy-efficient than mechanical systems. On the other hand, air-drying
systems require large amounts of land, and some require much more labor for cake
removal. Furthermore, winter weather and rainfall heavily influence the efficiency of
air-drying systems. Air-drying is considered for rural wastewater treatment plants
(WWTPs) with design flows less than 7500 m^3/d (2 mgd), which are located in warm,
dry-weather areas. In cold or wet climates, air-drying systems are less efficient and
should be covered or the plant will need to have an alternate disposal method avail-

able. Air-drying systems are intrusive to the public eye and the public nose. When they cause odors, there is no doubt where they are coming from. The following are characteristics of air-drying systems:

- The simplest beds are paved with asphalt or concrete;
- The sand bed consists of an impermeable liner and underdrains embedded in gravel, with graded sand above;
- Some beds are filled with reeds, which remove water by expiration;
- Most systems flocculate the biosolids with polymer to speed water break out (reed beds are an exception);
- Some have a provision for decanting the surface water;
- Some systems use wedge wires or plastic blocks in place of sand and gravel;
- Some use a vacuum to speed the separation;
- Some use a plastic or glass greenhouse to reduce odors and insulate from the weather; and
- Some have heating pipes to speed evaporation.

Attempts to get around the inefficiencies of air-drying are successful, but add considerable cost. Glass houses keep off the rain, but they also reduce the natural breeze so that the bed is more efficient in rainy weather but less efficient at other times. The underdrains allow some water to drain from the sludge, increasing efficiency, but they are subject to damage by heavy equipment and make solids removal more costly. Adding a grid strong enough to support light equipment helps with sludge removal, but the grid is a considerable expense and needs to be cleaned each time the bed is emptied. Polymer results in a free-draining sludge, which increases the loading, but polymer adds to the cost and complexity.

Air-drying systems are either reed beds or evaporative/drainage drying beds. All such beds must have a membrane lining to prevent contamination of groundwater; drainage into groundwater is not allowed.

Some also have an underdrain system and a sand layer above to allow removal of the sludge without damage to the membrane liner. The liquid that reaches the underdrain contains biosolids, so it needs to be recycled to the plant influent. The beds have to operate with some freeboard between the surface and the containment wall so that they will not overflow in wet weather (Figure 33.6).

PRINCIPLES OF OPERATION. This section describes the basic principles of operating the typical dewatering processes—reed beds, sand beds, vacuum-assisted beds, wedge-wire beds, and paved beds. All such systems will be out-of-service for substantial periods of time for cleaning or other maintenance, so several small systems

FIGURE 33.6   Reed harvesting in the spring.

are better than one large one. Depending on the local weather conditions, the capacity of the beds can be very low at times. Either substantial biosolids storage facilities or alternate methods of biosolids disposal must be available.

Reed Beds.   Reed drying beds attempt to replicate natural systems, in which reeds stabilize biosolids, and also provide an environment for worms and bacteria to further break down the biosolids (Garvey, 2002). The reeds absorb water through their roots and release it to the air through expiration. To the casual observer, reed beds look like a marsh and thus a natural area. Reed beds are lined with a flexible membrane, followed by an underdrain system set in sand and gravel. The reeds used are typically phragmites comminus reeds, which can be purchased commercially or are available for the digging at other WWTPs. Figure 33.6 shows the reeds growing in biosolids. Biosolids stabilized by aerobic or anaerobic digestion are periodically pumped into the bed, approximately 0.07-m (3-in.) deep. Some water drains through the sand into the underdrains, some water evaporates, and some is taken up by the reeds. Reeds grow best if the roots are wet. A good underdrain design allows water to pond in the bottom of the bed where the roots can reach it rather than draining or pumping the underdrains dry. The presence of other plants in the red bed suggests that the reeds are stressed and not getting a continuous supply of water. During a dry period, it may be necessary to pump plant effluent to water the beds. The transpiration of water by the reeds is one of the many advantages of this process; another is that the biosolids stay in the beds for 6 to 10 years, resulting in less volume to discard. Water that reaches the underdrains is returned to the plant as influent, but the drainage from reed beds is of better quality than the drainage from the other systems and will therefore have less effect on the plant. Biosolids are fed batchwise to the beds to a depth of 7 to 15 cm (3 to 5 in.) every 1 to 3 weeks. The reeds are harvested at the end of the growing season, as seen in Figure 33.6. They can be raked away from the edges and burned, in places where such practices are allowed, or removed, chopped, and composted or landfilled. Unfortunately, reporting requirements for reuse makes recycle of the harvested reeds an unattractive option. After several years, when the beds are 1-m (40-in.) deep, the top 78 to 80% is removed, and the process is begun again. The harvested material is typically land-applied. The reeds are not competitive with other plants outside of a wet marshy area, so reed growth in agriculture reuse has not been a problem. Some operators screen the reeds out, which results in a higher-grade product. Reed bed construction is very nearly a do-it-yourself dewatering operation. Reed beds typically require less capital, less land, and have much lower operating costs than the alternatives discussed below.

Sand Beds. Sand beds have been used since wastewater treatment became a recognized technology in the early 1900s. Dewatering on the sand bed occurs through decanting, gravity drainage of free water, and by evaporation until the desired solids concentration is reached. Figure 33.7 illustrates a typical sand bed. Generally, only stabilized biosolids are considered for air-drying. Sand bed construction is very similar to reed bed construction, except that the biosolids layer is generally no deeper than 150 to 300 mm (6 to 12 in.) and the bed is designed so that heavy equipment can operate on it.

Dewatering on sand beds is achieved by drainage by evaporation and by decanting. Initially, water drains from the biosolids through the sand and gravel to the underdrains. Biosolids from modern secondary plants do not drain very well without polymer, so beds that are intended to dewater by drainage need to use generous amounts of polymer. The drainage step requires several days, ending when the sand clogs with fine particles or all the free water is removed through the underdrains. If a supernatant layer forms, either from free water rising to the surface or from rain, it is typically decanted and returned to the plant influent. Decanting will likely be neces-

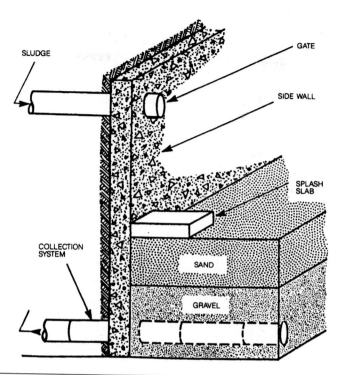

FIGURE 33.7   Sand bed details.

sary to remove the free water released by polymers added to the biosolids. Evaporation removes the water remaining after initial drainage and decanting. It is generally necessary to use tractor-mounted mixing devices to provide regular mixing and aeration. The mixing and aeration breaks up the surface crust that inhibits evaporation, allowing more rapid dewatering. When ready for harvest, front-end loaders go onto the beds and scoop up the dried biosolids, typically removing a small amount of sand. With the removal of the biosolids, more sand is spread, and the process repeats.

Vacuum-Assisted Drainage Beds.   An Internet search for vacuum-assisted drainage beds resulted in very few hits, suggesting that this design is mostly of historical interest. The high capital costs to build these beds preclude them from standing for weeks, allowing the biosolids to air-dry. They resemble air-drying beds, but actually operate by drainage. With this technology, heavily polymer-treated biosolids are flooded onto a surface of rigid, porous media plates to a depth of 900 mm (36 in.). A weak vacuum draws free water through the plates, until one section opens and allows air to break the vacuum. Figure 33.8 shows a plan view of a typical single-bed system. Quite a lot of

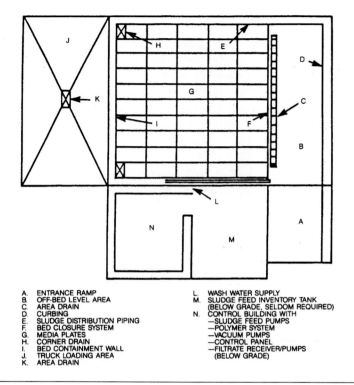

A. ENTRANCE RAMP
B. OFF-BED LEVEL AREA
C. AREA DRAIN
D. CURBING
E. SLUDGE DISTRIBUTION PIPING
F. BED CLOSURE SYSTEM
G. MEDIA PLATES
H. CORNER DRAIN
I. BED CONTAINMENT WALL
J. TRUCK LOADING AREA
K. AREA DRAIN

L. WASH WATER SUPPLY
M. SLUDGE FEED INVENTORY TANK
     (BELOW GRADE, SELDOM REQUIRED)
N. CONTROL BUILDING WITH
     —SLUDGE FEED PUMPS
     —POLYMER SYSTEM
     —VACUUM PUMPS
     —CONTROL PANEL
     —FILTRATE RECEIVER/PUMPS
      (BELOW GRADE)

FIGURE 33.8   Plant view of a vacuum-assisted drying bed system.

solids are recycled back to the plant. The filtration media requires cleaning, and the underdrains must be hosed down after each cycle.

Wedge-Wire Beds. The wedge-wire process is physically similar to the vacuum-assisted drying beds. The media consists of a filtration surface, called a *septum*, with wedge-shaped slots approximately 2.25-mm (0.01-in.) wide in place of the porous plates in vacuum-assisted beds. Alternate designs use porous plastic blocks. The septum supports the polymer flocculated cake and allows drainage through the slots (Figure 33.9). A controlled drainage process exerts a small hydrostatic suction on the bed to remove water from the biosolids. The synthetic media (plastic and wedge-wire) filtration drying beds require less land than the simpler air-drying systems, but they require more capital to build and maintain, more labor to operate, and about as much polymer as a BFP or centrifuge would require to dewater the biosolids to a truckable state.

Paved Beds. Paved beds are the simplest drying beds. They consist of a membrane liner, topped with concrete or asphalt, surrounded by a containment wall. Sludge is

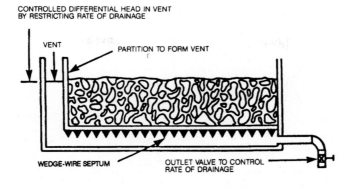

CONTROLLED DIFFERENTIAL HEAD IN VENT
BY RESTRICTING RATE OF DRAINAGE

VENT

PARTITION TO FORM VENT

WEDGE-WIRE SEPTUM

OUTLET VALVE TO CONTROL
RATE OF DRAINAGE

FIGURE 33.9   Cross-section of a wedge-wire drying bed.

added 150- to 300-mm (6- to 12-in.) deep. Occasionally, they benefit from polymer, but typically not. As the sludge crusts over, a tractor with a harrow breaks the crust, exposing wet sludge to the air. Operators who do this keep the mud flaps on the tractor in good repair. This is done several times, until the biosolids area is satisfactorily dry. Front-end loaders remove the cake and load it into trucks. Paved beds are the least efficient of the air-drying systems and require large amounts of land. They work well in desert conditions, and not so well elsewhere.

In an effort to improve efficiency and obtain some benefit from drainage, unpaved areas, constructed as sand drains, are placed around the perimeter or along the center of the bed to collect and convey drainage water under the bed, where it can be pumped back to the plant. The main advantage of paved beds is that heavy equipment can be used for dewatered cake removal, and the difficulties with sand and the underdrains is eliminated. On the negative side, pavement blocks drainage, requiring an increase in the total bed area to achieve the same results, and polymers may be needed to get most sludges to drain well. Figure 33.10 shows paved drying beds.

PROCESS VARIABLES. Similar process variables apply to all of the air-drying methods of dewatering. The performance of drying beds is affected by weather conditions, feed solids characteristics, system design, chemical conditioning, and the length of time before cake removal.

High temperatures, minimal rain, low humidity, and high wind velocities obviously improve drying. Conversely, cold, rain, high humidity, and little wind hinder drying. As a result of this weather dependency, either excess capacity must be built or the drying beds must be used to augment other disposal methods.

FIGURE 33.10   Paved beds.

FEED SOLIDS TYPE AND QUALITY. As with the mechanical dewatering process, the higher the proportion of secondary biosolids, the more difficult the biosolids will be to dry and the higher the chemical consumption will be. The process is broadly limited by the volume of water in the biosolids; the higher the solids concentration, the shorter the drainage and evaporation time to reach a given moisture level. Biosolids applied to drying systems are generally stabilized to class B solids to minimize odors.

SYSTEM DESIGN. The type of media, distribution of piping and drains, feed distribution system, and vacuum and pumping equipment can all affect the operation and maintenance of an air-drying system. The extent to which the system is isolated from the weather is also important.

CHEMICAL CONDITIONING. Polymer addition increases the drainage rate and volume, thus increasing the capacity of the beds and shortening the drying time. Other than reed beds and paved beds, most air-drying systems use at least some polymer. Polymer is always necessary for any filtration system.

PROCESS CONTROL—REED BEDS. A typical operating procedure (Fleurial, 2003) for reed beds is hugely simpler than other systems. Little is done from month to month. Once a year, the reeds are harvested, and, after 6 to 10 years of accumulation, approximately 0.6 to 0.9 m (2 to 3 ft) of accumulated biosolids are removed with an excavator. Polymers are typically not used. The following are characteristics of reed beds:

- A new reed bed must be planted with rhizomes and kept wet until they sprout. Biosolids can be used to a depth of 25 mm (1 in.). When the reeds sprout, the beds can then be flooded with biosolids to a depth of 75 to 100 mm (3 to 4 in.). The shoots must not be submerged.
- Biosolids are reapplied every 2 to 3 weeks, depending on the weather.
- The beds simulate wetlands, and the plant roots must be kept wet by drainage from the biosolids or by adding plant effluent. The presence of weeds is an indication that the reeds are stressed.
- Red worms, among other types, are helpful in reducing the organic material.
- The reeds are typically harvested once per year in the late winter. A cycle bar mower is excellent, but, if the bed is too soft to support the machinery, a weed whacker fitted with a blade (rather than string) is used.
- The reeds should be removed from the bed, either by raking back from the edges and burning, where allowed, or by physically removing them.

The beds' ability to accept biosolids is much reduced during cold weather and during wet weather. After several years, the solids level in the bed will have risen, and the beds must be emptied. The full bed should lie fallow for 4 to 6 months. The biosolids do not typically support heavy equipment, unless it is frozen. If heavy equipment bogs down, it can damage the underdrain system and the liner; an excavator on the edge of the reed bed can easily dig the sludge, purposely leaving the lowest 15 cm (6 in.). Typically, sufficient plant material remains so that the reeds sprout and grow without the need for further planting.

The residual biosolids can be screened for a high-quality soil amendment or spread on agricultural land. Reeds are a marsh-dwelling plant and do not grow when spread on farm land. Reed beds require less capital cost, less land area, and are much less labor-intensive than other evaporative systems. In addition to such budget-friendly attributes, they blend in with the natural landscape. As for capital cost, everything that is needed is available as generic items, typically available locally. As a result of all of these features, reed beds are becoming much more common.

TROUBLESHOOTING. *Sand Beds and Paved Beds.* The operation of a sand or paved drying bed depends on the condition of the sand bed and underdrain, feed solids concentration, depth to which the feed is applied, loss of water through the underdrain system, degree and type of digestion provided, evaporation rate, type of removal method used, and disposal method used. All these considerations determine the optimum loading for a given bed. The operator must be thoroughly familiar with the limitations of sand beds caused by the quality of the feed solids applied, conditioning effects on dewatering, and best methods of dealing with environmental conditions. Beds that use filtration as a mechanism are especially sensitive to the condition of the filtration media.

Table 33.4 presents a troubleshooting guide for sand beds that will enable the operator to identify problems and determine their solutions.

Sand Beds. The following are instructions for the use of sand beds:

- Remove weeds and debris;
- Add more sand if the beds are low;
- Rake and spread the sand evenly;
- Mix up polymer solution if used, and flood the bed to a depth of 150 to 300 mm (6 to 12 in.);
- Harrow the bed, if possible, to speed the drying;
- Allow to dry as desired, typically several weeks; and
- Carefully remove the dry sludge load into a truck.

Vacuum-Assisted Drying Beds. A typical operating cycle for a vacuum-assisted/ wedge-wire drying bed is similar to that above, except for the following:

- At the start of the cycle, the vacuum pumps are off, the bed closure system is in place, the filtrate pumps are set for automatic operation, sufficient polymer is mixed or otherwise available, and drains in the bed are covered and sealed.
- The bed is flooded with water to a depth of 2.5 cm (1 in.) to assist in distributing the sludge evenly.
- The biosolids and polymer pumps are started, and the polymer addition is adjusted to give a robust floc. Biosolids depth varies, but is typically 900 mm (36 in.).
- Gravity drainage is allowed to continue until the operator decides that the rate of filtrate collection is too slow. Gravity drainage time may range from 30 minutes to several hours after the end of the application.
- The operator starts the vacuum cycle at the end of the gravity drainage. The vacuum sequence typically proceeds in discrete steps. The highest vacuum level continues until the cake has dried sufficiently to crack, with the resultant loss of system vacuum.

TABLE 33.4   Troubleshooting guide for sand beds.

Indications/observations	Probable cause	Check or monitor	Solutions
Excessive dewatering time	Applied sludge depth is too great	Typically, 8 in.[a] of applied sludge is satisfactory	When bed has dried, remove sludge and clean; apply a smaller depth of sludge and measure the drawdown over a 3-day period; next application, apply twice the 3-day drawdown
	Sludge applied to improperly cleaned bed	Note condition of any empty beds	After sludge has dried, remove sludge; completely clean and rake surface of bed and replace with 0.5-1 in.[a] of clean sand, if necessary.
	Underdrain system has plugged or lines are broken		Backflush beds slowly by hooking clean water source to underdrain piping; check sand bed and replace media as needed; drain underdrain lines during cold weather to keep them from freezing
	Beds undersized	Effects of adding polymer	Typically 5–30 lb/ton[b] of dry solids of cationic polymer provides improved dewatering rates
	Weather conditions	Temperature, precipitation	Cover or enclose bed to protect from weather
Sludge feed lines are plugged	Accumulation of grit and solids in lines		Open valves fully at start of sludge application to clean lines; flush lines with water if necessary
Thin sludge being drawn from digester	Coning occuring in digester with water being pulled out and sludge left being		Reduce rate of withdrawal from digester
Flies breeding in sludge beds			Break sludge crust and use larvicides such as borax or calcium borate or kill adult flies with suitable insecticide
Odors when sludge is applied	Inadequate digestion of sludge	Operation of digestion process	Establish correct operation of digestion process
Sludge is dusty and crumbles	Excessive drying	Moisture content	Remove sludge from bed when it dries to 40–60% moisture content

[a]in. × 25.4 = mm.
[b]lb × 0.453 6 = kg; ton × 0.907 2 = Mg.

- An optional evaporation phase may be necessary to produce a liftable cake, typically requiring a minimum solids concentration of 10 to 12%. The time required for evaporation varies on a site-specific basis.
- A small tractor equipped with turf tires and front-loading bucket removes most of the cake. The front-end loader cannot completely remove all the cake. The small amount left on the bed requires manual removal with a shovel or scoop. Thorough removal is essential to prepare the media plates for final cleaning.
- Manual rinsing with a hose and nozzle begins at the end of the bed farthest from the drains and progresses toward the drains, recycling any remaining biosolids. Tile plates will have to be removed every so often to hose out the filtrate channels.
- A cycle without air-drying takes 8 hours and fully occupies one operator.

Two sources of problems are improper conditioning and improper plate cleaning. The wrong type of polymer, ineffective mixing of the polymer and biosolids, and incorrect dosage result in poor performance of the bed. Overdosing of polymer may lead to progressive plate clogging and the need for special cleaning procedures to regain plate permeability. Plate cleaning is critically important. If not performed regularly and properly, the media plates are certain to clog, and the beds will not perform as expected. Special cleaning measures, which are costly and time-consuming, are then required.

MECHANICAL DEWATERING EQUIPMENT

GENERAL. Before the development of organic coagulant polymers, vacuum and pressure filters were the only useful mechanical dewatering systems. They used large amounts of ferric chloride, lime, and other inorganic filter aids to condition the biosolids and improve their filterability. While a few advances have been made on such filtrations systems over the years, vacuum and pressure filters are only considered for peculiar plant requirements. They have largely been supplanted by BFPs and centrifuges. The end of dumps and the development of landfills raises the disposal cost to the point where it dominates the equipment selection. Better organic polymers have supported these changes. Generally speaking, as the polymer dosage increases, the cake dryness increases. Thus, the savings achieved by drier cake are partially offset by higher polymer costs. Likewise, higher feed rates result in wetter cake, so that higher capital costs are balanced by the benefits of drier cake. Nearly all dewatering systems are purchased based on a present-worth analysis, which seeks to find the lowest cost dewatering system for the planned design life.

DEWATERING CAPABILITY OF DEWATERING EQUIPMENT. Former editions of this book listed the performance claimed by various manufacturers for their dewatering equipment. Because the equipment varies so much and the biosolids were so ill-defined, the resulting numbers were misleading, at best. Operators interested in the capabilities of their present or future equipment are advised to request references from the various manufacturers. Check these references assiduously. Do not assume the engineer will do it. Be charitable if the phone numbers that are given are not current; area codes and telephone systems are constantly changing. Make a checklist of questions to ask each reference. Confirm who is being spoken to, and who their employer is. If possible, visit the actual plants. Unfortunately, the endorsement the plant operators tell a casual inquirer over the phone may be different from what is told to another operator one-on-one. Ask about the precise model of the equipment and especially about the biosolids process. If the equipment of interest is not what the reference has, it is a problem. If the biosolids and plant operation are not the same as your plant, then the references are different from the specific case and are suspect. It is much easier to avoid buying a poor performing piece of equipment than it is to get rid of a poor performing piece of equipment once it is installed in the plant.

INSTALLATION CONSIDERATIONS. Mechanical dewatering installations have a number of general requirements, which tend to be more important as the scale of the installation increases; for example, traveling bridges with electric hoists are very common on large installations and less often on small ones. Other features are universal. Following is an installations checklist:

- It is important that operators and maintenance people be able to safely move around the equipment. This requires a minimum of 0.9 to 1.2 m (3 to 4 ft), and, if the equipment is elevated, then catwalks on all sides are in order.
- A sink with handwashing facilities should be centrally located on each operating floor.
- Process sample points are important. A feed sample before polymer addition, a cake sample, an effluent sample, and a polymer solution sample are all necessary. It is very useful for the operator to be able to see the effluent conveniently. Sample lines that run continuously are better than those that do not. Long runs of sample piping will clog, and a flush system should be hard-piped. Avoid flat runs, and try to make the sample lines self-draining.
- The clarified effluent from the centrifuge (and sometimes filters) can contain large amounts of foam. Therefore, the lines must be oversized and designed to avoid traps.

- Centrifuges must be vented to work properly. Poor venting can cause solids buildup in the casing, catch fire, and/or erode the bowl shell. The centrifuge is automatically vented if the centrate and the cake chutes are open to atmospheric pressure. If one or both are enclosed, then both must be vented to the same pressure with a negative draft.
- Centrifuges typically vibrate excessively when they need service. A vibration meter is essential to monitor the vibration. It is much more useful than a vibration switch that only shuts the centrifuge down.
- Flow meters are very useful, on both the feed line and the polymer line. The flow meter cannot give an accurate reading unless it is upstream of the polymer-addition points. Each dewatering device should have its own flow meters.
- Most manufacturers recommend several polymer-addition points. The system should be set up so that the operator can close the valve to one addition point with one hand while opening the valve to another addition point with the other hand.
- From time to time, biosolids will spill on the floor and be hosed to the floor drains. With a bit of thought, the drains can be placed where it will be easy to direct the biosolids. The drains need to be of sufficient size (15 to 20 cm [6 to 8 in.]), so they will not plug readily.
- Centrifuges often spill liquid biosolids out of the solids end upon startup. The installation should use a knife gate or a reversing screw conveyor to contain the wet material.
- The best feed pumps for the feed biosolids and polymer are positive-displacement pumps. It is important to minimize the shear to both the polymer and the biosolids. Very large installations sometimes use a variable-speed centrifugal pump to charge a manifold with feed biosolids, with flow-control valves for each dewatering device. This works well as long as the pump speed is at a minimum and all of the flow control valves are 60 to 80% open. Pulsations in flow are not good and should be minimized.
- The control room for centrifuges does not need to be on the same floor as the centrifuges as there is nothing to see there. It is more convenient to be on the floor below where the effluent and cake are readily accessible.
- The control room for filters should be on the same floor as the filter because visual inspection of the filter is important.
- Biosolids often release hydrogen sulfide. In addition to the very serious health hazard to operators, even very low levels of hydrogen sulfide corrode copper electrical wires, contacts, and air conditioning coils. Both the operators and the controls need to be protected from hydrogen sulfide.

- Management should be sure safety devices are readily available and that operators use them. Ear plugs should be available in noisy environments. Nitrile, vinyl, or latex gloves are now standard protection when handling biosolids and need to be conveniently available if they are to be used.
- Lifting requirements. The instruction book typically gives the component weights of the equipment. The most versatile and most expensive is a traveling bridge crane. Sometimes, a monorail can be substituted if the principal lift points are in a row. Smaller installations may be well served with an A-frame. Anything over 0.9 to 1.8 Mg (1 to 2 tons) should have an electric hoist, especially if it will be lowered to the floor below. Fork lifts are not precise enough to be a good substitute for a hoist.
- The polymer area may require lifting equipment if drums or totes will be used.
- All plants should have flow meters on the polymer lines. This is especially important during polymer trials, where it is critical to measure the polymer rate.
- A facility that normally operates several dewatering units at once should have a provision to make-down and pump a trial polymer to just one unit.
- Maintenance should have a cabinet with dust-tight doors to store special tools for the disassembly of equipment.
- Consider having a storage room for spare parts in the dewatering building.
- Ventilation is important. Consider a two-mode operation—one level for operators who are only occasionally in the dewatering room, and a higher ventilation standard when someone is working in the room continually.

BIOSOLIDS FEED-PUMP CONSIDERATIONS. Aside from cost, there are two general considerations for pumps feeding dewatering devices.

1. Pumps can produce pulsations in flow and
2. Pumps can shear the biosolids.

The ideal pump would produce a steady, uniform flowrate, with little or no shear. Uniform flowrate is important because one is trying to add polymer at a constant dosage, which clearly is a problem if the feed flow is not uniform. A second problem with pulsing flow is that it causes surges in flow through the dewatering equipment. This aspect is less of a problem with filters, which have some buffering capacity in the thickening, gravity, and wedge sections, than with centrifuges that have no buffering. In the case of a piston pump, during the push stroke the flow to the dewatering device surges, but, when the piston returns, the flow stops completely. As a result, the peak flow is approximately double the average flow. This makes it impossible to meter polymer accurately to the biosolids. Generally, pumps with check valves (piston and diaphragm pumps) are the worst offenders because the pumps operate at low speeds. As the pump speed increases, the pulsation frequency also increases, and the difference between peak flow and minimum flow is easier to dampen.

The following evaluation of pumps for belt presses and centrifuges is based on the pulsation characteristics of the various pumps. It is not an absolute judgment.

- Centrifugal pumps recommended, except for shear (see below);
- Diaphragm pumps not recommended;
- Double disc pumps okay;
- Peristaltic pumps marginal;
- Piston pumps not recommended;
- Progressive cavity pumps okay; and
- Rotary lobe pumps okay.

Shear is another matter. The assumption is that shearing the biosolids before dewatering results in the break up of the natural floc, which, in turn, requires additional polymer to build the floc back up. Positive displacement pumps are generally low-shear devices and thus are generally recommended. Centrifugal pumps have a very uniform flowrate, but are more likely to shear the biosolids. The worse case is an oversized centrifugal pump operating at fixed speed, with a flow control valve to throttle the discharge, to maintain the flow set point; both the pump and the valve shear the biosolids. A better case would be to have a flow meter control the pump speed. In this manner, the flow control valve is eliminated, and the pump shear is at a minimum. Alternately, the pump speed could charge a manifold, such that at least one flow control valve off of the manifold is 85% open.

CONVEYANCE OF DEWATERED BIOSOLIDS. Once the biosolids leave the dewatering device, the designer is faced with transporting them to the next stage, towards ultimate disposal. The only advice is that simpler is better. A transport path that goes around several corners and rises 6 m (20 ft) in the air will be a difficult system to design and will therefore be expensive to build and costly to maintain. Some specific considerations are the following:

- Dewatered biosolids that free-fall from any significant height will splatter, at least some of the time. This is never a good thing.
- All dewatering devices will put out wet slop from time to time. The biosolids transport system must be designed to handle this.

- Avoid going around corners whenever possible.
- Avoid steep changes in elevation.
- Points of transfer from one conveyor to another are often problem areas.
- All moving parts will eventually need to be replaced. Plan for it in the design.
- Vertical screws and steeply inclined belts work most of the time.

The dewatering operation is at the end of the treatment process. Mistakes and problems upstream come to rest here and ensure that, at least some of the time, the cake volume and texture will be nowhere near that envisioned in the design. Versatility and overdesign is the only solution. Visit other plants and ask questions of the people who work with their design. Be vocal if trouble is foreseen as a result of a proposed design.

ADDITIVES. From time to time, various additives are used to improve the biosolids or mitigate problems. Shredded newsprint, coal, sand, and inorganic chemicals have all been added for one reason or another. Defoamers may be needed to reduce the foaming of the centrate/filtrate, odor control chemicals to reduce those complaints, and oxidants to reduce hydrogen sulfide. Perfumes attempt to conceal the underlying odors; oxidants, such as potassium permanganate and hydrogen peroxide, react chemically with hydrogen sulfide in the sludge and therefore make the environment safer for the operators and reduce the odor. The great majority of plants do not use additives. Many or most of the problems for which these additives are prescribed are really caused by poor operation (or design) in the plant. Obviously, it is preferable to correct the root problem rather than spend money daily to minimize the effect of the problem.

GENERAL CALCULATIONS FOR MECHANICAL DEWATERING. The general calculations for mechanical dewatering are based on analytical analysis of the various streams around the centrifuge. Sample-taking is addressed elsewhere, but one point should be emphasized here: If a sample of the feed biosolids is being sent to the laboratory, avoid sending one where the polymer has already been added. The laboratory must extract a representative sample of the biosolids from the jar full of floc and free water that was sent to them. It is very difficult to get a representative sample out of a biosolids–polymer mixture. Try to pull a valid sample of the feed upstream of the polymer addition.

Nomenclature. By custom, lowercase letters refer to concentrations, and uppercase letters refer to flowrates. For example, in U.S. customary units, a feed rate of 120 gpm of 2.5% solids biosolids would be expressed as $F = 120$ gpm and $f = 2.5\%$. The dry solids would be expressed as concentration times a flowrate, or

Solids loading $= f \times F \ 2.5\% \times 120$ gpm

$$\times \frac{1}{100\%} \times \frac{60 \text{ min}}{\text{hr}} \times \frac{8.34 \text{ lb}}{\text{gal}} \times \frac{1 \text{ ton}}{2000 \text{ lb}} = 0.75 \text{ tons/h DB (0.7 Mg)} \qquad (33.1)$$

Another convention is to refer to the clarified liquid as the effluent, rather than centrate, filtrate, or clarifier overflow. Around all unit processes, two mass balances can be drawn.

Conservation of dry solids; solids in equals solids out:

$$fF = sS + eE \qquad (33.2)$$

Conservation of mass; the weight of biosolids in must equal the weight of biosolids out:

$$F = S + E \qquad (33.3)$$

where (in U.S. customary units)
    F = feed rate, gpm (120 gpm);
    f = feed concentration, % (2.5%);
    S = solids or cake rate, gpm;
    s = solids or cake concentration, % (20%);
    E = effluent rate, gpm;
    e = effluent concentration, % (0.15%); and
    P = polymer rate, gpm (15 gpm).

Other units can be used for concentration and rate, as long as they are consistent. Further, *recovery* is defined as the percentage of dry solids removed in the dewatered solids, divided by the amount of solids in the feed. Or,

$$\text{Recovery} = \frac{sS}{fF} \times 100\% \qquad (33.4)$$

Solving Equations 33.2 and 33.3 for E, the effluent rate, and rearranging the result in terms of sS/fF, the percent recovery can be calculated entirely in terms of the feed, effluent, and solids analysis; the feed and other rates do not enter into the calculation at all.

$$\% \text{ Recovery} = \frac{s(f-e)}{f(s-e)} \times 100\% = \frac{20\%(2.5\% - 0.15\%)}{2.5\%(20\% - 0.15\%)} \times 100\% = 95\% \qquad (33.5)$$

As a quick logic check, if there are no solids in the effluent, then $e = 0$, and the recovery is 100%. At the other extreme, if the effluent concentration equals the feed concentration, the term $(f - e)$ in the numerator goes to zero, and therefore the recovery is zero. This formula is good for the capture across clarifiers, thickeners, and mechanical dewatering devices. It does not matter what the units for rate and concentration are, as long as they are the same.

Equations 33.2 and 33.3 can be combined with the feed rate and rearranged to allow the calculation of the effluent rate and the solids rate, as follows:

$$\text{The cake rate is } S = F\frac{(f - e)}{(s - e)} = 120 \text{ gpm}\frac{(2.5 - .015\%)}{(20\% - 0.15\%)} = 14 \text{ gpm of cake} \qquad (33.6)$$

$$\text{The centrate rate is } E = F\frac{(s - f)}{(s - e)} = 120 \text{ gpm}\frac{(2.5 - 2.5\%)}{(20\% - 0.15\%)} = 106 \text{ gpm of effluent} \quad (33.7)$$

Not mentioned here is the effect of polymer and rinse water. Commonly, if the volume of polymer or rinse water is less than 10% of the feed solids, the dilution effect is ignored. Similarly, the effect of polymer solids is also ignored, because it is very small compared with the feed solids. Where dilution resulting from polymer water or rinse water is important, the easiest way to compensate for it is to adjust the feed or effluent concentrations before calculating the percent recovery. For example, if F, the feed rate, is 120 gpm, and P, the polymer rate, is 15 gpm, the feed analysis, f_m, measured at 2.5%, can be corrected to take into account the dilution caused by the polymer and determine the calculated feed, f_c, as follows:

$$f_c = F_m\frac{F}{F + P} \qquad (33.8)$$

Using the numbers above, we obtain: $f_c = 2.5\%\dfrac{120 \text{ gpm}}{120 \text{ gpm} + 15 \text{ gpm}}$

where
   f_c = corrected feed concentration,
   f_m = measured feed concentration, and
   P = polymer rate (including any post-dilution).

For a given centrate concentration, the effect of any dilution water is to reduce the recovery level. Similarly, if the effluent is diluted by rinse water, as is commonly the case with a BFP, the measured effluent concentration will be lowered as a result of the dilution. If the press feed, F, is 120 gpm, and the belt wash water rate, W, is 100 gpm, with the feed concentration at 2.5%, dewatered cake at 20%, and effluent measured as 0.15%, then the formula to correct the effluent concentration for the effect of wash water is as follows:

$$e_c = e_m \frac{E+W}{E} = 0.15\% \frac{106 \text{ gpm} + 100 \text{ gpm}}{106 \text{ gpm}} = 0.29\% \qquad (33.9)$$

where

e_c = corrected effluent concentration,
e_m = measured effluent concentration,
E = effluent rate (Equation 33.6), gpm; and
W = wash water rate, gpm.

Using the numbers given above, taking the wash water into account nearly doubles the effluent concentration and therefore reduces the capture from 95 to 90%.

Polymer Dosage. The last formula is the calculation of the polymer dosage. While it makes some sense that the dosage should be calculated as the amount of polymer required to remove 0.9 Mg (1 ton) of solids, that is not how it is defined; it is the amount of polymer per Mg (ton) of feed solids. The formula is commonly expressed as two ratios, followed by a constant.

$$\text{Polymer dose} = \frac{\text{polymer concentration}}{\text{feed concentration}} \times \frac{\text{Polymer rate}}{\text{Feed rate}} \times 2000 \text{ lb/t} \qquad (33.10)$$

For metric calculation, substitute the metric 1000 kg/t for the U.S. customary term. The advantage of this formula is that any units can be used for rate (as long as the units are the same for both the feed and the polymer). Using the numbers given earlier (120 gpm of feed, 15 gpm of polymer, 2.5% feed solids, and assuming 0.12% polymer concentration), the following can be calculated using Equation 33.10:

$$\text{Polymer dose} = \frac{0.15\%}{2.5\%} \times \frac{15 \text{ gpm}}{120 \text{ gpm}} \times 2000 \text{ lb/t} = 15$$

The same formula is also used for jar testing, when the sample volume of feed and the sample volume of polymer are substituted for the rate terms. Supposing the same polymer and the same feed solids are used, 15 mL of polymer solution could be added to a 120-mL sample of biosolids, and the very same 7.5-g/kg (15-lb/ton) polymer dosage would be used as was being used on the machine.

BELT FILTER PRESS. This section describes the process variables that affect the dewatering process, operation, and maintenance required to run a BFP in optimum condition. Figure 33.11 is a typical process flow diagram of a BFP.

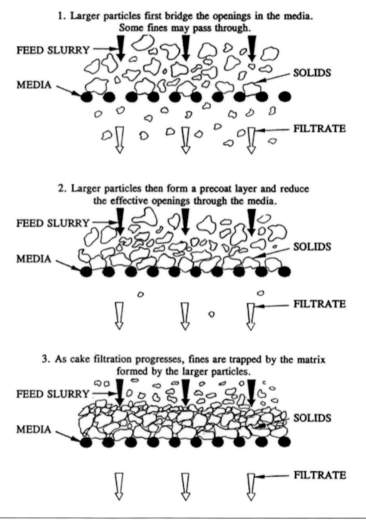

FIGURE 33.11   Filtration process.

Principles of Operation. The operation of a BFP is based on the principles of filtration and comprises the following zones:

- Gravity drainage zone, where the feed is thickened;
- Low-pressure zone; and
- High-pressure zone.

Figure 33.12 illustrates a BFP and the different zones encompassed in the system. In the gravity zone, most of the free water from the flocculation process drains through a porous belt. This is followed by the low-pressure zone, where the thickened feed is subjected to low pressure to further remove water and form a viscous biosolids matrix. In the high-pressure section, the matrix is sandwiched between porous belts passing through a series of decreasing diameter rollers. The roller arrangement progressively increases the pressure, filtering more water from the matrix. There are numerous design differences between the various manufacturers, mostly varying in the relative sizes of the different zones.

Generally, because the operator can see into the press, it is easy to determine what is happening, and the press is therefore easy to operate. Belt filter presses are very quiet,

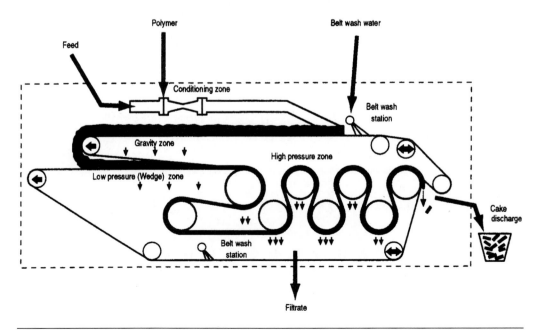

FIGURE 33.12   Belt filter press process flow diagram.

with low power consumption. Also, mechanical failure is easy to recognize, and, while the fix may be time-consuming, it is typically not difficult or expensive, as many parts can be obtained from local suppliers or manufactured locally, making presses relatively inexpensive for mechanics of average skills to repair or overhaul on-site. Negatively, they are messy, odorous, require significant operator attention, and are not very versatile.

Process Variables. There are several process variables that affect the performance of all dewatering systems. In general, dewatering devices must run at 95% capture or better, so capture really is not an operating variable. Of the remaining parameters—cake dryness, loading, and polymer dosage—within limits, the operator can take from one to give to another. For drier cake, one can reduce the loading and/or increase the polymer dosage.

Biosolids Type and Quality. As discussed earlier, the operation of the wet end of the plant determines the quality of the biosolids, which, in turn, greatly affects the dry end.

Polymer Activity. Polymer is selected once per year, but biosolids often change during the course of the year. If the polymer does not react well to the biosolids, performance suffers.

Polymer Addition and Mixing. Most presses have some sort of inline mixing system, which adjusts the amount of mixing given the sludge/polymer. Adding the polymer closer or further from the press can also affect performance.

Cake Dryness. Increased cake dryness comes at the price of lower capacity and/or higher polymer dosage. The ability to obtain higher cake dryness is, to a great extent, a function of the design of the press. Presses with extended gravity zones to better preconcentrate the feed and additional pressure rollers give longer sludge residence times and dryer cake.

Polymer Type and Dosage. Some polymers are designed to obtain drier cakes than others. Likewise, the dosage will increase and decrease with the cake dryness. Some polymers become less effective at higher dosages. This will be apparent from a quick jar test or observing that adding more polymer results in either poorer operation or the same operation.

Hydraulic Loading. Belt filter presses are limited by the volume of water that can go through the belts. As a result, thinner feed solids will result in less quantity of dry solids produced, with all else being equal.

Solids Loading. Likewise, more solids will result in less solids residence time inside of the press and therefore wetter solids, with all else being equal.

Capture. The solids capture is typically fixed by the plant management and is not an operating variable.

Mechanical Variables. Mechanical variables that affect BFP performance include the following:

- Belt composition and condition, speed, and tension;
- Size and number of rollers;
- Feed concentration;
- Polymer concentration;
- Polymer addition and mixing;
- Wash water flow and pressure; and
- Wash water suspended solids concentration,

Operation of the BFP should be monitored with data sheets, such as the one shown in Table 33.5 (BFP spreadsheet). If these data are taken frequently, it is easy for the operator to change one variable at a time to optimize the performance of the BFP. The spreadsheet calculates the cost of disposal, in U.S. dollars per dry ton. This keeps everyone informed of the goal of the dewatering operation. In addition, if further help is required, this is valuable information to aid in troubleshooting of the process.

Sequence of Operation. The sequence of operation for a BFP typically is set up in the following order:

1. Open wash water valve,
2. Start wash water pump,
3. Start pneumatic/hydraulic belt tension system,
4. Start belt drive and dewatered cake conveyor,
5. Start polymer solution feed pump, and
6. Start biosolids feed pump.

Modern presses typically have a one-button start system, so the operator only has to manually start the feed and polymer pumps. In any event, one benefit of filling out the operating log is that the operating conditions the last operator used are known, as are the conditions of the previous week and month.

The BFP is equipped with a series of alarms that shut down the unit if one of the following situations arise:

- Belt breakage,
- Belt misalignment,

TABLE 33.5 Belt filter press data sheet.[a]

Name of plant: _____	Telephone: _____
Address: _____	Date: _____
Sludge type: _____	Blend: _____
Contact name: _____	Press size: _____ (m)

Run #	1	2	3	4	5
Time (HH:MM)					
Feed pH					
Feed temperature, °F					
Feed alkalinity, mg/L					
Feed ash content, %					
Feed solids, % TS[b]					
Hydraulic loading (gpm)					
Solids loading, lb/hr					
Cake solids, % TS					
Cake thickness, in.					
Polymer type; dry, liquid					
Polymer number					
Polymer solution concentration, % TS					
Polymer flowrate, gpm					
Polymer dose, lb/ton					
Filtrate flow, gpm					
Filtrate TSS,[c] mg/L					
Inline mixer size, in.					
Retention time, sec					
Capture, %					
Belt type					
Belt speed, m/min					
Hydraulic pressure, psi					
Wash water pressure, psi					
Wash water flow, gpm					
Wash water TSS, mg/L					

[a]1 m³/h = 4.4 gpm; 1 kg = 2.204 lb; 1 ton = 2 000 lb; (°F − 32)0.555 6 = °C; 1 psi = 6 893 Pa; 1 lb/ton = 0.5 g/kg; 1 metric ton = 1.1 ton; 1 m = 39.37 in.
[b]TS = total solids.
[c]TSS = total suspended solids.

- Hydraulic or pneumatic unit fault,
- Low wash water pressure,
- Emergency cord surrounding the BFP is pulled,
- Stop button in the BFP panel is pressed,
- Solution polymer feed pump failure,
- Conveyor shutdown/storage bin overfilling,
- Biosolids feed pump failure, or
- Dilution water failure.

Belt Filter Press Optimization and Troubleshooting. After the cake solids, solids loading, and capture rate have been specified, adjustments are made to the BFP dewatering process and the polymer system to achieve those values. After the BFP is in operation, the approximate cake solids can be determined quickly (approximately 10 minutes) with a high-energy microwave oven or other such device.

If the belt wash water is included in the filtrate sample, then the actual filtrate analysis is diluted by the belt wash water and needs to be taken into account in calculating the capture. From a practical standpoint, the experienced operator can judge the cake and filtrate quality quite well by eye, and laboratory testing is used to confirm his or her estimate.

The computed values should be compared with the values previously achieved by the BFP operation. If low cake solids, low capture, low solids loading, or high polymer dosage are obtained, the following procedures illustrate some measures that can be taken to improve the process.

All filtration devices are sensitive to the amount of water that can filter through the cake. Generally, thicker feed dewaters better than thinner feed material. Increasing the polymer can sometimes compensate for poor process results caused by thinner feed solids.

As with other devices requiring polymer, problems in the polymer make-down system can also lead to problems in dewatering. Belt filters benefit from a polymer injection ring and some sort of mixing device. The injection ring is a spool piece with six or eight hoses adding the polymer around the circumference of the feed pipe. The hoses carrying the polymer to the ring occasionally plug with solids and need to be cleaned. When some of the hoses plug, the polymer is poorly distributed and, if uncorrected, require excess polymer. The mixing device is a weighted check valve, creating an adjustable amount of turbulence. More elaborate devices include static or mechanical mixing. Trial and error is necessary to determine the best setting of the mixing device. As with all optimization, the "best" setting often changes with the sludge, the polymer, or the dewatering goals.

Low Cake Solids. To improve this situation, the following steps can be taken:

- Adjust belt speed. The belt speed should be decreased in small increments (approximately 10%), to increase cake thickness. A thicker cake increases the shear rate exerted by the outer belt on the cake as it goes around the pressure rollers in the high-pressure zone. Let the BFP stabilize for 15 to 20 minutes, take a cake sample, and compare it with the previous cake solids reading.
- Adjust polymer flow/polymer dosage. A small increase in polymer dosage could result in faster drainage in the gravity zone, and less water to remove through the belts will produce a drier cake. This does increase the operating cost. Excessive polymer can slime the belts, halting dewatering, until the mess can be cleaned out. Thus, there is an upper limit to the effective use of polymer. Let the BFP stabilize for 15 to 20 minutes, take a cake sample, and compare it with the previous cake solids reading.
- Adjust hydraulic/pneumatic belt tension. Increase the belt tension in 10 to 15% increments. At some point, the higher pressure will press the biosolids into the belt weave, blinding the belts. Let the BFP stabilize for 15 to 20 minutes, take a cake sample, and compare it with the previous cake solids readings. Higher belt tension can shorten the belt life.
- Adjust the inline polymer mixer. The floc size is extremely important to the dewaterability of the sludge. The polymer dosage affects the mixing energy.

Low Capture Rate. This is typically caused by squeezing cake out from the lateral sides of the belts in the high-pressure zone. To correct this situation, the following steps can be taken:

- Adjust belt speed. Increase belt speed gradually until the cake ceases to squeeze out from the lateral sides of the belts. Let the BFP stabilize for 15 to 20 minutes, and take samples of the filtrate and cake solids.
- Adjust polymer flow/polymer dosage. A small increase (or sometimes decrease) in polymer dosage could result in faster drainage in the gravity zone, reducing the water content of the biosolids and allowing higher cake pressures in the high-pressure zone. Let the BFP stabilize for 15 to 20 minutes and take samples of the filtrate and cake solids.
- Adjust belt tension. Decrease belt tension gradually, until the cake stops squeezing out from the lateral sides of the belts. Let the BFP stabilize for 15 to 20 minutes, and take filtrate and cake solids samples.

Low Solids Loading.

- Make sure the gravity section support structure is clean. Solids built up on the wiper bars can lead to lower drainage rates.
- An obvious source of low solids loading is a decrease in solids concentration or flowrate of the biosolids feed. This can be corrected by taking a sample to determine the feed solids concentration and also checking the feed pump for wear or blockage.
- Less obvious is the gradual blinding of the belts. This can be the result of inadequate belt washing or chemical blinding. Manually cleaning the belts with a high-pressure hose may restore at least some of the drainage.

Maintenance. The following procedures will extend the life of the BFP and reduce its operating cost:

- Wash down the BFP every day after finishing the dewatering shift. This prevents cake from drying and accumulating in different sections of the BFP.
- Check to be sure all rollers are turning freely.
- Check the press weekly for damaged bearings.
- Check the grinder that prevents large particles from entering the press twice per year.
- Follow the manufacturer's operations and maintenance manual lubrication schedule. This extends the life of the roller bearings and belt drive motor.
- Clean the wash water nozzles as frequently as necessary (this depends on the quality of the wash water). This ensures proper cleaning of the belts.
- Inspect and change biosolids containment and washbox seals, as necessary.
- Inspect and clean doctor blades from any accumulated debris, hair, or any other foreign materials.
- Clean the chicanes (plows) in the gravity section after shutting down the press.

For any other maintenance of complex mechanical parts of the BFP, contact the manufacturer for advice.

Safety Considerations. The following guidelines should be implemented to avoid injuries:

- Watch out for spills of water, biosolids, and polymer; all are serious slip hazards.
- Do not remove solids or objects from the belts in the roller pressure section while the machine is in motion. If an object falls on the belts, try to stop the unit and remove it before it enters the roller pressure section.

- Never reach into an operating press with a hand.
- Adequate ventilation should be provided to avoid breathing hydrogen sulfide buildup. Also, the belt spray creates a lot of aerosols; the ventilation system should be maintained to keep them out of the room air. If the press is equipped with covers, keep them in place during operation. Some cover designs are better than others, and the cover design should be considered in any purchase decision.

DEWATERING CENTRIFUGES. This section describes the process variables that affect the dewatering process and the operation and maintenance required to run a centrifuge in optimum condition.

Principles of Operation. Centrifugation is the process of separating solids from liquids by the process of sedimentation, enhanced by centrifugal force. Thus, the plant thickeners and clarifiers, which also operate on the principle of sedimentation, are an accurate analogy to the centrifuge. In the clarifiers, feed flows across the surface of the clarifier to the discharge weirs. As feed flows across the surface, the solids settle out—first the larger and denser particles, and then the smaller and lighter ones. A rake turns slowly and pushes the solids to the discharge point. The rake is connected to a gear box, which, in turn, is driven by a motor. When the solids reach the discharge point, they accumulate, until the underflow pump removes them. The centrifuge is just the same.

The centrifuge (Figures 33.13 and 33.14) is a cylindrical drum that rotates at high speeds to develop centrifugal force. The force increases by the square of the bowl speed, so small changes in speed result in larger changes in centrifugal force. When a slurry enters the interior of a rotating centrifuge, it flows across the annular surface of the pool, to the discharge weirs (dams). As the liquid flows from the feed area to the weirs, the solids settle outward towards the bowl wall. The speed of the screw conveyor controls the discharge rate, which, in turn, determines the depth of the sludge blanket inside the centrifuge. The scroll drive motor is set to turn the scroll at a slightly different speed than the bowl and thus scrapes the solids along the bowl, up a conical section, and out of the centrifuge. This difference in revolutions per minute between the bowl and scroll is called the *differential speed* and serves the same purpose of controlling the biosolids blanket level in a centrifuge that the underflow pump does on a secondary clarifier. All centrifuges use some sort of device to drive the screw conveyor and also use some sort of variable-speed drive. The centrifuge has two motors—the larger one, called the *bowl drive*, which turns the bowl, and a smaller device, called the *scroll drive*, which controls the conveyor speed.

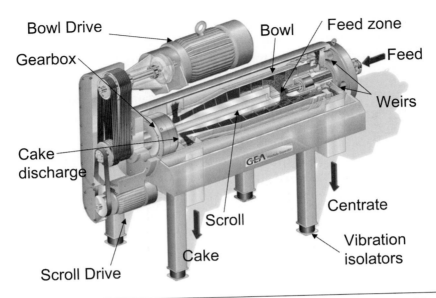

FIGURE 33.13   Centrifuge cutaway (courtesy of Westfalia Separator, Inc., Northvale, New Jersey).

FIGURE 33.14   Centrifuge disassembly (courtesy of Andritz, Arlington, Texas).

Traditionally, centrifuge illustrations suggest that the solids are scrolled out of the pool and drain on the beach area. This does not happen in wastewater processes. The pool level is at or above the solids discharge. There is no dry beach. Filtration is not a mechanism for dewatering in wastewater centrifuges—the only mechanism is sedimentation.

There are numerous design options within the centrifuge, as Figures 33.13 and 33.14 show. Manufacturers introduce these features to improve the product and provide product differentiation. Very few features are patented; therefore, the choice of design is open to all manufacturers. Manufacturing cost drives the design.

A centrifuge can thicken or dewater the biosolids with only a minor change in the weir setting (also called *pond setting*). Likewise, it can dewater biosolids to a moderate consistence at low polymer dose or produce very dry solids using higher polymer dosages. The centrifuge is a naturally sealed unit and is easy to connect to an odor control system. Centrifuges are unique in that the torque developed on the screw conveyor inside the centrifuge (when dewatering) is proportional to the cake viscosity plus assorted drag and mechanical inefficiencies. When the centrifuge is dewatering biosolids, the viscosity of the biosolids is large compared with the mechanical drag, so that, with most controllers, the operator can set a torque level and have the centrifuge operate at a fixed cake dryness. This makes automatic operation much simpler and reduces the need for supervision. Unfortunately, when the centrifuge is thickening a biosolids, the torque resulting from viscosity of the biosolids is small in comparison with the mechanical drag, so that autotorque does not work. An Andritz (Graz, Austria) centrifuge is depicted in Figure 33.14.

Centrifuges also have the reputation of being noisy and prone to vibrate, have weak gear reducers, and require frequent and expensive off-site repairs. To the extent that the manufacturer designs and manufactures them well, these problems are mitigated. Generally, it costs more to build a centrifuge with wear parts that can be easily removed when worn out, with new ones installed on-site. Similarly, hard surfacing materials with very long service lives cost more money than those that wear out quickly. It costs more to build a centrifuge that will operate for 20 000 hours or more with no mechanical failures than one that must be rebuilt every 6000 hours. On the other hand, dewatering will occur for 600 to 1000 hours per year, so a centrifuge that will operate for 6000 hours is not so bad.

Math Specific to Centrifuges. There are a few equations that are particularly useful with centrifuges. The centrifuges are rated by their manufacturer based on the specific gravity of the dewatered solids and the operating speed. The g force is a measure of the acceleration developed in the centrifuge and is proportional to the square of the speed. The g force at the bowl wall of a 74-cm (29-in.) diameter centrifuge rotating at 2600 r/min is as follows:

$$g \text{ force} = 0.0000142 \times (r/\text{min})^2 \times \text{diameter (in.)}$$
$$= 0.0000142 \times (2600)^2 \times 29 \text{ in.} = 2784 \, g \qquad (33.11)$$

This means that a 0.5-kg (1-lb) imbalance on the bowl wall would generate a force of 1263 kg (2784 lb) and create quite a bit of vibration.

Process Variables. There are several process variables that affect the performance of all dewatering systems. In general, dewatering devices must run at 95% capture or better, so capture really is not an operating variable. Of the remaining parameters—cake dryness, loading, and polymer dosage—the operator can take from one to give to another. For drier cake, one can reduce the loading and/or increase the polymer dosage.

Biosolids Type and Quality. As discussed earlier, the operation of the wet end of the plant determines the quality of the biosolids, which, in turn, greatly affects the dry end.

Polymer Activity and Mixing with the Biosolids. Polymer is selected once per year, but biosolids often change in the course of the year. If the polymer does not react well to the biosolids, performance suffers. Also, adding the polymer closer to or further from the centrifuge will affect performance.

Cake Dryness. Increased cake dryness comes at the price of lower capacity and/or higher polymer dosage. The ability to obtain higher cake dryness, to a great extent, is a function of the design of the centrifuge.

Polymer Type and Dosage. Some polymers are designed to obtain drier cakes than others. Likewise, the dosage will increase and decrease with the cake dryness. Some polymers become less effective at higher dosages. This will be apparent from a quick jar test or observing that adding more polymer results in either poorer operation or the same operation.

- Hydraulic loading. Centrifuges are less limited by the volume of water that passes through the centrifuge than filtration devices. As a result, thinner feed solids will have less effect on performance than in filtration devices.
- Solids loading. The solids residence time is important. If there are more biosolids to dewater, there will be less solids residence time and therefore wetter solids, with all else being equal.

- Capture. The solids capture is generally fixed by the plant management, and is not an operating variable.
- Sludge temperature. Centrifuge performance is linked to sludge temperature—warmer is better up to a limit of ~60 °C (140 °F).

Performance Level. Designing and building a centrifuge for high performance (i.e., dry cake and high hydraulic capacity) and long service life is expensive. There is a much greater probability that a more expensive centrifuge will have these attributes than will an inexpensive centrifuge.

Mechanical variables that affect the centrifuge performance include the following:

- Conveyor torque and/or differential speed,
- Bowl speed (within limits),
- Weir setting, and
- Polymer addition point.

Differential Speed—Scroll Torque. The scroll drive device allows the operator to change the differential speed between the bowl and the scroll (conveyor). Changing this differential speed changes the cake removal rate, which, in turn, changes the height of the solids blanket within the centrifuge, just as changing the underflow pumping rate changes the blanket level in a thickener. At low differentials, the biosolids blanket height increases, and the cake becomes drier. As the blanket rises, the solids may carry over into the centrate. The operator would respond by increasing the polymer rate to restore the centrate quality, or by increasing the differential speed to reduce the biosolids blanket. Similarly, increasing the differential speed increases the biosolids removal from the centrifuge and lowers the cake level. This results in wetter cake, but with cleaner centrate, allowing the operator to reduce the polymer rate.

Torque Control. In recent years, nearly all centrifuges have a controller that allows the operator to choose a scroll drive load or torque set point, and the controller then adjusts the differential speed to maintain that set point. In this manner, the torque and therefore the cake dryness is fixed. One way of looking at the centrifuge is that it is a very expensive viscometer. The conveyor is turning at a controlled speed immersed in biosolids. The effort or torque needed to turn the conveyor is measured by the scroll drive device. As the cake becomes drier, its viscosity increases, which, in turn, increases the torque or load on the scroll drive. When operating in load control, the controller automatically adjusts the differential revolutions per minute to maintain a constant torque level, and the operator's only task is to observe the centrate quality from time to time and adjust the polymer rate to maintain the desired centrate quality. This is a really simple control; one of its virtues is that the major operating cost—cake dryness—is fixed, and any operator error shows up in the centrate, which is easy to see and is not so costly if it is off slightly.

Bowl Speed. Centrifuges have a nameplate speed rating for a particular solids density. Each speed corresponds to a particular g force, and it is g force that drives the separation. Generally speaking, when the bowl speed is increased from 1800 to 2000 g, to 2500 g and sometimes 3000 g, and the pond is properly set, the process performance improves. Unfortunately, the power consumption, shear on the biosolids, erosion, noise, and vibration also increase with g force. The power consumption goes up proportionally to the g force. However, most centrifuges currently sold typically operate above 2000 g. Often, the centrifuge is operating at a lower speed than that for which it is rated because it is not designed to run continuously at the higher speed without excessive, noise, vibration, and maintenance. In theory, raising the g force should raise the centrifuge performance. Consult the manufacturer of the centrifuge for a recommendation on operating speed changes.

Weir Setting. The weir setting is a somewhat misunderstood operating variable. Considering the pool surface (the liquid–air interface within the centrifuge), raising the pool surface reduces pond radius and thus reduces the surface area of the pool where separation takes place. That this will give better centrate quality and the same or dryer cake is counterintuitive, but true. The drawback to operating with a deep pond is that liquid pours out of the centrifuge upon startup, and, the deeper the pond, the longer it takes to "make seal," which is the term for the discharge from the solids end to transition to dewatered cake. Do not assume that the pond setting on the centrifuge is the best one to use. Each year or two, try changing the pond up and down by approximately 4 mm to see which setting is best.

Polymer Addition Point. Most manufacturers recommend at least four polymer addition points. In general, if the polymer forms very weak floc, then closer to the centrifuge is better. The stronger the reactions are between the polymer and the biosolids, the further upstream (ahead) of the centrifuge the polymer should be added. As a general recommendation, addition points should be 20 to 30 m (60 to 75 ft) ahead of the centrifuge, 12 m (25 ft) ahead of the centrifuge, just in front of the flex coupling to the feed tube, and internal to the centrifuge. Some centrifuges have the feature of adding polymer inside the centrifuge, either in the feed zone or into the pool area. It is important to make the polymer additions convenient to use. The best design uses a manifold with valving for several addition points next to one another.

Operation. The operation of the centrifuge should be monitored with data sheets similar to that shown in Table 33.6. In a well-designed installation, the SCADA system

TABLE 33.6 Centrifuge data sheet.

Owner	Marion Haste WWTP
Sludge	Anaerobically Digested
Centrifuge	Bio Boogey 44
SERIAL NO:	93-DDN2500
Poly Type	Flocking Good FG1

Date	Run No.	Pond No.	Bowl Rpm	Delta Rpm	Feed Rate GPM	Feed Conc. %SS	Eff. Conc. %SS	Cake Conc. %TS	Feed tons/hr	Wet Cake tons/hr	Dry Cake tons/hr	Poly Conc. %	Load %	Poly Rate GPM	Poly Dose #/TDS	REC. %
3/4	1	14.7	2600	3.3	120	2.4	0.35	21	0.7	3.0	0.63	0.20	25	12.7	17.6	86.9
	2	14.7	2600	2.9	120	2.4	0.08	21	0.7	3.3	0.70	0.20	25	12.7	17.6	96.9
	3	14.7	2600	2.7	120	2.3	0.07	23.5	0.7	2.9	0.67	0.20	35	14.3	20.7	97.2
	4	14.7	2600	2.5	120	2.35	0.14	24.5	0.7	2.7	0.67	0.20	39	16.6	23.5	94.6
	5	14.7	2600	2.4	120	2.3	0.19	25.5	0.7	2.5	0.64	0.20	41	18.5	26.8	92.4
	6	14.7	2600	2.3	120	2.25	0.17	26	0.7	2.4	0.63	0.20	47	20.3	30.1	93.1
	7	14.7	2600	2.9	120	2.3	0.08	26.3	0.7	2.5	0.67	0.20	55	22.8	33.0	96.8
	8	14.7	2600	2.7	120	2.1	0.10	26.8	0.6	2.2	0.60	0.20	64	22.5	35.7	95.6

would store the data and automatically calculate the capture, polymer dosage, and de-watering cost. In addition, if further help is required, these sheets provide a ready record to aid in troubleshooting of the process.

Sequence of Operation. Because centrifuge construction and components vary, the manufacturer should be consulted for specific startup instructions. The following considerations apply to most units:

- Confirm operation of biosolids pumps, the polymer system, dewatered cake conveyor, and centrate return pumps.
- Open such valves as will enable polymer and feed biosolids.
- Press start button. Clear the alarm panel.
- When the centrifuge transitions from start to run, set the differential at minimum speed.
- Start the polymer and the feed at 25% of normal flow.
- When the centrifuge seals, increase the feed and polymer to 60% of normal flow, and change operation to load control.
- When the centrifuge reaches equilibrium, increase the feed and polymer to normal rates.
- In many installations, the centrifuge can be started in load control without causing torque trips. It is quicker to reach steady-state conditions in speed control.

The centrifuge is typically equipped with the following alarms:

- High torque alert and high torque alarm,
- High vibration alert and high vibration alarm,
- High bearing temperature,
- High motor temperature,
- Pump failure, and
- Biosolids conveying failure.

Furthermore, there is often an interlock to prevent the centrifuge from starting when the cover is open.

During operation, the operator should check for the following:

- On circulating oil systems, check the oil level and the flow of oil to the bearings;
- Flow of cooling water and oil temperature, to ensure it is operating in the proper range;

- Machine vibration;
- Ammeter reading on the bowl motor;
- Bearings for unusual noise;
- Bearing temperatures, by touching them;
- System for leaks;
- Centrate quality; and
- Scroll drive torque.

Because the centrifuge will shut itself down in the event of a fault, the operator typically only looks at the mechanical parameters once per shift.

Centrifuge Data Sheet. The centrifuge operation will be much better if the operators fill out the centrifuge data sheet (Table 33.6) once per shift. To be useful, an interactive spreadsheet is needed, to which operators enter raw data and the program performs calculations including disposal cost, polymer use cost, and total cost. There is value in the operators seeing the cost of dewatering for themselves and also for others.

Temperature Effects. Centrifuges operate on the density difference between solids and liquids. As the temperature rises, the density of water decreases. The density of the solids is not very affected by temperature. Numerous tests have shown that every increase of 5 °C (9 °F) raises the cake solids by 2 percentage points. The upper limit is approximately 60 to 65 °C (140 to 150 °F) when the polymer begins to break down. The BOD in the centrate edges up also, but not to the point of causing unnecessary concern. Considering that this has been known for years, few, if any, plants take advantage of it because it takes a lot of energy to heat the sludge and there are concerns that odors may increase.

Process Control. The following shutdown procedures are suggested:

- Stop biosolids and polymer feed to the centrifuge,
- Add flush with water (effluent water is acceptable) until the centrate is clear and the torque level begins to drop,
- Turn the centrifuge off,
- Continue flushing at 25% of normal feed flow until the centrifuge reaches 7 to 800 r/min, and
- Turn off the lubrication system and cooling water when the unit is completely stopped.

The purpose of flushing is to remove solids from the centrifuge. If no solids are coming out, then the flushing serves no purpose.

Troubleshooting. *Obtaining Drier Cakes.* Chances are, the polymer was chosen to achieve a given biosolids dryness. If a drier cake is needed, it will probably be necessary to change to a different polymer, so the polymer supplier will need to be involved. Likewise, the weirs may be set too low. Consider how long it takes to seal the centrifuge. If it seals very quickly (i.e., in 1 or 2 minutes), chances are the weir plates are set lower than is ideal for dry cake. Raise them in 5-mm (radial) increments, until it takes 5 to 8 minutes to seal the centrifuge. For the same torque level, the centrifuge should now operate at a lower differential, and the centrate will be cleaner. Increase the torque set point until more polymer will not clean it up or torque runs out.

Reducing the Polymer Consumption. Common causes of high polymer consumption are the following:

- Poor quality feed biosolids. See the "Operating Principles for Dewatering" section.
- If the biosolids is very viscous, add dilution water to the polymer. The polymer has to reach every biosolids particle to capture it in a floc, and lowering the viscosity helps.
- Change the polymer addition point. Most manufacturers recommend four addition points.
- Higher polymer consumption results in less aging time for the polymer solution, which, in turn, results in increased polymer usage. Check to see that the polymer aging time is sufficient.
- The biosolids do not react well with the polymer. Check the reference sample of polymer in the laboratory refrigerator to see that the polymer currently on-site reacts at least as well as that the vendor field tested. See the polymer supplier to determine if a different polymer will do better than the one being used.

Ferric salts will result in lower polymer dosages, although the polymer savings alone will not pay for the cost of the ferric salts.

Increasing Capacity.

- Increasing the feed rate or the feed solids rate will reduce the solids residence time. This will result in wetter cake.
- More polymer will offset this somewhat.
- As mentioned earlier, the difference in height between the pond level and the solids discharge level is important. When the feed rate changes, the crest over the dams changes as well. Higher cresting over the weirs may result in too high a ΔH, and the weirs need to be lowered a few millimeters (Figure 33.15).

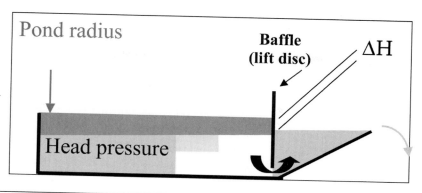

FIGURE 33.15   Differential head pressure.

- Increasing the feed solids concentration, thereby lowering the feed rate, helps as long as the feed is not too viscous and does not run the risk of going septic.
- Operating for more hours or operating more centrifuges will help.
- Add ferric chloride.

Table 33.7 presents a troubleshooting guide for centrifuges that enables the operator to identify problems early and determine their possible solutions.

VACUUM AND PRESSURE FILTERS. *Inorganic Chemical Conditioning.* Inorganic chemical conditioning is associated with rotary vacuum filters and pressure filtration dewatering. The chemicals typically are lime and ferric chloride. Ferrous sulfate, ferrous chloride, and aluminum sulfate are less commonly used.

Lime. Vacuum filters and filter presses commonly use lime and ferric chloride to make the sludge easier to filter and improve the release of the sludge from the filter media. Lime is available in two dry forms—quicklime (calcium oxide) and hydrated lime [$Ca(OH)_2$]. When using quicklime, it is first slurried with water and converted to calcium hydroxide, which is then used for conditioning. Because this process (known as *slaking*) generates heat, special equipment is required. Quicklime is typically available in three grades—high (88 to 96% calcium oxide), medium (75 to 88% calcium oxide), and low (50 to 75% calcium oxide). Because the grades can affect the slaking ability of the material, this should be considered before purchase. In general, only quicklime that is highly reactive and quick-slaking should be used for conditioning. Quicklime must be stored in a dry area, because it reacts with moisture in the air and can become unusable.

Hydrated lime is much easier to use than quicklime, because it does not require slaking, mixes easily with water with minimal heat generation, and does not require

TABLE 33.7   Troubleshooting for centrifuges.

Indications/ Observations	Probable cause	Action
Overall performance inadequate	The sludge no longer reacts well with the polymer	Check plant operations upstream, the polymer make-up system, and the polymer quality. If there is no obvious cause, see polymer vendor for help
	Excessive feed rate	Try lowering the feed rate
	Feed solids too thick	Conduct jar tests with diluted polymer try adding post dilution water to polymer, or thinning out the feed solids
	Raise bowl speed	Check with the centrifuge manufacturer before you do this.
	Pond depth too low	Check weir setting. Raise pond level by ±6 mm
High torque	Cake solids too viscous	Reduce torque setting to obtain wetter, less viscous cake
		Try a different polymer. Some polymers generate more torque than others
Centrifuge won't seal	The sludge no longer reacts well with the polymer	See above
	Pond setting too high	Lower pond by ±6 mm. This will buy time until you solve the problem
Erratic performance	Feed or polymer variations	Try to maintain uniform feedstock. Check polymer batches to see that they are uniform. It may be necessary to change the pond so as to function OK with all sludges, and great with none
High, intermittent vibration	Poor scrolling of the sludge inside the centrifuge	Lowering the bowl speed by several hundred rpm usually works. Alternately try deeper ponds, and/or higher feed rates
Vibration rising slowly over a week	Conveyor bearing failing	Disassemble and check bearings
	Solids blocking feed zone of clogging conveyor	Shut down, flush, and restart. If problem persits, try lowering the bowl speed.
Vibration rising quickly	Shut down immediately, and check bearing temperature	Call maintenance to see if the main bearings are bad
Continuous high vibration	Solids between the feed zone and rear conveyor bearing	Shut down, remove feed tube, and pressure wash
	Serous mechanical problems	Shut down and disassemble. Check for erosion, worn metal, bent flights, and bearings

any special storage conditions. Because hydrated lime is more expensive and less available than quicklime, quicklime should be used for applications that require more than 2 to 3 ton/d. This means that typically only larger WWTPs will use quicklime.

Lime typically is used in conjunction with ferric iron salts. Although lime has some slight dehydration effects on colloids, odor reduction, and disinfection, it is used because it improves filtration and release of the cake from the filter media. The lime reacts with bicarbonate to form a precipitate of calcium carbonate, which provides a granular structure that increases porosity and reduces compressibility of the biosolids.

Dosage Requirements. Iron salts are added at a dose rate of 20 to 62 kg/metric ton (40 to 125 lbs/ton) of dry feed solids, whether or not lime is used. Lime dosage varies from 75 to 277 kg/metric ton (150 to 550 lbs/ton) of dry solids dewatered. Table 33.8 lists typical ferric chloride and lime dosages. However, some organic polymers can be used with lime instead of ferric chloride.

Inorganic chemical conditioning greatly increases the mass of solids to be discarded. The operator should expect 0.5 kg (1 lb) of additional chemical solids for every 0.5 kg

TABLE 33.8 Typical conditioning dosages of ferric chloride and lime for municipal wastewater sludges.[a,b]

	Vacuum filter		Pressure filter	
Sludge	FeCl$_3$[c]	CaO[d]	FeCl$_3$	CaO
Raw sludges				
Primary	40–80	160–200	80–120	220–280
Waste activated	120–200	0–320	140–200	400–500
Primary/trickling filter	40–80	180–240		
Primary/WAS[e]	50–120	180–320		
Primary/WAS (septic)	50–80	240–300		
Elutriated anaerobically digested sludges				
Primary	50–80	0–100		
Primary/WAS	60–120	0–150		
Anaerobically digested sludges				
Primary	60–100	200–260		
Primary/WAS	60–120	300–420		
Primary/trickling filter	80–120	250–350		
Thermally conditioned sludges	None	None	None	None

[a]All values shown are for pounds of either FeCl$_3$ or CaO per ton of dry solids pumped to the dewatering unit.
[b]1 lb/ton = 0.5 kg/metric ton.
[c]Ferric chloride.
[d]Calcium oxide.
[e]Waste activated sludge.

(1 lb) of lime and ferric chloride added, and also more attendant water. This increases the total amount of biosolids for disposal and lowers its fuel value for incineration. With the development of polymers, the use of inorganic conditioning and related processes have declined severely.

Other Types of Conditioners. Other types of inorganic materials have been used as conditioning agents. Cement kiln dust fly ash, power plant ash, and biosolids incinerator ash have been used to increase the dewatering rate, improve cake release, increase cake solids, and, in some cases, reduce the dosage of other types of conditioning agents.

Pressure Filters. Principles of operation, process variables, process control, troubleshooting, startup and shutdown procedures, safety concerns, and maintenance considerations are described below. Figure 33.16 illustrates a plate and frame press.

FIGURE 33.16   Plate and frame press (courtesy of Siemens Water Technologies Corp.).

Principles of Operation. Filter presses for dewatering are generally either recessed plate filters or diaphragm filter presses. A typical pressure filter is illustrated in Figure 33.17. With the advent of better organic polymers, belt filters and centrifuges have largely displaced filter presses in the market. Filter presses can be attractive in unusual circumstances.

The fixed-volume recessed plate filter press (Figures 33.17, 33.18, and 33.19) consists of a series of plates, each with a recessed section that forms the volume into which the feed enters for dewatering. Figure 33.17 shows the general external mechanics of the basic filter press. Filter media or cloth, placed against each plate wall, retains the cake solids while permitting passage of the filtrate. The plate surface under the filter media is specifically designed with grooves between raised bumps to facilitate passage of the filtrate while holding the filter cloth.

Before pumping into the press, the feed must be chemically conditioned to flocculate the solids and release the water held within the solid mass. Most typical conditioning systems use inorganic chemicals and organic polymers.

Principles of Conditioning. Conditioning is performed to increase particle size by combining small particles into larger aggregates. Conditioning is a two-step process consisting of coagulation and flocculation. Coagulation involves the destabilization of the

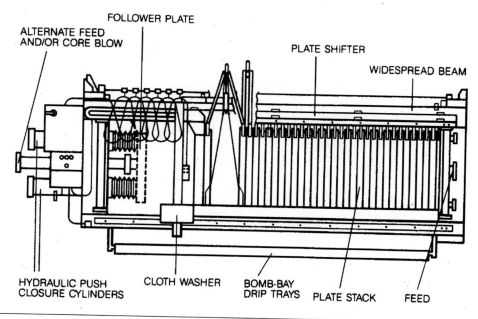

FIGURE 33.17 Fixed-volume recessed plate filter press.

FIGURE 33.18   Fixed-volume recessed plate filter press.

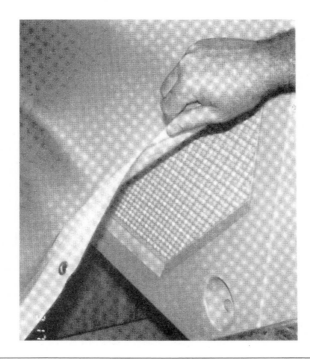

FIGURE 33.19   Plate drainage holes behind cloth on fixed-volume recessed plate filter press.

negatively charged particles by decreasing the magnitude of the repulsive electrostatic interactions between the particles.

The second step—flocculation—is the agglomeration of colloidal and finely divided suspended matter after coagulation by gentle mixing.

Inorganic Chemical Conditioning. Inorganic chemical conditioning is associated principally with vacuum and pressure filtration dewatering. The chemicals typically are lime and ferric chloride. Ferrous sulfate, ferrous chloride, and aluminum sulfate are also used, although less commonly.

Ferric Salts. Ferric chloride solutions typically are used at the concentration received from the supplier (30 to 40%); however, some WWTPs dilute the ferric chloride to approximately 10% to improve mixing and reduce the acidity and corrosivity of the material. This can be done in day tanks or inline. Dilution may lead to hydrolysis reactions and the precipitation of ferric chloride crystals.

An important consideration in the use of ferric chloride is its corrosive nature. It reacts with water to form hydrochloric acid, which attacks steel and stainless steel. When diluted with biosolids, the acidity is neutralized by the alkalinity of the biosolids and thoroughly diluted so that the end product is quite benign. Interlocks must be used to ensure that ferric chloride is always added to biosolids in the proper ratio and is never pumped into biosolids lines or process equipment by itself.

Special precautions must be taken when handling this chemical. The best materials are epoxy, rubber, ceramic, polyvinylchloride, and vinyl. Contact with the skin and eyes must be avoided. Rubber gloves, face shields, goggles, and rubber aprons must be used at all times. Ferric chloride can be stored indefinitely without deterioration. Customarily, it is stored in aboveground tanks constructed of resistant plastic and surrounded by a containment wall. Ferric chloride can crystallize at low temperatures, which means that the tanks must be kept indoors or heated.

Lime. Lime typically is used in conjunction with ferric iron salts, principally because it increases porosity and aids in the release of cake. Inorganic chemical conditioning increases the mass of solids to be discarded. The operator should expect 0.5 kg (1 lb) of additional chemical solids for every 0.5 kg (1 lb) of lime and ferric chloride added.

High-pressure pumps force the feed into the space between the two plates. The filtrate passes through the cake and the filter media and out of the press through special ports drilled in the plate. Figure 33.20 shows a schematic of a typical pressure filtration system.

Pumping continues up to a given pressure and is stopped when solids and water fill the void volume between the filter cloths and filtrate flow slows to a minimal rate. The press then opens mechanically and the cake is removed, one chamber at a time.

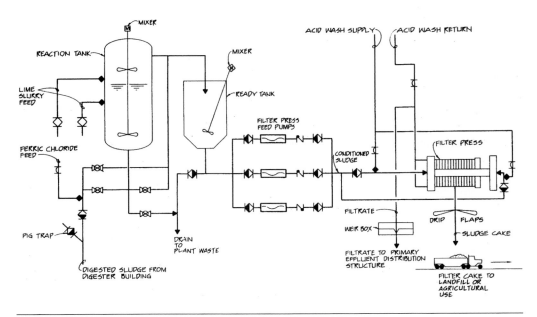

FIGURE 33.20   Schematic of a typical filter press system.

A variation of this process is the diaphragm press. After the press is full, the diaphragms are pressurized up to 1380 to 1730 kPa (200 to 250 psi) using either air or water, which expands the diaphragm and presses out more water.

Process Variables—Chemical Conditioning. Polymers have a narrow range of effective dosage. A dose that is too low or too high will result in a wet cake. Lime and ferric chloride have a broader range of effective dosage. While it is desirable for an operator to reduce the chemical usage to reduce costs, if erratic equipment operation or erratic feed qualities occur a higher lime dose typically will protect against a wet cake. Polymer conditioning requires much less chemical per unit mass of solids dewatered, which results in more room in the press for organic biosolids, and increased capacities.

- Ideally, the solids capture is 99%. A torn cloth immediately results in a filtrate flow that is very dirty and heavy with solids.
- Feed solids concentration. A very thin feed may blow out through the plate surfaces during the initial high-volume fill of the press because there would be too much filtrate flow for the drain capacity. A thin feed will at least require a longer filtration time and produce a wetter cake. A thick feed typically will produce a drier cake with a much shorter filtration time.

For a conventional filter press, the operator can control the following machine variables:

- Feed pressure. There are two differential operating ranges for filter presses— 656 to 897 kPa (95 to 130 psi) and 1380 to 1730 kPa (200 to 250 psi).
- Feed application rate, by pacing the flow to the filter press.
- Overall filtration time, including such variables as the time at each pressure level in multiple pressure level operations.
- Use and amounts of precoat or body feed. Typically, precoat is unnecessary when inorganic chemicals, such as lime and ferric chloride, are applied. Precoat may be needed if particle sizes are extremely small, filterability varies considerably, or a substantial loss of fine solids to and through the filter media is anticipated.
- Conditioning chemicals, type, dosage, location, and mixing efficiency. Polymer addition versus lime and ferric chloride conditioning typically are not interchangeable, as each chemical requires special mixing and flocculation energies and reaction times. Polymers only need a quick mix before injection to the press. Modifications to the piping and mixing systems are typically needed if a change in the type of chemical for conditioning is desired.
- Flocculation efficiency and energy vary with the type of chemical being used. Polymer floc shears easily and remains stable for only a few minutes. Lime floc is more durable and remains stable for a few hours.
- Filter media. Filter cloth media vary widely, with different filament composition, weave pattern, and weave tightness. A typical cloth for dewatering organic biosolids is made of polypropylene filaments woven to obtain 2.3 to 3.4 m^3/min (80 to 120 scfm) porosity.

Process Control. Typical operation of the filter press requires visual checks of equipment conditions before starting the automatic filter cycle. Most pressure filter installations have an automatic device that closes the filter plates and clamps them together. Before closing, the plates and filter cloth need to be inspected.

After the chemical solution has been prepared, and the day or ready tank contains sufficient feed stock, the filtration cycle can begin. Initially, a high feed rate is desired to fill the press evenly and completely in a relatively short time. Either multiple pumps or a high-volume, low-pressure pump is used for this initial quick fill. After the internal liquid pressure reaches approximately 380 kPa (55 psi), all feed pumps shut down, except for a small-volume, high-pressure pump that continues up to a given pressure. As the solids concentration in the filter press chambers increases, the filtration pressure will increase. After reaching the maximum pressure—typically 656 to 897 kPa (95 to 130 psi)—the pressure and the filtration cycle continue until a predetermined cake concentration is achieved.

Some presses are depressurized by initiating the core blow process, which removes wet solids that remain in the central feed core. If a core blow process is not included, some liquid will drain out the bottom when the press is opened.

1. Place the drip trays in the open position.
2. Retract the hydraulic ram.
3. Start the automatic plate shifter sequence to separate and jar the plates, one at a time, to drop the dewatered cake into the container below. Some of the cake may have to be removed with plastic scrapers. Note that poor cake release from the cloth may indicate improperly conditioned feed. A high lime dosage for the following filtration cycle may remove most of the residual wet solids from the cloth, or the filter cloth may need to be washed, as recommended in the manufacturer's literature.

Troubleshooting. Table 33.9 is a troubleshooting guide for pressure filters that will help operators to identify problems and determine their possible solutions.

Startup and Shutdown Procedures. Because each system has unique equipment with unique system interlocks, it is imperative that operators be thoroughly trained in the startup and shutdown procedures for their particular pressure filter system. Following are some general guidelines on startup and shutdown.

Whenever major changes occur to the chemical and biosolids feeds, the operator should perform a laboratory vacuum filtration test to confirm that the water will release from the solids. Additional testing is sometimes performed, such as a leaf test or a capillary suction test.

Also, before startup after major maintenance procedures and during each cake discharge cycle, the operator should check each filter plate and cloth as follows:

- Filter cloths should be attached to the plates without folds and be free of dirt or foreign objects;
- No objects should be on top of or between the plates; and
- The cloths should be observed for tears, especially at the center feed seams and at the boss (intermediary support) points.

Without these checks, the press can be damaged, and operation of the stem can be jeopardized. If lime is used, mix the proper concentration before biosolids transfer. Typi-

TABLE 33.9   Troubleshooting guide for pressure filtration.

Indicators/ observations	Probable cause	Check or monitor	Solutions
Plates fail to seal	Hydraulic pressure too low	Manufacturer's recommended pressure	Adjust to specifications
	Feed pressure too high	Feed pump pressure	Adjust to proper range
	Rags or solids on plate seal surfaces	Plate sealing surfaces	Inspect sealing surfaces during discharge and clean as needed
Cake discharge is difficult—wet cake	Inadequate precoat	Prevent feed	Increase precoat, feed at 172–276 kPa (25–40 psig)
	Improper conditioning	Conditioner type and dosage	Change conditioner type or dosage based on history or bench vacuum filter test
	Filtration time too short	Filtration time, filtrate flowrate, and feed pressure	Extend filtration cycle time
Filter cycle times excessive	Conditioned sludge too old	Time since conditioning	Do not save old conditioned sludge, or add extra lime before using
	Feed solids too low	Operation of thickening process	Improve solids thickening to increase solids concentration in press feed
	Improper precoat feed	Precoat feed	Decrease precoat feed substantially for a few cycles, then optimize
	Filter media plugged/ calcium buildup in media	Filter media	Wash filter media/acid wash (inhibited hydrochloric acid)
Dirty filtrate	Ripped cloth or cloth with too high a porosity	Inspect all cloths	Replace damaged or incorrect cloths
	High feed pressure	Check operating history	Adjust feed pressure within specifications
Frequent media binding	Precoat inadequate	Precoat feed	Increase precoat
	Initial feed rates too high	Pump rate	Develop initial cake more slowly
	Improper conditioning	Conditioning dosage	Change chemical dosage
	Cloth with too low a porosity	Inspect all cloths and consult with manufacturer	Replace incorrect cloths
Excessive moisture in cake	Improper conditioning	Conditioning dosage	Change chemical dosage
	Filter cycle too short	Correlate filtrate flowrate with cake moisture content	Lengthen filter cycle

TABLE 33.9   Troubleshooting guide for pressure filtration (*continued*).

Indications/observations	Probable cause	Check or monitor	Solutions
Sludge blowing out of press	Obstruction, such as rags, in the press forcing sludge between plates	Feed pressure	Shut down feed pump, hit press closure drive, restart feed pump; clean feed eyes of plates at end of cycle
	Cake complete	Feed pressure, filtrate flowrate, and filtration time	Shut down press and unload cake

cally, a 10% lime slurry solution is used, with a 1.06 specific gravity, as measured by a hydrometer.

The details of the press cycle vary greatly with the press design and will not be discussed here.

Safety Concerns. Following are some safety concerns for the operation of pressure filters. As always, complete site-specific safety procedures must be written, approved, and followed for each facility.

- Many presses are very noisy when in operation. Hearing protection may be needed.
- Never insert objects between the press plates as they are being discharged without first shutting the unit down by tripping the light curtain or flipping the emergency shutdown switch.
- Lime treatment results in considerable ammonia fumes being released during cake discharge. Make sure that adequate ventilation pulls these fumes away from the operator, preferably with a high-capacity, down-draft blower system. If an adequate ventilation system is not operational, then, if approved, short-term exposure may be allowed, if an approved ammonia respirator is worn by all personnel assisting with the cake discharge.
- Hydrochloric acid washing of the press releases volatile acid fumes, which should not be inhaled or exposed to moist body tissues, such as eyes and lungs. A high-capacity ventilation system, as previously noted, is essential. If approved, short-term exposure may be allowed with an approved respirator and complete coverage of all exposed skin.
- Lime powder is very caustic when it contacts moist body tissues. Therefore, an approved respirator and complete coverage of all exposed skin is necessary when working around lime.

Maintenance Considerations. Follow all equipment manufacturer's recommendations. Some typical areas that need special attention are as follows:

- The plate handles and the frame rails require frequent grease application to prevent binding and excessive wear.
- The plate shifter chain or other plate shifting devices require frequent lubrication.
- Even if a shredder is used before the conditioning step, rags will quickly accumulate on all mechanical mixer blades. These need to be removed frequently to prevent damaging the mixer gears and shaft bearings from operating out of balance. Some facilities have removed their mechanical mixers and converted to air mixing to avoid ragging problems.
- Lime systems scale up and plug over time and are unpleasant and potentially hazardous for operators to clean out. Some WWTPs have switched from quicklime to hydrated lime to avoid maintaining lime slakers.
- When lime is used, cloth and plate washing may require both an acid and a water wash because lime causes scaling.
- Ferric chloride, hydrochloric acid, lime, and ammonia cause considerable corrosion to metal surfaces, such as the plate handles with their retaining bolts, the shifter chain, and even steel plates under a hard rubber cover. Frequent cleaning and lubrication are necessary to reduce corrosion. Powder-coated steel handles, polypropylene plates, and polytetrafluoroethylene-coated frame rails have been used by some facilities to reduce corrosion problems. Also, adding an inhibitor to hydrochloric acid will reduce its metal corrosion properties.
- From time to time, the cloths and gaskets will have to be removed, the plates pressure cleaned, and new cloths and gaskets cut and installed.

Vacuum Filters. *Principles of Operation.* Until the mid-1970s, vacuum filtration was the most typical means of mechanical dewatering. The principal variations among filters was the filter medium itself. Some of those used were natural and synthetic fiber cloth, woven stainless steel mesh, and coil springs.

With the development of effective polymers, centrifuges and BFPs have replaced most vacuum filters. A vacuum filter consists of a horizontal cylindrical drum that rotates while partially submerged to approximately 20 to 35% of its depth in a vat of previously conditioned feed. The filter drum is partitioned into several compartments or sections (Figure 33.21). A pipe connects each compartment to a rotary valve. Bridge blocks in the valve divide the drum compartments into three zones—the cake formation zone, cake drying zone, and cake discharge zone.

The submerged zone is the cake formation zone. A vacuum applied to the submerged zone causes the filtrate to pass through the media and solids particles to be

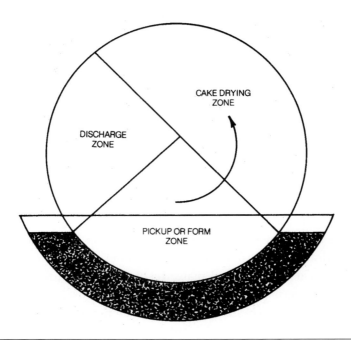

FIGURE 33.21   Operating zones in a rotary vacuum filter.

retained on the media. As the drum rotates, each section successively is carried through the cake formation zone to the cake drying zone. The cake drying zone begins when the filter drum emerges from the feed vat. The cake drying zone occupies 40 to 60% of the drum surface and ends at the point where the internal vacuum is shut off. At this point, the cake and drum section enter the cake discharge zone, where cake loosens and falls from the media. Figure 33.21 illustrates the various operating zones encountered during a complete revolution of the drum.

Process Variables. The principal variables that affect vacuum filter operation are as follows:

- Chemical conditioning,
- Filter media type and condition,
- Drum submergence,
- Drum speed, and
- Vacuum level.

Chemical conditioning most significantly influences dewaterability, because it changes the physical and chemical nature of the feed. Conditioning agents produce a feed that releases water more freely, thereby producing a drier cake. Conditioning agents also add to the solids to be discarded.

Drum submergence, the percentage of time the filter drum is submerged in the feed vat, typically varies between 15 and 25%. Increasing the submergence of the drum increases the filter cycle time devoted to cake formation, but reduces the cake drying ratio. This produces a thicker, but wetter, cake.

Drum speed affects the filter cycle time. Increasing the drum speed shortens both cake formation and cake dewatering cycles. Such action can produce a wetter cake, but increased filter yield, and vice versa. Drum speed also affects cloth cleaning. A higher speed decreases the time that the cloths are washed.

The vacuum is controlled by the drum design, drum speed, and vacuum pump size. The required vacuum level is also affected by the liquid level in the vat and the type of conditioning. For a dry cake, the vacuum typically ranges from 51 to 68 kPa (15 to 20 in. Hg). The lower the vacuum, the wetter the cake will be.

PREVENTIVE MAINTENANCE

At one time, labor was no problem at the plant, and most repairs were done in-house. Now, with a lot of pressure to reduce manpower, it is often more attractive to out-source the more time-consuming or specialized repairs. Management of the maintenance department becomes a key asset. They are the ones who supervise repairs and discuss repair methods with vendors. Repair of a million dollar centrifuge is much like repair of a car. Some jobs are for the dealer, some can be done by the owner, and some can be done by the local repair shop. Most repairs are more or less generic. Examples are broken fasteners, worn or damaged parts that can be removed and replaced, and worn or broken parts that can be repaired.

A good knowledge of mechanical things is essential. Likewise, knowledge of the capabilities of local and specialized repair firms is also valuable. It is hard to learn these things out of a book, but an intelligent person asks questions and considers the answers carefully.

As a rule, using the original manufacturer's parts, lubricants, repair services, and service intervals should result in better service life. Typically, it will also cost more. There are a number of alternatives for the operator to consider.

LUBRICATION. The manufacturer's instruction book may list oil change intervals for the worst-case scenario. However, if oil is sent out for analysis, and it meets new oil standards, then it does not need to be changed. Keep periodically testing the oil, to determine when it begins to break down, and adjust the lubrication schedule accordingly.

Consider buying lubricants in bulk directly from the lubricant manufacturer. Purchasing individual tubes of grease when 12 tubes are recommended to flush a bearing is not a good idea. Consider substituting generic lubricants. Most applications are standard, and a knowledgeable lubrication engineer should be able to select a suitable substitute.

BEARINGS. The same applies to bearing changes. When a bearing is removed, consider sending it out for analysis, and adjust the bearing change interval based on that analysis. When possible, buy the bearings from the bearing manufacturer rather than the equipment manufacturer. An internet search could locate other manufacturers who make a similarly rated bearing and will sell it at an even lower price. Or, if it is a common bearing application, it may be more convenient to go to the local bearing and drive store and pick one up. Unusual bearings and unusual applications involve a risk when substitutions are used. Ask other owners of the equipment for their experiences.

REPAIRS. The easiest repairs are simple replacements. When a part is worn out, a new one can be purchased and installed. Even specialized parts, such as grinder teeth and gearbox parts for specific brands and devices, are available from third-party parts suppliers. The trick is finding them and determining if their quality is adequate. One source of information is the annual WEFTEC conference. Each year, there are parts and repair specialists who set up booths at the convention, and attendees can meet with a dozen or more each day. It may be worth sending someone from the maintenance department to the conference to interview them.

DEALING WITH THE REPAIR SHOP. Many repairs are more or less standard. Shafts become grooved, metal wears away, and bearing surfaces warp to become egg-shaped (out of round). Unless there is something special about these sorts of repairs, a local repair shop can probably do a very good job with them. To begin with, tell the repair shop staff specifically what they need to do. Do not send in a piece of hardware and simply say "fix it." If there is something unusual about the failure, tell them that also. A pump with a bad bearing will get a different treatment than a pump with bearings that fail repeatedly. The repair person will not know the history of the piece until they are told. Solicit bids from several sources. Take digital photos, and send them to the following:

- The manufacturer of the equipment,
- Manufacturers who make similar equipment, and
- Independent repair shops that are dependable.

Take advantage of expertise. Ask the following questions:

- What repair options are there?
- Why did it fail, and is there something that can be done to prevent it from happening again so soon?
- Is there something that can be done now to reduce the repair cost next time?

For example, a typical seal surface can be chrome-plated and ground back to dimension; it can be built up with weld and ground back; or it can be machined out and a replaceable sleeve installed. It is important to consider which is the best fix in a particular case.

If, in the end, it is determined that the problem is something only the original manufacturer should work on, a price can be negotiated. When the equipment is sent to someone for big repairs, demand a full report of what was done, including the return of replaced parts. The report should document the damage found, with digital photos, and a description of the repair method. The report should list, by name and part number, all of the parts used, the price paid for them, and the name of the company who made the parts. While this might seem excessive, all of this is information that the repair shop needed to calculate the price of the job.

DATA COLLECTION AND LABORATORY CONTROL. Process control is the keystone for optimum performance of a dewatering system. The operator needs to keep records of all dewatering performance parameters. Sample points must be located at various places throughout the system. For mechanical dewatering systems, at least twice per shift, a sample should be taken of the feed biosolids, cake discharge, and filtrate or centrate. However, composite sampling at these three locations is better than grab sampling because a composite sample will give a better picture of performance. Total solids should be determined for the feed and cake samples and total suspended solids determined for the filtrate.

Standard Methods for the Examination of Water and Wastewater (APHA et al., 2005) is the principal source for all laboratory protocol. The operator needs to make the following types of calculations:

- Calculation of dilute polymer concentration from a water solution of polymer, liquid polymers, and dry polymers; and
- Calculation of polymer use and cost of solids dewatered.

Successful dewatering depends on periodic, frequent laboratory tests to check the polymer dosages. Regular tests are critical. Furthermore, it is vital that the operator understands the conditioning process. By observing the feed, filtrate, and cake, and by properly adjusting polymer dosages, the operator can ensure optimum performance of the dewatering system.

SAFETY AND HOUSEKEEPING. The phrase "accidents do not happen, they are caused" may be overused, but it is true. Common sense and a thorough understanding of the equipment and processes are essential to a safely operating WWTP. Before operating any conditioning or dewatering system, the best safety precaution is that all personnel involved read and thoroughly understand the operation and maintenance manuals provided by the process equipment suppliers.

Good WWTP housekeeping can prevent many accidents. Equipment should always be kept in first-class condition. Unsafe actions and conditions should be corrected.

Employees have the responsibility of safety to themselves, their fellow workers, and the WWTP. Cutting corners or taking chances while operating or maintaining equipment is not advisable. Refer to Chapter 5 for a thorough discussion of safety procedures and a listing of WWTP safety references.

REFERENCES

American Public Health Association; American Water Works Association; Water Environment Federation (2005) *Standard Methods for the Examination of Water and Wastewater*, 21st ed.; American Public Health Association: Washington, D.C.

Fleurial, D. (2003) Personal communication, Shelburne Falls WWTP, Shelburne Falls, Massachusetts.

Garvey, D. D. (2002) Personal communication, Keystone Water Quality Manager.

Gillette, R. A.; Scott, J. D. (2001) Dewatering System Automation: Dream or Reality? *Water Environ. Technol.*, **13** (5), 44–50.

Lambert, S. D.; McGrath, K. E. (2000) Can Stored Sludge Cake Be Deodorized by Chemical or Biological Treatment? *Water Sci. Technol.*, **41** (6), 133–139.

Macray, J. (2000) Personal communication, City of Windsor, Ontario, Canada.

Minneapolis Wastewater Treatment Plant (1999) Full-Scale Centrifuge Demonstration Project Centrifuge Test Program Final Report, June 30; Minneapolis Wastewater Treatment Plant: Minneapolis, Minnesota.

Pramanik, A.; LaMontagne, P.; Brady, P. (2002) Automatic Improvements, Installing an Integrated Control System Can Improve Sludge Dewatering Performance and Cut Costs. *Water Environ. Technol.*, **14** (10), 46–50.

Rudolph, D. J. (1994) Solution to Odor Problems Gives Unexpected Savings. *Water Eng. Manage.*, 9/94.

Vesilind, P. A. (1994) The Role of Water in Sludge Dewatering. *Water Environ. Res.*, **66,** 4–11.

Water Environment Federation (1995) *Engineered Equipment Procurement Options to Ensure Project Quality*; Water Environment Federation: Alexandria, Virginia.

Water Environment Federation (1998) *Design of Municipal Wastewater Treatment Plants,* 4th ed.; Manual of Practice No. 8; Water Environment Federation: Alexandria, Virginia.

Water Environment Research Foundation (1993) *Guidance Manual for Polymer Selection in Wastewater Treatment Plants;* Water Environment Research Foundation: Alexandria, Virginia.

Water Environment Research Foundation (1995) *Polymer Characterization and Control in Biosolids Management*; Water Environment Research Foundation: Alexandria, Virginia.

Water Environment Research Foundation (2001) Thickening and Dewatering Processes: How to Evaluate and Implement an Automation, Package 2001 Final Report, Project 98-REM-3; Water Environment Research Foundation: Alexandria, Virginia.

Symbols and Acronyms

A	Area of transmission loss
A^2/O	Anaerobic/anoxic/oxic
AC	Alternating current
A/O	Anoxic/oxic
ABF	Activated biofilter
ACOE	Army Corps of Engineers
AD	Apparent density
ADA	Americans with Disabilities Act
AD/AT	Factor representing number of diffusers per unit area of basin
ADF	Average daily flow
AE	Aeration (electrical) efficiency
AIDS	Acquired immune deficiency syndrome
AIHA	American Industrial Hygiene Association
AlCl$_3$	Aluminum chloride
Al$_2$(SO$_4$)$_3$	Aluminum sulfate
AMA	American Management Association
AMB	Approximate mass balance
ANSI	American National Standards Institute
APHA	American Public Health Association
APWA	American Public Works Association
ASA	American Standards Institute
ASCE	American Society of Civil Engineers
ASD	Adjustable speed drives
ASHRAE	American Society of Heating, Refrigerating and Air-Conditioning Engineers
ASM	Activated sludge models
ASM2	Activated sludge models, second model

ASM3	Activated sludge models, third model
ASP	Application service provider
ATAD	Autothermal thermophilic aerobic digestion
AST	Activated sludge treatment
ATS	Automatic transfer switch
AWWA	American Water Works Association
AWWARF	American Water Works Association Research Foundation
Ax	Anoxic
AxSF	Anoxic step feed
B	Y-axis intercept
BACT	Best available control technology
BBP	Blood-borne pathogen
BEP	Best engineering practices
BEPs	Best efficiency points
BETX	benzene, ethyl benzene, toluene, and xylene isomers
BF/AS	Biofilter/activated sludge
BFP	Belt filter press
BHC	Benzenehexachloride
BioP	*Acinetobacter*
BioP	Biological phosphorus
BMPs	Best management practices
BMR	Baseline monitoring report
BNR	Biological nutrient removal
BOD	Biochemical oxygen demand
BOD_5	Five-day biochemical oxygen demand
BOD_{2INF}	Influent BOD
BOT	build–own–transfer
BOOT	build–own–operate–transfer
BPR	Biological phosphorus removal
BrCl	Bromine chloride
BTU	British thermal unit
CH_3COOH	Acetic acid
$C_2H_4O_3$	Peracetic acid
C	Carbon
C	Constant: 0.25 for fecal coliform and 0.9 for total coliform
CF	Float solids concentration (percent) at the top of the unit
C_s	Specific heat of sludge
C:N	Carbon-to-nitrogen ratio
CAC	Combined available chlorine
CAD	Computer-aided design
CaO	Calcium oxide
$CaCO_3$	Calcium carbonate
$Ca(OH_2)$	Calcium hydroxide
$Ca(OCl)_2 \cdot 4 H_2O$	Calcium hypochlorite

CAS	Chemical Abstracts Service
CAS	Conventional activated sludge
CBOD	Carbonaceous biochemical oxygen demand
$CBOD_5$	Five-day carbonaceous biochemical oxygen demand
$CBOD_{5,IN}$	Influent carbonaceous biochemical oxygen demand
$CBOD_{5,SE}$	Secondary effluent carbonaceous biochemical oxygen demand
CCP	Composite Correction Program
CDs	Computer disks
CD-ROM	Compact disc read-only memory
CDX	Central data exchange
CERCLA	Comprehensive Environmental Response, Compensation, and Liability Act
CFD	Computational fluid dynamics
CFM	Computational fluid mixing
CFR	*Code of Federal Regulations*
CFU	Colony forming units
CH_4	Methane
CHIP	Chemicals Hazard Information and Packaging Regulations
CIUs	Categorical industrial users
Cl_2	Chlorine
ClO_2	Chlorine dioxide
CM	Corrective maintenance
CMMS	Computerized maintenance management systems
CMOM	Capacity, management, operation, and maintenance
C:N	carbon-to-nitrogen ratio
CPE	Comprehensive performance evaluations
CPU	Central processing unit
CO	Carbon monoxide
CO_2	Carbon dioxide
COD	Chemical oxygen demand
COTS	Commercial off-the-shelf
CPR	Cardiopulmonary resuscitation
CPVC	Chlorinated polyvinyl chloride
CSF	Clarifier safety factor
CSOs	Combined sewer overflows
CST	Capillary suction time
CSTRs	Completely stirred tank reactors
CT	Contact time
CWA	Clean Water Act
CWF	Combined waste stream formula
CWSRF	Clean Water State Revolving Fund
D	Diameter
D	UV dose
dB	Depth of float below the water level

dW	Height of float above water level
dB + dW	Total float depth
DAF	Dissolved air flotation
dB	Decibels
DBO	Design–build–operate
DBPs	Disinfection byproducts
DC	Direct current
DCS	Distributed control systems
DDD	dichlorodiphenyldichloroethane
DDE	dichlorodiphenyldichloroethylene
DDT	dichlorodiphenyltrichloroethane
DMRs	Discharge monitoring reports
dN	Denitrification
DNA	Deoxyribonucleic acid
dN/dt	Rate of change of the number of organisms population
DO	Dissolved oxygen
DOB	Depth of blanket
DOT	Department of Transportation
DPAM	Dry polyacrylamide
DPIV	Digital particle image velocimetry
DV	Digester volume
DWP	Dynamic wet pressure
E	Effluent rate
E	Effluent concentration
e_c	Corrected effluent concentration
e_m	Measured effluent concentration
EBPR	Enhanced biological phosphorus removal
ED_{50}	Effective dose at the 50% level
EDMS	Electronic document management system
EHS	Extremely hazardous substances
EH&S	Environmental Health and Safety
EM	Electromotive force
EMP	Energy management plan
EMS	Environmental management system
EPAct	Energy Policy Act
EPAM	Emulsion polyacrylamide
EPCRA	Emergency Planning and Community Right-to-Know Act
EQ	Exceptional quality
ERG	Enforcement response guide
ERP	Enforcement response plan
U.S. EPA	U.S. Environmental Protection Agency
ES	Extended Service
ESD	Egg-shaped digesters
ESEERCo.	Empire State Electrical Energy Research Corporation

E_x	Ultraviolet reactor dispersion coefficient
F	Force
F	Feed rate
f	Feed concentration
F:M	Food-to-microorganism ratio
FAC	Free available chlorine
f_c	Corrected feed concentration
Fe	Iron
$FeCl_2$	Ferrous chloride
$FeCl_3$	Ferric chloride
FEMA	Federal Emergency Management Agency
f_m	Measured feed concentration
FF/SG	Fixed-film and suspended growth processes
FFT	Fast Fourier transform
FIS	Financial information system
FOG	Fats, oil, and grease
FR	Federal Regulations
FRP	Fiberglass-reinforced plastic
FSS	Fixed suspended solids
FWA	Flow-weighted average
G	Balance quality grade number
GAC	Granular activated carbon
GALPs	Good automated laboratory practices
GASB	Government Accounting Standards Board
GBT	Gravity belt thickener
GC	Gas chromatography
GFI	Ground fault interrupter
GISs	Geographic information systems
GPS-X	General purpose simulator
GS	General service
GW&PCA	Georgia Water and Pollution Control Association
H	Head
H_2	Hydrogen
HAZCOM	Hazardous Communication Standard
HID	High-intensity discharge
HO	High output
H/O/A	Hand-off-auto
HCO_3^-	Biocarbonate
HS^-	Hydrosulfide
H_2CO_3	Carbonic acid
H_2O	Water
H_2O_2	Hydrogen peroxide
H_2SO_4	Sulfuric acid
H_2S	Hydrogen sulfide

H_3PO_4	Phosphoric Acid
HCl	Hydrochloric acid
HDPE	High-density polyethylene
HGL	Hydraulic grade line
HIV	Human immunodeficiency virus
h_L	Headloss
HLR	Hydraulic loading rate
HOA	Hand-off automatic
H/O/A	Hands/off/automatic
HOBr	Hypobromous acid
HOCl	Hypochlorous acid
HP	Horsepower
HPO	High-purity oxygen
$HS\text{-}1^{A/M}$	Start/stop handswitch
$HS\text{-}2^{S/S}$	Automatic/manual handswitch
HRMS	Human resources management system
HRR	High resolution redox
HRT	Hydraulic residence (retention) time;
HVAC	Heating, ventilation, and air conditioning
I	Intensity of UV radiation
I&C	Instrumentation and control
ICMA	International City and County Management Association
IFAS	Integrated fixed-film activated sludge
IMLR	Internal mixed-liquor recirculation
IRS	U.S. Internal Revenue Service
ISO	International Organization for Standardization
ISV	Initial settling velocity
IT	Information technology
$iTSS_{2INF}$	Inert total suspended solids in secondary influent
IUs	Industrial users
IWA	International Water Association
JHB	Johannesburg
K	Disinfection rate constant
k	Organism die-off rate constant
k	Rate constant
$KMnO_4$	Potassium permanganate
KVAR	Kilovar
L	Concentration of organisms after ultraviolet exposure
L	Effective length of rectangular draft
L_0	Concentration of organisms before ultraviolet exposure
LAN	Local area network
LiCl	Lithium chloride
LIF	Laser-induced fluorescence
LC_{50}	Lethal concentration at the 50% level

LDA	Laser Doppler anemometry
LE	Ludzack–Ettinger
LEL	Lower explosivity limit
LIMS	Laboratory information management systems
LMMSS	Lockheed Martin Michoud Space Systems
LP–LI	Low pressure–low intensity
μ	Biomass growth rate
μ_{max}	Maximum growth rate
m	Slope of straight line
M	Mass
M_{Ox}	Mass of oxic zone
MAs	Member associations
MAHL	Maximuum allowable headworks loading
MBRs	Membrane bioreactors
MCCs	Motor control center
MCRT	Mean cell residence time
$MCRT_{MIN}$	Minimum mean cell residence time
$MCRT_{Ox}$	Oxic zone mean cell residence time
$MgNH_4PO_4$	Magnesium ammonium phosphate
MIL	Moisture indicator light
MIS	Management information system
ML	Mixed liquor
mLE	Modified Ludzack–Ettinger
MLR	Mixed-liquor recycle
MLR_{Ax}	Denitrified mixed-liquor recycle
MLR_{Ax}	Mixed-liquor recycle anoxic zone
MLR_{ox}	Mixed-liquor recycle oxic zone
MLSS	Mixed-liquor suspended solids
MLVSS	Mixed-liquor volatile suspended solids
MMR	Mumps, Measles, and Rubella
MOP	Manual of Practice
MPN	Most probable number
MS	Mass spectrometry
MSDS	Material safety data sheet
MW	Medium wave
MWWTP	Minneapolis (Minnesota) Wastewater Treatment Plant
N	Effluent bacteria density after ultraviolet radiation
N	Rotating speed
N	Number of surviving organisms per unit volume at any given time
N	Speed
N/dN	Nitrify/denitrify
N_0	Influent bacteria density before ultraviolet radiation
N_2	Nitrogen
NaCl	Sodium chloride

$Na_2Al_2O_4$	Sodium aluminate
Na_2CO_3	Sodium carbonate
Na_2SO_3	Sodium sulfite
$NaClO_2$	Sodium chlorite
$NaHCO_3$	Sodium bicarbonate
$NaOCl$	Sodium hypochlorite
$NaOH$	Sodium hydroxide
NaS_2O_5	Sodium metabisulfite
NACWA	National Association of Clean Water Agencies
NASA	National Aeronautics and Space Administration
NBP	National Biosolids Partnership
NBOD	Nitrogenous biochemical oxygen demand
NEC	National Electric Code
NEMA	National Electrical Manufacturers Association
NFPA	National Fire Protection Association
NH_3	Ammonia
NH_3-N	Ammonia as nitrogen
NH_3-N_{IN}	Influent total ammonia-nitrogen
NH_3-N_{SE}	Influent total Kjeldahl nitrogen
NH_4^+	Ammonia ion
NH_4HCO_3	Ammonium bicarbonate
NH_4OH	Ammonium hydroxide
NIOSH	National Institute for Occupational Safety and Health
NO_2^-	Nitrite
NO_3^-	Nitrate
$NO_{3, IN}$	Influent nitrate concentration
$NO_{3, SE}$	Secondary effluent nitrate concentration
$NO, NO_2,$ and NO_x	Nitrogen oxides
NOD	Nitrogenous oxygen demand
NO_x	Nitrogen oxide
NOx_{2EFF}	Secondary effluent nitrogen oxide concentration
NPDES	National Pollutant Discharge Elimination System
NPSH	Net positive suction head
NRC	National Research Council
NTU	Nephelometric turbidity unit
NWRI	National Water Research Institution
NYSERDA	New York State Energy Research and Development Authority
O&M	Operations and maintenance
O_2	Molecular or free oxygen
O_3	Ozone
OC	Oxygenation capacity
OCl^-	Hypochlorite ion
OCP	Operator control panel
OD	Oxygen demand

OH^-	Hydroxyl
OMS	Operations management system
ORP	Oxidation–reduction potential
OSHA	Occupational Safety and Health Act
OTE	Oxygen transfer efficiency
OTR	Oxygen transfer rate
OTMAX	Oxygen transferred to the mixed liquor
OUR	Oxygen uptake rate
Ox	Aerobic or oxic
$\alpha Fk_L a$	Oxygen mass transfer coefficient
P	Brake horsepower
P	Phosphorus
P	Polymer rate
P	Predictive
P	True power
P3	Public–private partnerships
P_{atm}	Absolute atmospheric pressure
P&IDs	Process and instrumentation drawings
PAM	Polyacrylamide
PASS	Pull, aim, squeeze, sweep
P_B	*Brake power*
P_M	Wire or motor power
P_{atm}	Absolute atmospheric pressure
P_{vp}	Fluid vapor pressure
P_W	Water power
PAOs	Phosphorus-accumulating organisms
PAC	Powdered activated carbon
PCs	Personal computers
PCBs	Polychlorinated biphenyls
PCS	Process control system
PD	Positive displacement
PDA	Personal digital assistant
PDM	Predictive maintenance
PELs	Permissible exposure limits
PFC	Power factor correction
PFRP	Processes to further reduce pathogens
PLCs	Programmable logic controllers
PM	Preventive maintenance
PMP	Planned maintenance program
PMS	Personnel management system
PO_4^-	Phosphate
POs	Purchase orders
POTWs	Publicly owned treatment works
PP	Polypropylene

PPE	Personal protective equipment
P/PM	Predictive/preventive maintenance
PR	Pumping rate
PSA	Pressure swing adsorption
PSRPs	Processes to significantly reduce pathogens
PRV	Pressure-reducing valve
PTFE	Polytetrafluoroethylene
PUCs	Public utilities commissions
PVC	Polyvinyl chloride
PVDF	Fluorinated polyvinylidene
Q	Digester transmission loss
Q	Flow capacity
Q	Sludge heating requirement
Q	Reactive power
$Q_{digested\ sludge}$	Volumetric flowrate of the digested sludge
Q_{feed}	Volumetric flowrate of the feed sludge
Q_{2INF}	Secondary influent flow
Q_{IN}	Influent wastewater flowrate
Q_S	Solids loading rate
Q_{WW}	Wastewater flow
QA	Quality assurance
QAP	Quality assurance plan
QAPP	Quality assurance project plan
QAU	Quality assurance unit
QC	Quality control
Q_{RAS}	Return activated sludge flow
Q_{WAS}	Waste activated sludge flow
RAS	Return activated sludge
RBCs	Rotating biological contactors
RCM	Reliability centered maintenance
RCFA	Root-cause failure analysis
RCRA	Resource Conservation and Recovery Act
RDT	Rotary drum thickeners
RF/AS	Roughing filter activated sludge
RFP	Request for proposals
RFQ	Request for qualifications
RMP	Risk management program
RNA	Ribonucleic acid
RTDs	Resistance temperature detectors
S	Solids or sludge mass flowrate
S	Solids or cake rate
S	Apparent power
s	Solids or cake concentration
SA	Surface area

SAE	Standard aeration (electrical) efficiency
SAR	Sodium adsorption ratio
SARA	Superfund Amendments and Reauthorization Act
SARS	Severe acute respiratory syndrome
SBOD	Soluble biochemical oxygen demand
SBR	Sequencing batch reactors
SCADA	Supervisory control and data acquisition
SCBA	Self-contained breathing apparatus
SDNRs	Super digital noise reductions
SGT-HEM	Silica gel treated n-hexane extractable material
SIK	Timer, indicator, and control
SIU	Significant industrial user
SK	Spulkraft
SLA	Service level agreement
SLR	Solids loading rate
SMR	Self-monitoring report
SNDN	Simultaneous nitrification/denitrification
SO_2	Sulfur dioxide
SOL	Soluble organic loading
SOPs	Standard operating procedures
SOTR	Standard oxygen transfer rate
SOUR	Specific oxygen uptake rate
SP	Soluble phosphorus
SPAM	Solution polyacrylamide
SPCC	Spill prevention and countermeasures control
SPDES	State Pollution Discharge Elimination System
SRF	State Revolving Fund
SRT	Solids residence (retention) time
SS	Suspended solids
SSO	Sanitary sewer overflow
SSV	Settled sludge volume
STOWA	Foundation for Applied Water Research
SVI	Sludge volume index
SWD	Side-water depth
t	Time of exposure
T_i	Temperature of influent
T_o	Digester operating temperature
T_s	Temperature of surroundings
T	Turbidity
t_C	Float scraper cycle time
t_{OFF}	Time that the scraper is off during a given cycle
t_{ON}	Time that the scraper is on during a given cycle
TCDD	tetrachlorodibenzo-p-dioxin
TD	Tetanus and diphtheria

TDH	Total dynamic head
TDS	Total dissolved solids
TF	Trickling filter
TF/AS	Trickling filter/activated sludge
TF/RBC	Trickling filter/rotating biological contactor
TF/SC	Trickling filter/solids contact
THMs	Trihalomethanes
TIMS	Training information management system
TKN	Total Kjeldahl nitrogen
TKN_{SE}	Secondary effluent TKN
TOC	Total organic carbon
TOD	Total oxygen demand
TOL	Total organic loading
TP	Total phosphorus
TPAD	Temperature-phased anaerobic digestion
TR	Turnover rate
TRE	Toxicity reduction evaluation
TSCA	Toxic Substances Control Act
TS	Total solids
TSS	Total suspended solids
TSS_{2EFF}	Total suspended solids in secondary effluent
TSS_{WAS}	Total suspended solids in waste activated sludge
TSU	Total sludge units
TTO	Total toxic organics
TWAS	Total waste activated sludge
TWUA	Texas Water Utilities Association
u	Fluid velocity
u_{per}	Permissible unbalance
U	Heat transfer coefficient
UCT	University of Capetown
UEL	Upper explosive limit
U.K.	United Kingdom
UL	Underwriters Laboratory
UPSs	Uninterruptible Power Systems
U.S.	United States
USD	United States dollar
USDA	United States Department of Agriculture
U.S.S.R	Union of Soviet Socialist Republics
UV	Ultraviolet
V	Velocity of travel of scrapers
V	Voltage
V	Volume of reactor
V_{AX}	Anoxic volume
V_{AB}	Volume of aeration basin

V_B	Basin volume
vL	Downflow rate
$v^2/2g$	Velocity head
VA/ALK	Volatile acid-to-alkalinity ratio
VAR	Vector attraction reduction
VFA	Volatile fatty acid
VHO	Very high output
VIP	Virginia Initiative Plant
VOC	Volatile organic compound
VFDs	Variable frequency drives
VS	Volatile solids
$VS_{digested\ sludge}$	Volatile solids concentration of the digested sludge
VS_{feed}	Volatile solids concentration of the feed sludge
VSA	Vacuum swing adsorption
VSR	Volatile solids reduction
VSS	Volatile suspended solids
VSS_{2INF}	Volatile suspended solids in secondary influent
W	Watts
W	Weight
W	Wash water rate
W/WW ITA	Water and Wastewater Instrumentation Testing Association
ω	Rotational speed
WA	Weight alarm
WAN	Wide area network
WAS	Waste activated sludge
WEAO	Water Environment Association of Ontario
WEF	Water Environment Federation
WEFTEC	Water Environment Federation Technical Exposition and Conference
WERF	Water Environment Research Foundation
WPCF	Water Pollution Control Federation
WRFs	Water reclamation facilities
WRRC	Water Resources Research Center
WRS	White data medium reference signal
WS	Weight switch
WWTPs	Wastewater treatment plants
x	Forward distance traveled during exposure
Z_W	Vertical distance

Glossary

absorption   (1) Taking up of matter in bulk by other matter, as in dissolving of a gas by a liquid. (2) Penetration of substances into the bulk of the solid or liquid. See also *adsorption*.

absorption capacity   A measure of the quantity of a soluble substance that can be absorbed by a given quantity of a solid substance.

acclimation   The dynamic response of a system to the addition or deletion of a substance until equilibrium is reached; adjustment to a change in the environment.

accuracy   The absolute nearness to the truth. In physical measurements, it is the degree of agreement between the quantity measured and the actual quantity. It should not be confused with "precision," which denotes the reproducibility of the measurement.

acid   (1) A substance that tends to lose a proton. (2) A substance that dissolves in water with the formation of hydrogen ions. (3) A substance containing hydrogen which may be replaced by metals to form salts.

acid-forming bacteria   Microorganisms that can metabolize complex organic compounds under anaerobic conditions. This metabolic activity is the first step in the two-step anaerobic fermentation process leading to the production of methane.

acidity   The quantitative capacity of aqueous solutions to neutralize a base; measured by titration with a standard solution of a base to a specified end point; usually expressed as milligrams of equivalent calcium carbonate per liter (mg/L $CaCO_3$); not to be confused with pH. Water does not have to have a low pH to have high acidity.

acre-foot (ac-ft)   A volume of water 1-ft deep and 1 ac in area, or 43 560 cu ft (1 233.5 m^3).

activated carbon   Adsorptive particles or granules usually obtained by heating carbonaceous material in the absence of air or in steam and possessing a high capacity to selectively remove trace and soluble components from solution.

activated carbon adsorption   Removal of soluble components from aqueous solution by contact with highly adsorptive granular or powdered carbon.

activated carbon treatment   Treatment process in which water is brought into contact with highly adsorptive granular or powdered carbon to remove soluble components; process may be applied to raw water, primary effluent, or chemically clarified wastewater for nonspecific removal of organics, or to secondary effluent as a polishing process to remove specific organics.

activated sludge   Sludge particles produced by the growth of organisms in the aeration tank in the presence of dissolved oxygen.

activated-sludge loading   The pounds (kilograms) of BOD in the applied liquid per unit volume of aeration capacity or per pound (kilogram) of activated sludge per day.

activated-sludge process   A biological wastewater treatment process that converts nonsettleable (suspended, dissolved, and colloidal solids) organic materials to a settleable product using aerobic and facultative microorganisms.

adsorption   The adherence of a gas, liquid, or dissolved material to the surface of a solid or liquid. It should not be confused with *absorption*.

adsorption water   Water held on the surface of solid particles by molecular forces with the emission of heat (heat of wetting).

advanced waste treatment   Any physical, chemical, or biological treatment process used to accomplish a degree of treatment greater than that achieved by secondary treatment.

aerated contact bed   A biological treatment unit consisting of stone, cement-asbestos, or other surfaces supported in an aeration tank, in which air is diffused up and around the surfaces and settled wastewater flows through the tank; also called a *contact aerator*.

aerated pond   A natural or artificial wastewater treatment pond in which mechanical or diffused air aeration is used to supplement the oxygen supply.

aeration   (1) The bringing about of intimate contact between air and a liquid by one or more of the following methods: (a) spraying the liquid in the air; (b) bubbling air through the liquid; and (c) agitating the liquid to promote surface absorption of air. (2) The supplying of air to confined spaces under nappes, downstream from gates in conduits, and so on, to relieve low pressures and to replenish air entrained and removed from such confined spaces by flowing water. (3) Relief of the effects of cavitation by admitting air to the affected section.

aeration period   (1) The theoretical time, usually expressed in hours, during which mixed liquor is subjected to aeration in an aeration tank while undergoing activated-sludge treatment. It is equal to the volume of the tank divided by the volumetric rate of flow of the wastewater and return sludge. (2) The theoretical time during which water is subjected to aeration.

aeration tank   A tank in which wastewater or other liquid is aerated.

aerator   A device that brings air and a liquid into intimate contact. See *diffuser*.

aerobic   Requiring, or not destroyed by, the presence of free or dissolved oxygen in an aqueous environment.

aerobic bacteria   Bacteria that require free elemental oxygen to sustain life.

aerobic digestion   The breakdown of suspended and dissolved organic matter in the presence of dissolved oxygen. An extension of the activated-sludge process, waste sludge is stored in an aerated tank where aerobic microorganisms break down the material.

aerobic lagoon   An oxygen-containing lagoon, often equipped with mechanical aerators, in which wastewater is partially stabilized by the metabolic activities of bacteria and algae. Small lagoons (less than 0.5 ac [0.2 ha] and less than 3-ft [0.9-m] deep) may remain aerobic without mechanical aeration. See also *anaerobic lagoon*.

aerosol   Colloidal particles dispersed in a gas, smoke, or fog.

agglomeration   Coalescence of dispersed suspended matter into larger flocs or particles.

agitator   Mechanical apparatus for mixing or aerating. A device for creating turbulence.

air-bound   Obstructed, as to the free flow of water, because of air entrapped in a high point; used to describe a pipeline or pump in such condition.

air chamber   A closed pipe chamber installed on the discharge line of a reciprocating pump to take up irregularities in hydraulic conditions, induce a uniform flow in suction and discharge lines, and relieve the pump of shocks caused by pulsating flow.

air-chamber pump   A displacement pump equipped with an air chamber in which the air is alternately compressed and expanded by the water displaced by the pump, resulting in the water being discharged at a more even rate.

air diffuser   Devices of varied design that transfer oxygen from air into a liquid.

air diffusion   The transfer of air into a liquid through an oxygen-transfer device. See *diffusion*.

air gap   The unobstructed vertical distance through the free atmosphere between the lowest opening from any pipe or outlet supplying water to a tank, plumbing fixture, or other device, and the flood-level rim of the receptacle.

air lift   A device for raising liquid by injecting air in and near the bottom of a riser pipe submerged in the liquid to be raised.

air-lift pump   A pump, used largely for lifting water from wells, from which fine pressured air bubbles are discharged into the water at the bottom of the well. The bubbles reduce the density of the water at the bottom, allowing the denser surrounding water to push it up in the discharge pipe to the outlet. Also called an *air lift*.

air stripping   A technique for removal of volatile substances from a solution; employs the principles of Henry's Law to transfer volatile pollutants from a solution of high concentration into an air stream of lower concentration. The process ordinarily is designed so that the solution containing the volatile pollutant contacts large volumes of air. The method is used to remove ammonia in advanced waste treatment.

algae   Photosynthetic microscopic plants that contain chlorophyll that float or are suspended in water. They may also be attached to structures, rocks, etc. In high concentrations, algae may deplete dissolved oxygen in receiving waters.

algal assay   An analytical procedure that uses specified nutrients and algal inoculums to identify the limiting algal nutrient in water bodies.

algal bloom   Large masses of microscopic and macroscopic plant life, such as green algae, occurring in bodies of water.

alkali   Generally, any substance that has highly basic properties; used particularly with reference to the soluble salts of sodium, potassium, calcium, and magnesium.

alkaline   The condition of water, wastewater, or soil that contains a sufficient amount of alkali substances to raise the pH above 7.0.

alkalinity   The capacity of water to neutralize acids; a property imparted by carbonates, bicarbonates, hydroxides, and occasionally borates, silicates, and phosphates. It is expressed in milligrams of equivalent calcium carbonate per liter (mg/L $CaCO_3$).

alkyl benzene sulfonate (ABS)   A type of surfactant, or surface active agent, present in synthetic detergents in the United States prior to 1965. ABS was troublesome because of its foam-producing characteristics and resistance to breakdown by biological action. ABS has been replaced in detergents by linear alkyl sulfonate, which is biodegradable.

alum, aluminum sulfate   [$Al_2(SO_4)_3 \times 18H_2O$] Used as a coagulant in filtration. Dissolved in water, it hydrolyzes into $Al(OH)_2$ and sulfuric acid (H_2SO_4). To precipitate the hydroxide, as needed for coagulation, the water must be alkaline.

ambient   Generally refers to the prevailing dynamic environmental conditions in a given area.

ammonia, ammonium   (NH_3, NH_4^+) Urea and proteins are degraded into dissolved ammonia and ammonium in raw wastewaters. Typically, raw wastewater contains

30 to 50 mg/L of NH_3. Reactions between chlorine and ammonia are important in disinfection.

ammonia nitrogen   The quantity of elemental nitrogen present in the form of ammonia (NH_3).

ammoniator   Apparatus used for applying ammonia or ammonium compounds to water.

ammonification   Bacterial decomposition of organic nitrogen to ammonia.

amoeba   A group of simple protozoans, some of which produce diseases such as dysentery in humans.

ampere   The unit of measurement of electrical current. It is proportional to the quantity of electrons flowing through a conductor past a given point in one second and is analogous to cubic feet of water flowing per second. It is the current produced in a circuit by one volt acting through a resistance of one ohm.

amperometric   Pertaining to measurement of electric current flowing or generated, rather than by voltage.

anaerobic   (1) A condition in which free and dissolved oxygen are unavailable. (2) Requiring or not destroyed by the absence of air or free oxygen.

anaerobic bacteria   Bacteria that grow only in the absence of free and dissolved oxygen.

anaerobic digestion   The degradation of concentrated wastewater solids, during which anaerobic bacteria break down the organic material into inert solids, water, carbon dioxide, and methane.

anaerobic lagoon   A wastewater or sludge treatment process that involves retention under anaerobic conditions.

anion   A negatively charged ion attracted to the anode under the influence of electrical potential.

anionic flocculant   A polyelectrolyte with a net negative electrical charge.

anoxic   Condition in which oxygen is available in the combined form only; there is no free oxygen. Anoxic sections in an activated-sludge plant may be used for denitrification.

antagonism   Detrimental interaction between two entities. See also *synergism*.

antichlors   Reagents, such as sulfur dioxide, sodium bisulfite, and sodium thiosulfate, that can be used to remove excess chlorine residuals from water or watery wastes by conversion to an inert salt.

anticorrosion treatment   Treatment to reduce or eliminate corrosion-producing qualities of a water.

appurtenances   Machinery, appliances, or auxiliary structures attached to a main structure enabling it to function, but not considered an integral part of it.

aqueous vapor   The gaseous form of water. See *water vapor*.

area drain   A drain installed to collect surface or stormwater from an open area of a building.

automatic recording gauge   An automatic instrument for measuring and recording graphically and continuously. Also called *register*.

automatic sampling   Collecting of samples of prescribed volume over a defined time period by an apparatus designed to operate remotely without direct manual control. See also *composite sample*.

autothermal thermophilic aerobic digestion   As part of the aerobic digestion process, heat is evolved. In a contained vessel, the sufficient heat is generated to maintain temperatures in the thermophilic range. At higher temperatures, detention time requirements are reduced for a given solids reduction resulting in an end product that is relatively pathogen free.

autotrophic organisms   Organisms including nitrifying bacteria and algae that use carbon dioxide as a source of carbon for cell synthesis. They can consume dissolved nitrates and ammonium salts.

available chlorine   A measure of the total oxidizing power of chlorinated lime, hypochlorites, and other materials used as a source of chlorine as compared with that of elemental chlorine.

average   An arithmetic mean obtained by adding quantities and dividing the sum by the number of quantities.

average daily flow   (1) The total quantity of liquid tributary to a point divided by the number of days of flow measurement. (2) In water and wastewater applications, the total flow past a point over a period of time divided by the number of days in that period.

average flow   Arithmetic average of flows measured at a given point.

average velocity   The average velocity of a stream flowing in a channel or conduit at a given cross section or in a given reach. It is equal to the discharge divided by the cross-sectional area of the section or the average cross-sectional area of the reach. Also called *mean velocity*.

axis   A line about which a figure or a body is symmetrically arranged, or about which such a figure or body rotates.

backflow connection   In plumbing, any arrangement whereby backflow can occur. Also called *interconnection, cross connection*.

backflow preventer   A device on a water supply pipe to prevent the backflow of water into the water supply system from the connections on its outlet end. See also *vacuum breaker* and *air gap*.

backflushing   The action of reversing the flow through a conduit for the purpose of cleaning the conduit of deposits.

back-pressure valve   A valve provided with a disk hinged on the edge so that it opens in the direction of normal flow and closes with reverse flow; a check valve.

backwashing   The operation of cleaning a filter by reversing the flow of liquid through it and washing out matter previously captured in it. Filters include true filters such as sand and diatomaceous earth-types, but not other treatment units such as trickling filters.

bacteria   A group of universally distributed, rigid, essentially unicellular microscopic organisms lacking chlorophyll. They perform a variety of biological treatment processes including biological oxidation, sludge digestion, nitrification, and denitrification.

bacterial analysis   The examination of water and wastewater to determine the presence, number, and identity of bacteria; more commonly called *bacterial examination*.

bacterial examination   The examination of water and wastewater to determine the presence, number, and identity of bacteria. Also called *bacterial analysis*. See also *bacteriological count*.

bacteriological count   A means for quantifying numbers of organisms. See also *most probable number*.

baffles   Deflector vanes, guides, grids, gratings, or similar devices constructed or placed in flowing water, wastewater, or slurry systems as a check or to produce a more uniform distribution of velocities; absorb energy; divert, guide, or agitate the liquids; and check eddies.

barminutor   A bar screen of standard design fitted with an electrically operated shredding device that sweeps vertically up and down the screen cutting up material retained on the screen.

bar screen   A screen composed of parallel bars, either vertical or inclined, placed in a waterway to catch debris. The screenings are raked from it either manually or automatically. Also called *bar rack, rack*.

base   A compound that dissociates in aqueous solution to yield hydroxyl ions.

basic data   Records of observations and measurements of physical facts, occurrences, and conditions, as they have occurred, excluding any material or information developed by means of computation or estimate. In the strictest sense, basic data include only the recorded notes of observations and measurements, although in general use it is taken to include computations or estimates necessary to present a clear statement of facts, occurrences, and conditions.

Beggiatoa   A filamentous organism whose growth is stimulated by H_2S.

belt screen   A continuous band or belt of wire mesh, bars, plates, or other screening medium that passes around upper and lower rollers and from which the material caught on the screen is usually removed by gravity, brushes, or other means.

bicarbonate alkalinity   Alkalinity caused by bicarbonate ions.

bioassay   (1) An assay method using a change in biological activity as a qualitative or quantitative means of analyzing a material's response to biological treatment. (2) A method of determining the toxic effects of industrial wastes and other wastewaters by using viable organisms; exposure of fish to various levels of a chemical under controlled conditions to determine safe and toxic levels of that chemical.

biochemical   (1) Pertaining to chemical change resulting from biological action. (2) A chemical compound resulting from fermentation. (3) Pertaining to the chemistry of plant and animal life.

biochemical oxidation   Oxidation brought about by biological activity resulting in the chemical combination of oxygen with organic matter. See *oxidized wastewater*.

biochemical oxygen demand (BOD)   A measure of the quantity of oxygen used in the biochemical oxidation of organic matter in a specified time, at a specific temperature, and under specified conditions.

biochemical oxygen demand (BOD) load   The BOD content (usually expressed in mass per unit of time) of wastewater passing into a waste treatment system or to a body of water.

biodegradation   The destruction of organic materials by microorganisms, soils, natural bodies of water, or wastewater treatment systems.

biofilm   Accumulation of microbial growth on the surface of a support material.

biological contactors   Inert surfaces engineered to provide a high specific surface area on which a biofilm can develop; usually designed so that the surface is cyclically moved through the medium to be biologically oxidized and through the open air so that oxygen transfer occurs.

biological denitrification   The transformation of nitrate nitrogen to inert nitrogen gas by microorganisms in an anoxic environment in the presence of an electron donor to drive the reaction.

biological filter   A bed of sand, gravel, broken stone, or other medium through which wastewater flows or trickles. It depends on biological action for its effectiveness.

biological filtration   The process of passing a liquid through a biological filter containing fixed media on the surfaces of which develop zoogleal films that absorb and adsorb fine suspended, colloidal, and dissolved solids and release end products of biochemical action.

biological oxidation   The process by which living organisms in the presence of oxygen convert organic matter into a more stable or mineral form.

biological process   (1) The process by which metabolic activities of bacteria and other microorganisms break down complex organic materials into simple, more stable substances. Self-purification of polluted streams, sludge digestion, and all the so-called secondary wastewater treatments depend on this process. (2) Process involving living organisms and their life activities. Also called *biochemical process*.

biomass   The mass of biological material contained in a system.

biosolids   The organic product of municipal wastewater treatment that can be beneficially used.

bleed   (1) To drain a liquid or gas, as to vent accumulated air from a water line or to drain a trap or a container of accumulated water. (2) The exuding, percolation, or seeping of a liquid through a surface.

blinding   (1) Clogging of the filter cloth of a vacuum filter, belt press, belt thickener, or pressure filter. (2) Obstruction of the fine media of a sand filter.

blowdown   (1) The removal of a portion of any process flow to maintain the constituents of the flow within desired levels. The process may be intermittent or continuous. (2) The water discharged from a boiler or cooling tower to dispose of accumulated dissolved solids.

bottom contraction   The reduction in the area of overflowing water caused by the crest of a weir contracting the nappe.

bottom ventilation   Movement of air through the medium of a wastewater filter facilitated by vent stacks or provisions for the entrance or exit of air at the base of the filter.

bound water   (1) Water held strongly on the surface or in the interior of colloidal particles. (2) Water associated with the hydration of crystalline compounds.

branch circuit   That portion of the wiring system between the final overcurrent device that protects the circuit and the outlet.

breakpoint chlorination   Addition of chlorine to water or wastewater until the chlorine demand has been satisfied, with further additions resulting in a residual that is directly proportional to the amount added beyond the breakpoint.

brush aerator   A surface aerator that rotates about a horizontal shaft with metal blades attached to it; commonly used in oxidation ditches.

buffer   A substance that resists a change in pH.

bulking   Inability of activated-sludge solids to separate from the liquid under quiescent conditions; may be associated with the growth of filamentous organisms, low DO, or high sludge loading rates. Bulking sludge typically has an SVI > 150 mL/g.

bus   An electrical conductor that serves as a common connection for two or more electrical circuits. A bus may be in the form of rigid bars, either circular or rectangular in cross section, or in the form of stranded conductor cables held under tension.

butterfly valve   A valve in which the disk, as it opens or closes, rotates about a spindle supported by the frame of the valve. The valve is opened at a stem. At full opening, the disk is in a position parallel to the axis of the conduit.

bypass   An arrangement of pipes, conduits, gates, and valves by which the flow may be passed around a hydraulic structure appurtenance or treatment process; a controlled diversion.

cake   Wastewater solids that have been sufficiently dewatered to form a semisolid mass.

calcium hypochlorite   [$Ca(OCl)_2 \cdot 4H_2O$] A solid that, when mixed with water, liberates the hypochlorite ion OCl^- and can be used for disinfection.

calibration   (1) The determination, checking, or rectifying of the graduation of any instrument giving quantitative measurements. (2) The process of taking measurements or of making observations to establish the relationship between two quantities.

calorie   The amount of heat necessary to raise the temperature of 1 g of water at 15 °C by 1 °C.

capacitor (condenser)   A device to provide capacitance, which is the property of a system of conductors and dielectrics that permit the storage of electrically separated charges when potential differences exist between the conductors. A dielectric is an insulator.

capacity   (1) The quantity that can be contained exactly, or the rate of flow that can be carried out exactly. (2) The load for which an electrical apparatus is rated either by the user or manufacturer.

carbon (C)   (1) A chemical element essential for growth. (2) A solid material used for adsorption of pollutants.

carbonaceous biochemical oxygen demand (CBOD)   A quantitative measure of the amount of dissolved oxygen required for the biological oxidation of carbon-containing compounds in a sample. See *BOD*.

carbon adsorption   The use of either granular or powdered carbon to remove organic compounds from wastewater or effluents. Organic molecules in solution are drawn to the highly porous surface of the carbon by intermolecular attraction forces.

carbonate hardness   Hardness caused by the presence of carbonates and bicarbonates of calcium and magnesium in water. Such hardness may be removed to the limit of solubility by boiling the water. When the hardness is numerically greater than the sum of the carbonate alkalinity and bicarbonate alkalinity, the amount of hardness is equivalent to the total alkalinity and is called *carbonate hardness*. It is expressed in milligrams of equivalent calcium carbonate per liter (mg/L $CaCO_3$). See also *hardness*.

carbonation   The diffusion of carbon dioxide gas through a liquid to render the liquid stable with respect to precipitation or dissolution of alkaline constituents. See also *recarbonation*.

carcinogen   A material that induces excessive or abnormal cellular growth in an organism.

carrying capacity   The maximum rate of flow that a conduit, channel, or other hydraulic structure is capable of passing.

cascade aerator   An aerating device built in the form of steps or an inclined plane on which are placed staggered projections arranged to break up the water and bring it into contact with air.

cathodic protection   An electrical system for prevention of rust, corrosion, and pitting of steel and iron surfaces in contact with water. A low-voltage current is made to flow through a liquid or a soil in contact with the metal in such a manner that the external electromotive force renders the metal structure cathodic and concentrates corrosion on auxiliary anodic parts used for that purpose.

cation   A positively charged ion attracted to the cathode under the influence of electrical potential.

cationic flocculant   A polyelectrolyte with a net positive electrical charge.

caustic alkalinity   The alkalinity caused by hydroxyl ions. See also *alkalinity*.

cavitation   (1) The action, resulting from forcing a flow stream to change direction, in which reduced internal pressure causes dissolved gases to expand, creating negative pressure. Cavitation frequently causes pitting of the hydraulic structure affected. (2) The formation of a cavity between the downstream surface of a moving body (e.g., the blade of a propeller) and a liquid normally in contact with it. (3) Describing the action of an operating centrifugal pump when it is attempting to discharge more water than suction can provide.

Celsius   The international name for the centigrade scale of temperature, on which the freezing point and boiling point of water are 0 °C and 100 °C, respectively, at a barometric pressure of 1.013×10^5 Pa (760-mm Hg).

centigrade   A thermometer temperature scale in which 0° marks the freezing point and 100° the boiling point of water at 760-mm Hg barometric pressure. Also called *Celsius*. To convert temperature on this scale to Fahrenheit, multiply by 1.8 and add 32.

centrate   Liquid removed by a centrifuge; typically contains high concentrations of suspended, nonsettling solids.

centrifugal pump   A pump consisting of an impeller fixed on a rotating shaft and enclosed in a casing having an inlet and a discharge connection. The rotating impeller creates pressure in the liquid by the velocity derived from centrifugal force.

centrifugal screw pump   A centrifugal pump having a screw-type impeller; may be of axial flow or combined axial and radial flow.

centrifugation   Imposition of a centrifugal force to separate solids from liquids based on density differences. In sludge dewatering, the separated solids commonly are called *cake* and the liquid is called *centrate*.

centrifuge   A mechanical device in which centrifugal force is used to separate solids from liquids or to separate liquids of different densities.

certification   A program to substantiate the capabilities of personnel by documentation of experience and learning in a defined area of endeavor.

cfs (cu ft/sec)   The rate of flow of a material in cubic feet per second; used for measurement of water, wastewater, or gas; equals 2.832×10^{-2} m^3/s.

chain bucket   A continuous chain equipped with buckets and mounted on a scow. Also called a *ladder dredge*.

chamber   Any space enclosed by walls or a compartment; often prefixed by a descriptive word indicating its function, such as *grit* chamber, *screen* chamber, *discharge* chamber, or *flushing* chamber.

change of state   The process by which a substance passes from one to another of the solid, the liquid, and the gaseous states, and in which marked changes in its physical properties and molecular structure occur.

channel   (1) A perceptible natural or artificial waterway that periodically or continuously contains moving water or forms a connecting link between two bodies of water. It has a definite bed and banks that confine the water. (2) The deep portion of a river or waterway where the main current flows. (3) The part of a body of water deep enough to be used for navigation through an area otherwise too shallow for navigation. (4) Informally, a more or less linear conduit of substantial size in cavernous lime-stones or lava rocks. See also *open channel*.

channel roughness   That roughness of a channel including the extra roughness owing to local expansion or contraction and obstacles, as well as the roughness of the stream bed proper; that is, friction offered to the flow by the surface of the bed of the channel in contact with the water. It is expressed as the roughness coefficient in velocity formulas.

check valve   A valve with a disk hinged on one edge so that it opens in the direction of normal flow and closes with reverse flow. An approved check valve is of substantial construction and suitable materials, is positive in closing, and permits no leakage in a direction opposite to normal flow.

chemical   Commonly, any substance used in or produced by a chemical process. Certain chemicals may be added to water or wastewater to improve treatment efficiency; others are pollutants that require removal.

chemical analysis   Analysis by chemical methods to show the composition and concentration of substances.

chemical coagulation   The destabilization and initial aggregation of colloidal and finely divided suspended matter by the addition of an inorganic coagulant. See also *flocculation*.

chemical conditioning   Mixing chemicals with a sludge prior to dewatering to improve the solids separation characteristics. Typical conditioners include polyelectrolytes, iron salts, and lime.

chemical dose   A specific quantity of chemical applied to a specific quantity of fluid for a specific purpose.

chemical equilibrium   The condition that exists when there is no net transfer of mass or energy between the components of a system. This is the condition in a reversible chemical reaction when the rate of the forward reaction equals the rate of the reverse reaction.

chemical equivalent   The weight (in grams) of a substance that combines with or displaces 1 g of hydrogen. It is found by dividing the formula weight by its valence.

chemical feeder   A device for dispensing a chemical at a predetermined rate for the treatment of water or wastewater. The change in rate of feed may be effected manually or automatically by flowrate changes. Feeders are designed for solids, liquids, or gases.

chemical gas feeder   A feeder for dispensing a chemical in the gaseous state. The rate is usually graduated in gravimetric terms. Such devices may have proprietary names.

chemical oxidation   The oxidation of compounds in wastewater or water by chemical means. Typical oxidants include ozone, chlorine, and potassium permanganate.

chemical oxygen demand (COD)   A quantitative measure of the amount of oxygen required for the chemical oxidation of carbonaceous (organic) material in wastewater using inorganic dichromate or permanganate salts as oxidants in a 2-hour test.

chemical precipitation   (1) Formation of particulates by the addition of chemicals. (2) The process of softening water by the addition of lime or lime and soda to form insoluble compounds; usually followed by sedimentation or filtration to remove the newly created suspended solids.

chemical reaction   A transformation of one or more chemical species into other species resulting in the evolution of heat or gas, color formation, or precipitation. It may be initiated by a physical process such as heating, by the addition of a chemical reagent, or it may occur spontaneously.

chemical reagent   A chemical added to a system to induce a chemical reaction.

chemical sludge   Sludge obtained by treatment of water or wastewater with inorganic coagulants.

chemical solution tank   A tank in which chemicals are added in solution before they are used in a water or wastewater treatment process.

chemical tank   A tank in which chemicals are stored before they are used in a water or wastewater treatment process.

chemical treatment   Any treatment process involving the addition of chemicals to obtain a desired result such as precipitation, coagulation, flocculation, sludge conditioning, disinfection, or odor control.

chloramines   Compounds of organic or inorganic nitrogen formed during the addition of chlorine to wastewater. See *breakpoint chlorination*.

chlorination   The application of chlorine or chlorine compounds to water or wastewater, generally for the purpose of disinfection, but frequently for chemical oxidation and odor control.

chlorinator   Any metering device used to add chlorine to water or wastewater.

chlorine (Cl_2)   An element ordinarily existing as a greenish-yellow gas about 2.5 times heavier than air. At atmospheric pressure and a temperature of -30.1 °F (-48 °C), the gas becomes an amber liquid about 1.5 times heavier than water. Its atomic weight is 35.457, and its molecular weight is 70.914.

chlorine contact chamber   A detention basin provided to diffuse chlorine through water or wastewater and to provide adequate contact time for disinfection. Also called a *chlorination chamber* or *chlorination basin*.

chlorine demand   The difference between the amount of chlorine added to a wastewater and the amount of chlorine remaining after a given contact time. Chlorine dosage is a function of the substances present in the water, temperature, and contact time.

chlorine dose   The amount of chlorine applied to a wastewater, usually expressed in milligrams per liter (mg/L) or pounds per million gallons (lb/mil. gal).

chlorine ice   A yellowish ice formed in a chlorinator when chlorine gas comes in contact with water at 49 °F (9 °C) or lower. Chlorine ice is frequently detrimental to the performance of a chlorinator if it is formed in quantities sufficient to interfere with the safe operation of float controls or to cause plugging of openings essential to flow indication, control, or rate of application.

chlorine residual   The amount of chlorine in all forms remaining in water after treatment to ensure disinfection for a period of time.

chlorine room   A separate room or building for housing chlorine and chlorination equipment, with arrangements for protecting personnel and plant equipment.

chlorine toxicity   The detrimental effects on biota caused by the inherent properties of chlorine.

chromatography   The generic name of a group of separation processes that depend on the redistribution of the molecules of a mixture between a gas or liquid phase in contact with one or more bulk phases. The types of chromatography are adsorption, column, gas, gel, liquid, thin-layer, and paper.

ciliated protozoa   Protozoans with cilia (hair-like appendages) that assist in movement; common in trickling filters and healthy activated sludge. Free-swimming ciliates are present in the bulk liquid, stalked ciliates are commonly attached to solids matter in the liquid.

circuit   A conductor or a system of conductors through which an electrical current flows or is intended to flow.

circuit breaker   A device designed to open or close a circuit by nonautomatic means and open the circuit automatically on a predetermined overload of current without injury to itself.

clarification   Any process or combination of processes whose primary purpose is to reduce the concentration of suspended matter in a liquid; formerly used as a synonym for *settling* or *sedimentation*. In recent years, the latter terms are preferable when describing settling processes.

clarifier   Any large circular or rectangular sedimentation tank used to remove settleable solids in water or wastewater. A special type of clarifier, called an *upflow clarifier*, uses flotation rather than sedimentation to remove solids.

clear-water basin   A reservoir for the storage of filtered water of sufficient capacity to prevent the necessity of frequent variations in the rate of filtration with variations in demands. Also called *filtered-water reservoir, clear-water reservoir, clear well.*

closed centrifugal pump   A centrifugal pump having its impeller built with the vanes enclosed within circular disks.

closed conduit   Any closed artificial or natural duct for conveying fluids.

closed impeller   An impeller having the side walls extended from the outer circumference of the suction opening to the vane tips.

coagulant   A simple electrolyte, usually an inorganic salt containing a multivalent cation of iron, aluminum, or calcium [for example, $FeCl_3$, $FeCl_2$, $Al_2(SO_4)_3$, and CaO]. Also, an inorganic acid or base that induces coagulation of suspended solids. See also *flocculant*.

coagulant **or** *flocculant aid*   An insoluble particulate used to enhance solid–liquid separation by providing nucleating sites or acting as a weighting agent or sorbent; also used colloquially to describe the action of flocculents in water treatment.

coagulation   The conversion of colloidal (<0.001 mm) or dispersed (0.001 to 0.1 mm) particles into small visible coagulated particles (0.1 to 1 mm) by the addition of a coagulant, compressing the electrical double layer surrounding each suspended particle, decreasing the magnitude of repulsive electrostatic interactions between particles, and thereby destabilizing the particle. See also *flocculation*.

coagulation basin   A basin used for the coagulation of suspended or colloidal matter, with or without the addition of a coagulant, in which the liquid is mixed gently to induce agglomeration with a consequent increase in the settling velocity of particulates.

coating   A material applied to the inside or outside of a pipe, valve, or other fixture to protect it primarily against corrosion. Coatings may be of various materials.

Cocci   Sphere-shaped bacteria.

codisposal   Joint disposal of wastewater sludge and municipal refuse in one process or facility. Disposal can be intermediate, as with incineration or composting, or final, as with placement in a sanitary landfill.

coefficient   A numerical quantity, determined by experimental or analytical methods, interposed in a formula that expresses the relationship between two or more variables to include the effect of special conditions or to correct a theoretical relationship to one found by experiment or actual practice.

coefficient of viscosity   A numerical factor that is a measure of the internal resistance of a fluid to flow; the greater the resistance to flow, the larger the coefficient. It is equal to the shearing force in dynes per square centimeter ($dyne/cm^2$) transmitted from one fluid plane to another parallel plane 1 cm distant, and is generated by a difference in fluid velocities in the two planes of 1 cm/s in the direction of the force. The coefficient varies with temperature. Also called *absolute viscosity*. The unit of measure is the poise, a force of 1 dyne/cm^2.

cohesion   The force of molecular attraction between the particles of any substance that tends to hold them together.

coil   A set of windings with or without an iron core, shaped to produce a magnetic force when current flows through the windings. This force is used in relays and other electrical equipment to pull contacts together or to separate them.

coliform-group bacteria   A group of bacteria predominantly inhabiting the intestines of man or animal, but also occasionally found elsewhere. It includes all aerobic and facultative anaerobic, Gram-negative, non-spore-forming, rod-shaped bacteria that ferment lactose with the production of gas. Also included are all bacteria that produce a dark, purplish-green metallic sheen by the membrane filter technique used for coliform identification. The two groups are not always identified, but they are generally of equal sanitary significance.

collection system   In wastewater, a system of conduits, generally underground pipes, that receives and conveys sanitary wastewater or stormwater; in water supply, a system of conduits or canals used to capture a water supply and convey it to a common point.

colloids   Finely divided solids (less than 0.002 mm and greater than 0.000 001 mm) that will not settle but may be removed by coagulation, biochemical action, or membrane filtration; they are intermediate between true solutions and suspensions.

colony   A discrete clump of microorganisms on a surface as opposed to dispersed growth throughout a liquid culture medium.

color   Any dissolved solids that impart a visible hue to water.

colorimeter   An instrument that quantitatively measures the amount of light of a specific wavelength absorbed by a solution.

combined available chlorine   The concentration of chlorine that is combined with ammonia as chloramine or as other chloro derivatives, yet is still available to oxidize organic matter.

combined available residual chlorine   That portion of the total residual chlorine remaining in water or wastewater at the end of a specified contact period that will react chemically and biologically as chloramines.

combined residual chlorination   The application of chlorine to water or wastewater to produce, with natural or added ammonia or with certain organic nitrogen compounds, a combined chlorine residual.

combined sewer   A sewer intended to receive both wastewater and storm or surface water.

combustible-gas indicator   An explosimeter; a device for measuring the concentration of potentially explosive fumes. The measurement is based on the catalytic oxidation of a combustible gas on a heated platinum filament that is part of a Whetstone bridge.

commercially dry sludge   Sludge containing not more than 10% moisture by weight; the limit is 5% in the fertilizer trade.

comminution   An in-stream process of cutting and screening solids contained in wastewater flow.

comminutor   A shredding or grinding device that reduces the size of gross suspended materials in wastewater without removing them from the liquid.

complete mix   Activated sludge process whereby wastewater is rapidly and evenly distributed throughout the aeration tank.

composite sample   A combination of individual samples of water or wastewater taken at preselected intervals to minimize the effect of the variability of the individual sample. Individual samples may be of equal volume or may be proportional to the flow at the time of sampling.

compost   The product of the thermophilic biological oxidation of sludge or other materials.

concentration   (1) The amount of a given substance dissolved in a discrete unit volume of solution or applied to a unit weight of solid. (2) The process of increasing the dissolved solids per unit volume of solution, usually by evaporation of the liquid. (3) The process of increasing the suspended solids per unit volume of sludge as by sedimentation or dewatering.

concentrator   A solids contact unit used to decrease the water content of sludge or slurry.

condensate   Condensed steam from any heat exchanger.

condensation   The process by which a substance changes from the vapor state to the liquid or solid state. Water that falls as precipitation from the atmosphere has condensed from the vapor state to rain or snow. Dew and frost are also forms of condensation.

condenser   Any device for reducing gases or vapors to liquid or solid form.

conditioning   The chemical, physical, or biological treatment of sludges to improve their dewaterability.

conductor   A material that offers very little resistance to the flow of current and is, therefore, used to carry current or conduct electricity.

conduit (duct) bank   A length of one or more conduits or ducts (which may be enclosed in concrete) that is designed to contain cables.

contacts   Any set of points that may be joined manually or automatically to complete a circuit. Contacts are found in breakers, switches, relays, and starters.

contact stabilization   Modification of the activated-sludge process involving a short period of contact between wastewater and sludge for rapid removal of soluble BOD by adsorption, followed by a longer period of aeration in a separate tank where sludge is oxidized and new sludge synthesized.

contact tank   A tank used in water or wastewater treatment to promote contact between treatment chemicals or other materials and the liquid treated.

contact time   The time that the material processed is exposed to another substance (such as activated sludge or activated carbon) for completion of the desired reaction. See also *detention time*.

contamination   The introduction into water of microorganisms, chemicals, wastes, or wastewater in a concentration that makes the water unfit for its intended use.

continuous-flow pump   A displacement pump within which the direction of flow of the water is not changed or reversed.

continuous-flow tank   A tank through which liquid flows continuously at its normal rate of flow, as distinguished from a fill-and-draw or batch system.

continuous load   A load where the maximum current is expected to continue for 3 hours or more.

contracted weir   A rectangular notched weir with a crest width narrower than the channel across which it is installed and with vertical sides extending above the upstream water level producing a contraction in the stream of water as it leaves the notch.

controlled discharge   Regulation of effluent flowrates to correspond with flow variations in receiving waters to maintain established water quality.

controller   A device or group of devices, that serve to govern, in some predetermined measure, the electrical power delivered to the apparatus to which it is connected.

convection   (1) In physics, mass motions within a fluid resulting in the transport and mixing of the properties of that fluid, caused by the force of gravity and by differences in density resulting from nonuniform temperature. (2) In meteorology, atmospheric motions that are predominantly vertical, resulting in vertical transport and mixing of atmospheric properties; sometimes caused when large masses of air are heated by contact with a warm land surface.

conventional aeration   Process design configuration whereby the aeration tank organic loading is higher at the influent end than at the effluent end. Flow passes through a serpentine tank system, typically side-by-side, before passing on to the secondary clarifier. Also called *plug flow*.

conventional treatment   Well-known or well-established water or wastewater treatment processes, excluding advanced or tertiary treatment; it generally consists of primary and secondary treatment.

conversion factor   A numerical constant by which a quantity with its value expressed in units of one kind is multiplied to express the value in units of another kind.

cooling coil   A coil of pipe or tubing containing a stream of hot fluid that is cooled by heat transfer to a cold fluid outside. Conversely, the coil may contain a cold fluid to cool a hot fluid in which the coil is immersed.

core sampler   A long, slender pole with a foot valve at the bottom end that allows the depth of the sludge blanket to be measured. Sometimes called a *sludge judge*.

correlation   (1) A mutual relationship or connection. (2) The degree of relative correspondence, as between two sets of data.

corrosion   The gradual deterioration or destruction of a substance or material by chemical action, frequently induced by electrochemical processes. The action proceeds inward from the surface.

corrosion control   (1) In water treatment, any method that keeps the metallic ions of a conduit from going into solution, such as increasing the pH of the water, removing free oxygen from the water, or controlling the carbonate balance of the water. (2) The sequestration of metallic ions and the formation of protective films on metal surfaces by chemical treatment.

critical depth   The depth of water flowing in an open channel or partially filled conduit corresponding to one of the recognized critical velocities.

critical flow   (1) A condition of flow in which the mean velocity is at one of the critical values, ordinarily at Belanger's critical depth and velocity; also used in reference to Reynolds' critical velocities, which define the point at which the flow changes from streamline or nonturbulent flows. (2) The maximum discharge of a conduit that has a free outlet and has the water ponded at the inlet.

cross connection   (1) A physical connection through which a supply of potable water could be contaminated or polluted. (2) A connection between a supervised potable water supply and an unsupervised supply of unknown potability.

culture   Any organic growth that has been developed intentionally by providing suitable nutrients and environment.

culture media   Substances used to support the growth of microorganisms in analytical procedures.

cyclone separator   A conical unit used for separating particles by centrifugal force.

data   Records of observations and measurements of physical facts, occurrences, and conditions reduced to written, graphical, or tabular form.

debris   Generally, solid wastes from natural and man-made sources deposited indiscriminately on land and water.

decantation   Separation of a liquid from solids or from a liquid of higher density by drawing off the upper layer after the heavier material has settled.

dechlorination   The partial or complete reduction of residual chlorine by any chemical or physical process. Sulfur dioxide is frequently used for this purpose.

declining growth phase   Period of time between the log-growth phase and the endogenous phase, where the amount of food is in short supply, leading to ever-slowing bacterial growth rates.

decomposition   The breakdown of complex material into simpler substances by chemical or biological processes.

decomposition of wastewater   (1) The breakdown of organic matter in wastewater by bacterial action, either aerobic or anaerobic. (2) Chemical or biological transformation of the organic or inorganic materials contained in wastewater.

defoamer   A material having low compatibility with foam and a low surface tension. Defoamers are used to control, prevent, or destroy various types of foam, the most widely used being silicone defoamers. A droplet of silicone defoamer contacting a bubble of foam will cause the bubble to undergo a local and drastic reduction in film strength, thereby breaking the film. Unchanged, the defoamer continues to contact other bubbles, thus breaking up the foam. A valuable property of most defoamers is their effectiveness in extremely low concentration. In addition to silicones, defoamers for special purposes are based on polyamides, vegetable oils, and stearic acid.

degasification   (1) The removal of a gas from a liquid medium. (2) In water treatment, the removal of oxygen from water to inhibit corrosion. It may be accomplished by mechanical methods, chemical methods, or a combination of both.

degreasing   (1) The process of removing greases and oils from waste, wastewater, sludge, or solids. (2) The industrial process of removing grease and oils from machine parts or iron products.

degree   (1) On the centigrade or Celsius thermometer scale, 1/100 of the interval from the freezing point to the boiling point of water under standard conditions; on the

Fahrenheit scale, 1/180 of this interval. (2) A unit of angular measure; the central angle subtended by 1/360 of the circumference of a circle.

demand   The rate at which electrical energy is delivered to a piece of power-consuming equipment or system.

demand average   The demand on an electrical system or any of its parts over an interval of time, as determined by dividing the total number of watt-hours by the number of hours (units of time) in the interval.

demand coincident   The sum of two or more demands that occurs in the same demand interval.

demand factor   The ratio of the maximum demand of the system or part of a system to the total connected load of the system or part of the system under consideration.

demand instantaneous peak   The maximum demand at the instant of greatest load.

demand interval   The period of time that electrical energy flows is averaged to determine demand, such as 60 minutes, 15 minutes, or instantaneous.

demand maximum   The greatest of all demands of the load under consideration that occurs during a specified period of time.

demand noncoincident   The sum of two or more individual demands that do not occur in the same demand interval, which is meaningful only when considering demands within a limited period of time, such as a day, week, month, and a heating or cooling season.

denitrification   The anaerobic biological reduction of nitrate nitrogen to nitrogen gas; also, removal of total nitrogen from a system. See also *nitrification*.

density current   A flow of water through a large body of water that retains its unmixed identity because of a difference in density.

deoxygenation   The depletion of the dissolved oxygen in a liquid either under natural conditions associated with the biochemical oxidation of the organic matter present or by addition of chemical reducing agents.

deoxygenation constant   A constant that expresses the rate of the biochemical oxidation of organic matter under aerobic conditions. Its value depends on the time unit involved (usually 1 day) and varies with temperature and other test conditions.

departure   The difference between any single observation and the normal.

deposition   The act or process of settling solid material from a fluid suspension.

depth of blanket   Level of sludge in the bottom of a secondary clarifier, typically measured in feet.

design criteria   (1) Engineering guidelines specifying construction details and materials. (2) Objectives, results, or limits that must be met by a facility, structure, or process in performance of its intended functions.

design flow   Engineering guidelines that typically specify the amount of influent flow that can be expected on a daily basis over the course of a year. Other design flows can be set for monthly or peak flows.

design loadings   Flowrates and constituent concentrations that determine the design of a process unit or facility necessary for proper operation.

design voltage   The nominal voltage for which a line or piece of equipment is designed. This is a reference level of voltage for identification and not necessarily the precise level at which it operates.

detention time   The period of time that a water or wastewater flow is retained in a basin, tank, or reservoir for storage or completion of physical, chemical, or biological reaction. See also *contact time, retention time.*

detergent   (1) Any of a group of synthetic, organic, liquid, or water-soluble cleaning agents that are inactivated by hard water and have wetting and emulsifying properties but, unlike soap, are not prepared from fats and oils. (2) A substance that reduces the surface tension of water.

detoxification   Treatment to modify or remove a toxic material.

dewater   (1) To extract a portion of the water present in a sludge or slurry. (2) To drain or remove water from an enclosure. A river bed may be dewatered so that a dam can be built; a structure may be dewatered so that it can be inspected or repaired.

dewatered sludge   The solid residue remaining after removal of water from a wet sludge by draining or filtering. Dewatering is distinguished from thickening in that dewatered sludge may be transported by solids handling procedures.

dewatering   The process of partially removing water; may refer to removal of water from a basin, tank, reservoir, or other storage unit, or the separation of water from solid material.

dewpoint   The temperature to which air with a given concentration of water vapor must be cooled to cause condensation of the vapor.

dialysis   The selective separation of dissolved or colloidal solids on the basis of molecular size by diffusion through a semipermeable membrane. See also *reverse osmosis.*

differential plunger pump   A reciprocating pump with a plunger so designed that it draws the liquid into the cylinder on the upward stroke but is double-acting on the discharge stroke.

diffused aeration   Injection of air under pressure through submerged porous plates, perforated pipes, or other devices to form small air bubbles from which oxygen is transferred to the liquid as the bubbles rise to the water surface.

diffused air   Small air bubbles formed below the surface of a liquid to transfer oxygen to the liquid.

diffuser   A porous plate, tube, or other device through which air is forced and divided into minute bubbles for diffusion in liquids. In the activated sludge process, it is a device for dissolving air into mixed liquor. It is also used to mix chemicals such as chlorine through perforated holes.

diffusion   (1) The transfer of mass from one fluid phase to another across an interface, for example liquid to solid or gas to liquid. (2) The spatial equalization of one material throughout another.

diffusion aerator   An aerator that blows air under low pressure through submerged porous plates, perforated pipes, or other devices so that small air bubbles rise continuously through the water or wastewater.

digested solids   Solids digested under either aerobic or anaerobic conditions until the volatile content has been reduced to the point at which the solids are relatively nonputrescible and inoffensive.

digester   A tank or other vessel for the storage and anaerobic or aerobic decomposition of organic matter present in the sludge. See also *anaerobic digestion*.

digester coils   A system of hot water or steam pipes installed in a digestion tank to heat the digester contents.

digestion   (1) The biological decomposition of the organic matter in sludge, resulting in partial liquefaction, mineralization, and volume reduction. (2) The process carried out in a digester.

discharge   The flow or rate of flow from a canal, conduit, pump, stack, tank, or treatment process. See also *effluent*.

discharge area   The cross-sectional area of a waterway. Used to compute the discharge of a stream, pipe, conduit, or other carrying system.

discharge capacity   The maximum rate of flow that a conduit, channel, or other hydraulic structure is capable of passing.

discharge head   A measure of the pressure exerted by a fluid at the point of discharge, usually from a pump.

discharge rate   (1) The determination of the quantity of water flowing per unit of time in a stream channel, conduit, or orifice at a given point by means of a current meter, rod float, weir, pitot tube, or other measuring device or method. The operation includes not only the measurement of velocity of water and the cross-sectional area of the stream of water, but also the necessary subsequent computations. (2) The numerical results of a measurement of discharge, expressed in appropriate units.

disconnecting means   A device, group of devices, or other means whereby the conductors of a circuit can be disconnected from the source of power.

discrete sedimentation   Sedimentation in which removal of suspended solids is a function of terminal settling velocity.

disinfectant   A substance used for disinfection and in which disinfection has been accomplished.

disinfected wastewater   Wastewater to which a disinfecting agent has been added.

disinfection   (1) The killing of waterborne fecal and pathogenic bacteria and viruses in potable water supplies or wastewater effluents with a disinfectant; an operational term that must be defined within limits, such as achieving an effluent with no more than 200 colonies fecal coliform/100 mL. (2) The killing of the larger portion of microorganisms, excluding bacterial spores, in or on a substance with the probability that all pathogenic forms are killed, inactivated, or otherwise rendered nonvirulent.

dispersion   (1) Scattering and mixing. (2) The mixing of polluted fluids with a large volume of water in a stream or other body of water. (3) The repelling action of an electric potential on fine particles in suspension in water, as in a stream carrying clay. This dispersion usually is ended by contact with ocean water causing flocculation and precipitation of the clay, a common cause of shoaling in harbors. (4) In a continuous-flow treatment unit, the phenomenon of short-circuiting.

displacement pump   A type of pump in which the water is induced to flow from the source of supply through an inlet pipe and valve and into the pump chamber by a vacuum created therein by the withdrawal of a piston or piston-like device which, on its return, displaces a certain volume of the water contained in the chamber and forces it to flow through the discharge valves and discharge pipes.

disposal   Release to the environment. See also *ultimate disposal*.

dissolved air flotation (DAF)   A separation process in which air bubbles emerging from a supersaturated solution become attached to suspended solids in the liquid undergoing treatment and float them up to the surface. See also *diffused air*.

dissolved oxygen (DO)   The oxygen dissolved in liquid, usually expressed in milligrams per liter (mg/L) or percent saturation.

dissolved solids   Solids in solution that cannot be removed by filtration; for example, NaCl and other salts that must be removed by evaporation. See also *total dissolved solids*.

distributor   A device used to apply liquid to the surface of a filter or contact bed. Distributors are of two general types: fixed and movable. The fixed type consists of perforated pipes, notched troughs, sloping boards, or sprinkler nozzles. The movable type consists of rotating, reciprocating, or traveling perforated pipes or troughs applying a spray or a thin sheet of liquid.

diurnal   (1) Occurring during a 24-hour period; diurnal variation. (2) Occurring during the day (as opposed to night). (3) In tidal hydraulics, having a period or cycle of approximately 1 tidal day.

diversity   The characteristic or variety of electrical loads whereby individual maximum demands usually occur at different times. Diversity among equipment loads results in diversity among the loads of transformers, feeders, and substations.

diversity factor   The ratio of the sum of the coincident demands of two or more loads to their maximum demands for the same period.

domestic wastewater   Wastewater derived principally from dwellings, business buildings, institutions, and the like. It may or may not contain groundwater, surface water, or stormwater.

dosing tank   Any tank used in applying a dose; specifically used for intermittent application of wastewater to subsequent processes.

double-suction impeller   An impeller with two suction inlets, one on each side of the impeller.

double-suction pump   A centrifugal pump with suction pipes connected to the casing from both sides.

DPD method   An analytical method for determining chlorine residual using the reagent DPD (n-diethyl-p-phenylenediamine). This is the most commonly and officially recognized test for free chlorine residual.

drag   The resistance offered by a liquid to the settlement or deposition of a suspended particle.

drag coefficient   A measure of the resistance to sedimentation or flotation of a suspended particle as influenced by its size, shape, density, and terminal velocity. It is the ratio of the force per unit area to the stagnation pressure and is dimensionless. See also *friction factor*.

drain   (1) A conduit or channel constructed to carry off, by gravity, liquids other than wastewater, including surplus underground, storm, or surface water. It may be an open ditch, lined or unlined, or a buried pipe. (2) In plumbing, any pipe that carries water or wastewater in a building drainage system.

drawdown   (1) The magnitude of the change in surface elevation of a body of water as a result of the withdrawal of water. (2) The magnitude of the lowering of the water surface in a well, and of the water table or piezometric surface adjacent to the well, resulting from the withdrawal of water from the well by pumping. (3) In a continuous water surface with accelerating flow, the difference in elevation between downstream and upstream points.

drum screen   A screen in the form of a cylinder or truncated cone that rotates on its axis.

dry-bulb temperature   The temperature of air measured by a conventional thermometer.

dry feeder   A feeder for dispensing a chemical or other fine material to water or wastewater at a rate controlled manually or automatically by the rate of flow. The constant rate may be either volumetric or gravimetric.

drying beds   Confined, shallow layers of sand or gravel on which wet sludge is distributed for draining and air drying; also applied to underdrained, shallow, dyked, earthen structures used for drying sludge.

dry suspended solids   The weight of the suspended matter in a sample after drying for a specified time at a specific temperature.

dry weather flow   (1) The flow of wastewater in a combined sewer during dry weather. Such flow consists mainly of wastewater, with no stormwater included. (2) The flow of water in a stream during dry weather, usually contributed entirely by groundwater.

dual-media filters   Deep-bed filters using discrete layers of dissimilar media, such as anthracite and sand, placed one on top of the other.

duplex pump   A reciprocating pump consisting of two cylinders placed side by side and connected to the same suction and discharge pipe; the pistons move so that one exerts suction while the other exerts pressure resulting in continuous discharge from the pump.

dynamic equilibrium   See *population dynamics*.

dynamic head   (1) When there is flow, (a) the head at the top of a waterwheel; (b) the height of the hydraulic grade line above the top of a waterwheel; and (c) the head against which a pump works. (2) That head of fluid that would produce statically the pressure of a moving fluid.

dynamic suction head   The reading of a gauge on the suction line of a pump corrected for the distance of the pump below the free surface of the body of liquid being pumped; exists only when the pump is below the free surface. When pumping proceeds at the required capacity, the vertical distance from the source of supply to the center of the pump minus velocity head and entrance and friction losses. Internal pump losses are not subtracted.

dynamic suction lift   When pumping proceeds at the required capacity, the vertical distance from the source of supply to the center of the suction end of a pump, plus velocity head and entrance and friction losses. Internal pump losses are not added.

E. coli   See *Escherichia coli.*

eductor   A device for mixing air with water; a liquid pump operating under a jet principle, using liquid under pressure as the operating medium to entrain air in the liquid. See also *ejector.*

effective size   The diameter of the particles, spherical in shape, equal in size, and arranged in a given manner, of a hypothetical sample of granular material that would have the same transmission constant as the actual material under consideration. There are a number of methods for determining effective size, the most common being that developed by Allen Hazen, which consists of passing the granular material through sieves with varying dimensions of mesh. In this method, the effective size is determined from the dimensions of that mesh, which permits 10% of the sample to pass and will retain the remaining 90%; in other words, the effective size is that for which 10% of the grains are smaller and 90% larger.

effervescence   The vigorous escape of small gas bubbles from a liquid, especially as a result of chemical action.

efficiency   The relative results obtained in any operation in relation to the energy or effort required to achieve such results. It is the ratio of the total output to the total input, expressed as a percentage.

effluent   Wastewater or other liquid, partially or completely treated or in its natural state, flowing out of a reservoir, basin, treatment plant, or industrial treatment plant, or part thereof.

effluent quality   The physical, biological, and chemical characteristics of a wastewater or other liquid flowing out of a basin, reservoir, pipe, or treatment plant.

ejector   A device for moving a fluid or solid by entraining it in a high-velocity stream or air or water jet.

elbow   A pipe fitting that connects two pipes at an angle. The angle is always 90 deg unless another angle is stated. Also called an *ell.*

electromotive force   The property of a physical device that tends to produce an electrical current in a circuit. It is the moving force that causes current to flow (see *volt*).

elevation head   The energy possessed per unit weight of a fluid because of its elevation above some point. Also called *position head* or *potential head.*

elutriation   A process of sludge conditioning whereby the sludge is washed with either fresh water or plant effluent to reduce the demand for conditioning chemicals and to improve the settling or filtering characteristics of the solids. Excessive alkalinity is removed in this process.

emission   Discharge of a liquid, solid, or gaseous material.

emulsifying agent   An agent capable of modifying the surface tension of emulsion droplets to prevent coalescence. Examples are soap and other surface-active agents, certain proteins and gums, water-soluble cellulose derivatives, and polyhydric alcohol esters and ethers.

emulsion   A heterogeneous liquid mixture of two or more liquids not normally dissolved in one another, but held in suspension one in the other by forceful agitation or by emulsifiers that modify the surface tension of the droplets to prevent coalescence.

endogenous respiration   Autooxidation by organisms in biological processes.

energy (electrical)   As commonly used in the utility industry, electrical energy means kilowatt-hours.

Enterococci   A group of Cocci that normally inhabit the intestines of man and animals. Incorrectly used interchangeably with *fecal Streptococci*.

entrainment   The carryover of drops of liquid during processes such as distillation. The trapping of bubbles in a liquid produced either mechanically through turbulence or chemically through a reaction.

enzyme   A catalyst produced by living cells. All enzymes are proteins, but not all proteins are enzymes.

epidemic   A disease that occurs simultaneously in a large fraction of the community.

equalization   In wastewater systems, the storage and controlled release of wastewaters to treatment processes at a controlled rate determined by the capacity of the processes, or at a rate proportional to the flow in the receiving stream; used to smooth out variations in temperature and composition as well as flow.

equalizing basin   A holding basin in which variations in flow and composition of a liquid are averaged. Such basins are used to provide a flow of reasonably uniform volume and composition to a treatment unit. Also called *balancing reservoir*.

equilibrium   A condition of balance in which the rate of formation and the rate of consumption or degradation of various constituents are equal. See also *chemical equilibrium*.

equilibrium constant   A value that describes the quantitative relationship between chemical species in a system at equilibrium.

equivalent calcium carbonate   A common form of expressing hardness, the acidity, or the carbon dioxide, carbonate, bicarbonate, noncarbonate, hydroxide, or total alkalinity of water; expressed in milligrams per liter (mg/L). It is calculated by multiplying the number of chemical equivalents of any of these constituents present in 1 L by 50, the equivalent weight of calcium carbonate. See also *chemical equivalent*.

Escherichia coli (E. coli)   One of the species of bacteria in the fecal coliform group. It is found in large numbers in the gastrointestinal tract and feces of warm-blooded animals and man. Its presence is considered indicative of fresh fecal contamination, and it is used as an indicator organism for the presence of less easily detected pathogenic bacteria.

eutrophication   Nutrient enrichment of a lake or other water body, typically characterized by increased growth of planktonic algae and rooted plants. It can be accelerated by wastewater discharges and polluted runoff.

evaporation   (1) The process by which water becomes a vapor. (2) The quantity of water that is evaporated; the rate is expressed in depth of water, measured as liquid water removed from a specified surface per unit of time, generally in inches or centimeters per day, month, or year. (3) The concentration of dissolved solids by driving off water through the application of heat.

evaporation opportunity   The ratio of the rate of evaporation from a land or water surface in contact with the atmosphere to evaporation under existing atmospheric conditions; that is, the ratio of the actual to the potential rate of evaporation. Also called *relative evaporation*.

evaporation rate   The quantity of water, expressed in terms of depth of liquid water, evaporated from a given water surface per unit of time. It is usually expressed in inches or millimeters per day, month, or year.

evapotranspiration   Water withdrawn from soil by evaporation or plant transpiration; considered synonymous with consumptive use.

evapotranspiration potential   Water loss that would occur if there was never a deficiency of water in the soil for use by vegetation.

explosimeter   A device for measuring the concentration of potentially explosive fumes. Also called a *combustible-gas indicator*.

extended aeration   A modification of the activated-sludge process using long aeration periods to promote aerobic digestion of the biological mass by endogenous respiration. The process includes stabilization of organic matter under aerobic conditions and disposal of the gaseous end products into the air. Effluent contains finely divided suspended matter and soluble matter.

extended aeration process   A modification of the activated-sludge process. See *extended aeration*.

extraction   The process of dissolving and separating out particular constituents of a liquid by treatment with solvents specific for those constituents. Extraction may be liquid–solid or liquid–liquid.

facultative   Having the ability to live under different conditions; for example, with or without free oxygen.

facultative bacteria   Bacteria that can grow and metabolize in the presence, as well as in the absence, of dissolved oxygen.

facultative lagoon   A lagoon or treatment pond with an aerobic upper section and an anaerobic bottom section so that both aerobic and anaerobic biological processes occur simultaneously.

Fahrenheit   A temperature scale in which 32° marks the freezing point and 212° the boiling point of water at 760-mm Hg. To convert to centigrade (Celsius), subtract 32 and multiply by 0.5556.

false filter bottom   A type of underdrainage system consisting of a porous or perforated floor suspended above the true bottom of the filter. See also *underdrain*.

fats (wastes)   Triglyceride esters of fatty acids; erroneously used as a synonym for *grease*.

fecal coliform   Aerobic and facultative, Gram-negative, non-spore-forming, rod-shaped bacteria capable of growth at 44.5 °C (112 °F), and associated with fecal matter of warm-blooded animals.

fecal indicators   Fecal coliform, fecal Streptococci, and other bacterial groups originating in human or other warm-blooded animals, indicating contamination by fecal matter.

fecal Streptococci   The subgroup of enterococci that is of particular concern in water and wastewater. See also *Enterococci*.

feeder   A circuit conductor between the service equipment or switchboard and the branch circuit overcurrent device.

fermentation   Changes in organic matter or organic wastes brought about by anaerobic microorganisms and leading to the formation of carbon dioxide, organic acids, or other simple products. See also *biological oxidation*.

ferric chloride ($FeCl_3$)   A soluble iron salt often used as a sludge conditioner to enhance precipitation or bind up sulfur compounds in wastewater treatment. See also *coagulant*.

ferric sulfate [$Fe_2(SO_4)_3$]   A water-soluble iron salt formed by reaction of ferric hydroxide and sulfuric acid or by reaction of iron and hot concentrated sulfuric acid; also obtainable in solution by reaction of chlorine and ferrous sulfate; used in conjunction with lime as a sludge conditioner to enhance precipitation.

ferrous chloride (FeCl₂)   A soluble iron salt used as a sludge conditioner to enhance precipitation or bind up sulfur. See also *coagulant*.

ferrous sulfate (FeSO₄·7H₂O)   A water-soluble iron salt, sometimes called *copperas*; used in conjunction with lime as a sludge conditioner to enhance precipitation.

field groundwater velocity   The actual or field velocity of groundwater percolating through water-bearing material. It is measured by the volume of groundwater passing through a unit cross-sectional area in unit time divided by the effective porosity. Also called *effective groundwater velocity, true groundwater velocity, actual groundwater velocity*.

field moisture capacity   The approximate quantity of water that can be permanently retained in the soil in opposition to the downward pull of gravity. It may be expressed as a percentage of dry weight or in inches for a given depth of soil. The length of time required for a soil to reach field moisture capacity varies considerably with various soils, being approximately 24 to 48 hours for sandy soils, 5 to 10 days for silt clay soils, and longer for clays. Also called *capillary capacity, field carrying capacity, maximum water-holding capacity, moisture-holding capacity, normal moisture capacity*.

field permeability coefficient   The rate of flow of water, in gallons per day (gpd) or liters per second (L/s), under prevailing conditions, through each 1 ft (0.3 m) of thickness of a given aquifer in a width of 1 mile (1.6 km), for each 1 ft/mile (0.19 m/km) of hydraulic gradient. Also called *hydraulic conductivity*.

filamentous growth   Intertwined, thread-like biological growths characteristic of some species of bacteria, fungi, and algae. Such growths reduce sludge settleability and dewaterability.

filamentous organisms   Bacterial, fungal, and algal species that grow in thread-like colonies resulting in a biological mass that will not settle and may interfere with drainage through a filter.

filter   A device or structure for removing solid or colloidal material, usually of a type that cannot be removed by sedimentation, from water, wastewater, or other liquid. The liquid is passed through a filtering medium, usually a granular material but sometimes finely woven cloth, unglazed porcelain, or specially prepared paper. There are many types of filters used in water and wastewater treatment. See also *pressure filter*.

filter aid   Solid particulate media (for example, diatomaceous earth) added to a filter to improve the rate of filtration; also used colloquially to describe flocculents in water treatment; same as filtration aid. See also *coagulant* or *flocculent aid*.

filter bed   (1) A type of bank revetment consisting of layers of filtering medium of which the particles gradually increase in size from the bottom upward. Such a filter allows the groundwater to flow freely, but it prevents even the smallest soil particles from being washed out. (2) A tank for water filtration that has a false bottom covered with sand, such as a rapid sand filter. (3) A pond with sand bedding, as a sand filter or slow sand filter. (4) The media that comprise a trickling filter.

filter bottom   (1) The underdrainage system for collecting the water that has passed through a rapid sand filter and for distributing the wash water that cleans the filtering medium. (2) The underdrainage system supporting the graded gravel of a biological bed. It may consist of specially fabricated tile or concrete blocks containing waterways and slots in the top for conveying the underdrainage, or it may consist of inverted half tile.

filter cake   The solids collected on the surface of a mechanical filter. It also applies to spent cake removed from a diatomaceous earth filter.

filter clogging   The effect occurring when fine particles fill the voids of a sand filter or biological bed, or when growths form surface mats that retard the normal passage of liquid through the filter.

filter cloth   A fabric stretched around the drum of a vacuum filter.

filtered wastewater   Wastewater that has passed through a mechanical filtering process but not through a trickling filter bed.

filter efficiency   The operating results of a filter as measured by various criteria such as percentage reduction in suspended matter, total solids, BOD, bacteria, or color.

filter flooding   The filling of a trickling filter to an elevation above the top of the medium by closing all outlets in order to reduce or control filter flies.

filter gallery   A gallery provided in a treatment plant for the installation of conduits and valves and used as a passageway to provide access to them. See also *pipe gallery*.

filter loading   Organically, the pounds (kilograms) of BOD in the applied liquid per unit of filter bed area or volume per day. Hydraulically, the quantity of liquid applied per unit of filter bed area or volume per day.

filter media   (1) Material through which water, wastewater, or other liquid is passed for the purpose of purification, treatment, or conditioning. (2) A cloth or metal material of some appropriate design used to intercept sludge solids in sludge filtration. (3) Particulate (sand, gravel, or diatomaceous earth) or fibrous (cloth) material placed within a filter to collect suspended particles.

filter ponding   The formation of ponds on the surface of trickling filters, caused by excessive biofilm growth, media degradation, or inadequate ventilation. Sometimes called *filter pooling*.

filter press   A plate and frame press operated mechanically to produce a semisolid cake from a slurry. See also *plate press*.

filter rate   The rate of application of material to some process involving filtration, for example, application of wastewater sludge to a vacuum filter, wastewater flow to a trickling filter, or water flow to a rapid sand filter.

filter run   (1) The interval between the cleaning and washing operation of a rapid sand filter. (2) The interval between the changes of the filter medium on a sludge dewatering filter.

filter strainer   A perforated device inserted in the underdrain of a rapid sand filter through which the filtered water is collected and through which the wash water is distributed when the filter is washed. Also called a strainer head.

filter underdrain   A system of subsurface drains to collect water that passes through a sand filter or biological bed. See also *filter bottom*.

filter wash   The reversal of flow through a rapid sand filter to wash clogged material out of the filtering medium and relieve conditions causing loss of head. Also called *backwash*.

filtrate   The liquid that has passed through a filter.

filtration   The process of contacting a dilute liquid suspension with filter media for the removal of suspended or colloidal matter, or for the dewatering of concentrated sludge.

final effluent   The effluent from the final treatment unit of a wastewater treatment plant.

final sedimentation   The separation of solids from wastewater in the last settling tank of a treatment plant.

fire flow   The rate of flow, usually expressed in gallons per minute (gpm) or cubic meters per second (m^3/s), that can be delivered from a water distribution system at a specified residual pressure for fire fighting. When delivery is to fire department pumpers, the specified residual pressure is generally 20 psi (138 kPa).

fire pressure   The pressure necessary in water mains when water is used for fire fighting; applied to cases in which the pressure for fire fighting is increased above that normally maintained for general use.

fire-service connection   A pipe extending from a main to supply a sprinkler, standpipe, yard main, or other fire protection system.

fire system   A separate system of water pipes or mains and their appurtenances installed solely to furnish water for extinguishing fires.

first-stage BOD   That part of oxygen demand associated with biochemical oxidation of carbonaceous material. Usually, the greater part of the carbonaceous material is oxidized before the second stage (active oxidation of the nitrogenous material) takes place.

five-day biochemical oxygen demand (BOD$_5$)   A standard test to assess wastewater pollution due to organic substances, measuring the oxygen used under controlled conditions of temperature (20 °C) and time (5 days).

fixed distributor   A distributor consisting of perforated pipes or notched troughs, sloping boards, or sprinkler nozzles that remain stationary when the distributor is operating. See also *distributor*.

fixed solids   The residue remaining after ignition of suspended or dissolved matter.

flame arrester   (1) A device incorporating a fine-mesh wire screen or tube bundle inserted in a vent or pipe and designed to resist the flashback of flame. (2) Device consisting of a multiple number of corrugated stamped sheets in a gas-tight housing. As a flame passes through the sheets, it is cooled below the ignition point.

flange   A projecting rim, edge, lip, or rib.

flap gate   A gate that opens and closes by rotation around a hinge or hinges at the top side of the gate.

flap valve   A valve that is hinged at one edge and opens and shuts by rotating about the hinges. See also *check valve*.

flash dryer   A device for vaporizing water from partly dewatered and finely divided sludge through contact with a current of hot gas or superheated vapor. It includes a squirrel-cage mill for separating the sludge cake into fine particles.

flash mixer   A device for uniform, quick dispersal of chemicals throughout a liquid.

flash point   The temperature at which a gas, volatile liquid, or other substance ignites.

flat-crested weir   A weir with a horizontal crest in the direction of flow and of appreciable length when compared with the depth of water passing over it.

flight   A scraper in a rectangular sedimentation tank with blades that move sludge along the bottom of the tank to a collection point. As the flights return, scum is collected on the surface of the tank and pushed to an outlet point.

float control   A float device that is triggered by changing liquid levels that activates, deactivates, or alternates process equipment operation.

float gauge   A device for measuring the elevation of the liquid, the actuating element of which is a buoyant float that rests on the surface of the liquid and rises or falls with it. The elevation of the surface is measured by a chain or tape attached to the float.

floating cover   A gas-tight metal cover floating on the sludge in a digestion tank, with guides to assist in smooth vertical travel as the sludge level changes.

float switch   An electrical switch operated by a float in a tank or reservoir and usually controlling the motor of a pump.

float valve   A valve, such as a plug or gate, that is actuated by a float to control the flow into a tank.

floc   Collections of smaller particles agglomerated into larger, more easily settleable particles through chemical, physical, or biological treatment. See also *flocculation*.

flocculant   Water-soluble organic polyelectrolytes that are used alone or in conjunction with inorganic coagulants, such as aluminum or iron salts, to agglomerate the solids present to form large, dense floc particles that settle rapidly.

flocculating tank   A tank used for the formation of floc by the gentle agitation of liquid suspensions, with or without the aid of chemicals.

flocculation   In water and wastewater treatment, the agglomeration of colloidal and finely divided suspended matter after coagulation by gentle stirring by either mechanical or hydraulic means. For biological wastewater treatment in which coagulation is not used, agglomeration may be accomplished biologically.

flocculation agent   A coagulating substance that, when added to water, forms a flocculent precipitate that will entrain suspended matter and expedite sedimentation; examples are alum, ferrous sulfate, and lime.

flocculator   (1) A mechanical device to enhance the formation of floc in a liquid. (2) An apparatus for the formation of floc in water and wastewater.

flood flow   The discharge of a stream during periods of flood.

flood frequency   The frequency with which the maximum flood may be expected to occur at a site in any average interval of years. Frequency analysis defines the "n-year flood" as being the flood that will, over a long period of time, be equaled or exceeded on the average once every n years. Thus, the 10-year flood would be expected to occur approximately 100 times in a period of 1 000 years, and of these, 10 would be expected to reach the 100-year magnitude. Sometimes expressed in terms of percentage of probability; for example, a probability of 1% would be 100-year flood; a probability of 10% would be a 10-year flood.

flood-protection works   Structures built to protect lands and property from damage by floods.

flotation   (1) Separation of suspended particles, or oil and grease, from solution by naturally or artificially raising them to the surface, usually with air. (2) Thickening of waste activated sludge by injecting air into it and introducing the mixture into a tank where the air buoys the sludge to the surface.

flow   (1) The movement of a stream of water or other fluid from place to place; the movement of silt, water, sand, or other material. (2) The fluid that is in motion. (3) The quantity or rate of movement of a fluid discharge; the total quantity carried by a stream. (4) To issue forth or discharge. (5) The liquid or amount of liquid per unit time passing a given point.

flow-control valve   A device that controls the rate of flow of a fluid.

flow equalization   Transient storage of wastewater for release to a sewer system or wastewater treatment plant at a controlled rate to provide a reasonably uniform flow for treatment.

flowrate   The volume or mass of a gas, liquid, or solid material that passes through a cross section of conduit in a given time; measured in such units as kilograms per hour (kg/h), cubic meters per second (m^3/s), liters per day (L/d), or gallons per day (gpd).

flow recording   Documentation of the rate of flow of a fluid past a given point. The recording is normally accomplished automatically.

flow regulator   A structure installed in a canal, conduit, or channel to control the flow of water or wastewater at the intake or to control the water level in a canal, channel, or treatment unit. See also *rate-of-flow controller*.

flow sheet   A diagrammatic representation of the progression of steps in a process showing their sequence and interdependence.

fluidized bed reactor   A pressure vessel or tank that is designed for liquid–solid or gas–solid reaction. The liquid or gas moves upward through the solids particles at a velocity sufficient to suspend the individual particles in the fluid. Applications include ion exchange, granular activated carbon adsorbers, and some types of furnaces, kilns, and biological contactors.

flushing   The flow of water under pressure in a conduit or well to remove clogged material.

foam   (1) A collection of minute bubbles formed on the surface of a liquid by agitation, fermentation, and so on. (2) The frothy substance composed of an aggregation of bubbles on the surface of liquids and created by violent agitation or by the admission of air bubbles to liquid containing surface-active materials, solid particles, or both. Also called *froth*.

food-to-microorganism (F:M) ratio   In the activated-sludge process, the loading rate expressed as pounds of BOD$_5$ per pound of mixed liquor or mixed liquor volatile suspended solids per day (lb BOD$_5$/d/lb MLSS or MLVSS).

foot valve   (1) A valve placed at the bottom of the suction pipe of a pump that opens to allow water to enter the suction pipe, but closes to prevent water from passing out of it at the bottom end. (2) A valve with the reverse action attached to the drainage pipe of a vacuum chamber. It allows water to drain out, but closes to hold the vacuum.

forced aeration   The bringing about of intimate contact between air and liquid where the air, under pressure, is applied below the surface of the liquid through diffusers or other devices that promote the formation of small bubbles.

force main   A pressure pipe joining the pump discharge at a water or wastewater pumping station with a point of gravity flow.

formazine turbidity unit (FTU)   A standard unit of turbidity based on a known chemical reaction that produces insoluble particulates of uniform size. The FTU has largely replaced the JTU. Also known as *nephelometric turbidity unit*.

fouling   A gelatinous, slimy accumulation resulting from the activity of organisms in the water. Fouling may be found on concrete, masonry, or metal surfaces, but tuberculation is found only on metal surfaces.

Francis turbine   A reaction turbine of the radial inward-flow type.

free available chlorine   The amount of chlorine available as dissolved gas, hypochlorous acid, or hypochlorite ion that is not combined with an amine or other organic compound.

free available residual chlorine   That portion of the total residual chlorine remaining in water or wastewater at the end of a specified contact period that will react chemically and biologically as hypochlorous acid or hypochlorite ion.

freeboard   The vertical distance between the normal maximum level of the surface of the liquid in a conduit, reservoir, tank, or canal and the top of the sides of an open conduit or the top of a dam or levee, which is provided so that waves and other movements of the liquid will not overflow the confining structure.

free flow   A condition of flow through or over a structure where such flow is not affected by submergence or the existence of tailwater.

free oxygen   Elemental oxygen (O_2).

free-swimming ciliate   Mobile, one-celled organisms using cilia (hair-like projections) for movement.

free water   Suspended water constituting films covering the surface of solid particles or the walls of fractures, but in excess of pellicular water; mobile water is free to move in any direction under the pull of the force of gravity and unbalanced film pressure.

frequency   (1) The time rate of vibration or the number of complex cycles per unit time. (2) The number of occurrences of a certain phenomenon in a given time. (3) The number of occasions on which the same numerical measure of a particular quantity has occurred between definite limits. (4) The number of cycles through which an alternating current passes per second. Frequency has been generally standardized in the electrical utility industry in the United States at 60 cycles per second (60 Hz).

fresh-air inlet   A specially constructed opening usually provided with a perforated cover to facilitate ventilation of a wastewater line.

fresh sludge   Sludge in which decomposition is little advanced.

fresh wastewater   Wastewater of recent origin containing dissolved oxygen.

friction factor   A measure of the resistance to flow of fluid in a conduit as influenced by wall roughness.

friction head   The head loss resulting from water flowing in a stream or conduit as the result of the disturbances set up by the contact between the moving water and its containing conduit and by intermolecular friction. In laminar flow, the head loss is approximately proportional to the first power of the velocity; in turbulent flow to a higher power, approximately the square of the velocity. While, strictly speaking, head losses such as those caused by bends, expansions, obstructions, and impact are not included in this term, the usual practice is to include all such head losses under this term.

friction loss   The head loss resulting from water flowing in a stream or conduit as the result of the disturbances set up by the contact between the moving water and its containing conduit and by intermolecular friction. See also *friction head*.

fungi   Small, nonchlorophyll-bearing plants that lack roots, stems, or leaves; occur (among other places) in water, wastewater, or wastewater effluents; and grow best in the absence of light. Their decomposition may cause disagreeable tastes and odors in water; in some wastewater treatment processes they are helpful and in others they are detrimental.

fuse   A protective device that carries the full current of a circuit. If the current is higher than the fuse rating, it contains a substance that will melt and break the current. Fuses cannot be reset but must be replaced.

gas chromatography   A method of separating a mixture of compounds into its constituents so they can be identified. The sample is vaporized into a gas-filled column, fractionated by being swept over a solid adsorbent, selectively eluted, and identified.

gas chromatography-mass spectrometry (GC-MS)   An analytical technique involving the use of both gas chromatography and mass spectrometry, the former to separate a complex mixture into its components and the latter to deduce the atomic and molecular weights of those components. It is particularly useful in identifying organic compounds.

gas dome   In sludge digestion tanks, usually a steel cover floating entirely or in part on the liquid sludge.

gasification   The transformation of soluble and suspended organic materials into gas during waste decomposition.

gas production   The creation of a gas by chemical or biological means.

gate valve   A valve in which the closing element consists of a disk that slides over the opening or cross-sectional area through which water passes.

gauge   (1) A device for indicating the magnitude or position of an element in specific units when such magnitude or position is subject to change; examples of such elements are the elevation of a water surface, the velocity of flowing water, the pressure of water, the amount or intensity of precipitation, and the depth of snowfall. (2) The act or operation of registering or measuring the magnitude or position of a thing when these characteristics are subject to change. (3) The operation of determining the discharge in a waterway by using both discharge measurements and a record of stage.

globe valve   A valve having a round, ball-like shell and horizontal disk.

gpcd   The rate of water, wastewater, or other flow measured in U.S. gallons (liters) per capita of served population per day.

gpd   The rate of water, wastewater, or other flow measured in U.S. gallons (liters) per day.

gpm   The rate of water, wastewater, or other flow measured in U.S. gallons (liters) per minute.

grab sample   A sample taken at a given place and time. It may be representative of the flow. See also *composite sample*.

gradient   The rate of change of any characteristic per unit of length or slope. The term is usually applied to such things as elevation, velocity, or pressure. See *slope*.

grease and oil   In wastewater, a group of substances including fats, waxes, free fatty acids, calcium and magnesium soaps, mineral oils, and certain other nonfatty materials; water-insoluble organic compounds of plant and animal origins or industrial wastes that can be removed by natural flotation skimming.

grease skimmer   A device for removing floating grease or scum from the surface of wastewater in a tank.

grit   The heavy suspended mineral matter present in water or wastewater, such as sand, gravel, or cinders. It is removed in a pretreatment unit called a *grit chamber* to avoid abrasion and wearing of subsequent treatment devices.

grit chamber   A detention chamber or an enlargement of a sewer designed to reduce the velocity of flow of the liquid to permit the separation of mineral (grit) from organic solids by differential sedimentation.

grit collector   A device placed in a grit chamber to convey deposited grit to a point of collection.

grit separator   Any process or device designed to separate grit from a water or wastewater stream.

grit washer   A device for washing organic matter out of grit.

ground   A conducting connection, whether intentional or accidental, between an electrical circuit or equipment and earth, or to some conducting body that serves in place of earth.

grounded   Connected to earth or to some conducting body that serves in place of the earth.

hardness   A characteristic of water imparted primarily by salts of calcium and magnesium, such as bicarbonates, carbonates, sulfates, chlorides, and nitrates, that causes curdling and increased consumption of soap, deposition of scale in boilers, damage in some industrial processes, and sometimes objectionable taste. It may be determined by a standard laboratory titration procedure or computed from the amounts of calcium and magnesium expressed as equivalent calcium carbonate. See also *carbonate hardness*.

hazardous waste   Any waste that is potentially damaging to environmental health because of toxicity, ignitability, corrosivity, chemical reactivity, or other reasons.

head   (1) The height of the free surface of fluid above any point in a hydraulic system; a measure of the pressure or force exerted by the fluid. (2) The energy, either kinetic or potential, possessed by each unit weight of a liquid, expressed as the vertical height through which a unit weight would have to fall to release the average energy possessed. It is used in various compound terms such as *pressure head, velocity head,* and *loss of head*. (3) The upper end of anything, such as a headworks. (4) The source of anything, such as a head-water. (5) A comparatively high promontory with either a cliff or steep face extending into a large body of water, such as a sea or lake. An unnamed head is usually called a *headland*.

header   (1) A structure installed at the head or upper end of a gully to prevent overfall cutting. (2) A supply ditch for the irrigation of a field. (3) A large pipe installed to intercept the ends of a series of pipes; a manifold. (4) The closing plate on the end of a sewer lateral that will not be used immediately.

head gate   A gate at the entrance to a conduit such as a pipeline, penstock, or canal.

head loss   Energy losses resulting from the resistance of flow of fluids; may be classified into conduit surface and conduit form losses.

headworks   (1) All the structures and devices located at the head or diversion point of a conduit or canal. The term as used is practically synonymous with *diversion works*; an intake heading. (2) The initial structures and devices of a water or wastewater treatment plant.

heat exchanger   A device providing for the transfer of heat between two fluids.

heat treatment   A sludge conditioning process combining high temperature, time, and pressure to improve the dewaterability of organic sludge.

heavy metals   Metals that can be precipitated by hydrogen sulfide in acid solution, for example, lead, silver, gold, mercury, bismuth, and copper.

high-purity oxygen   A modification of the activated-sludge process using relatively pure oxygen and covered aeration tanks in a conventional flow arrangement.

high-rate aeration   A modification of the activated-sludge process whereby the mixed liquor suspended solids loadings are kept high, allowing high food-to-microorganism (F:M) ratios and shorter detention times.

humus sludge   Sloughed particles of biomass from trickling media that are removed in the secondary clarifier.

hydrated lime   Limestone that has been "burned" and treated with water under controlled conditions until the calcium oxide portion has been converted to calcium hydroxide.

hydraulic loading   The amount of water applied to a given treatment process, usually expressed as volume per unit time, or volume per unit time per unit surface area.

hydraulic radius   The cross-sectional area of a stream of water divided by the length of that part of its periphery in contact with its containing conduit; the ratio of area to wetted perimeter. Also called *hydraulic mean depth*.

hydrocarbon   Any of the class of compounds consisting solely of carbon and hydrogen. Usually derived from petroleum.

hydrogen-ion concentration   The concentration of hydrogen ions in moles per liter of solution (moles/L). Commonly expressed as the pH value, which is the logarithm of the reciprocal of the hydrogen-ion concentration. See also *pH*.

hydrogen sulfide (H_2S)   A toxic and lethal gas produced in sewers and digesters by anaerobic decomposition. Detectable in low concentrations (%) by its characteristic "rotten egg" odor. It deadens the sense of smell in higher concentrations or after prolonged exposure. Respiratory paralysis and death may occur quickly at concentrations as low as 0.07% by volume in air.

hydrostatic level   The level or elevation to which the top of a column of water would rise from an artesian aquifer or basin, or from a conduit under pressure.

hypochlorination   The use of sodium hypochlorite ($NaOCl_2$) for disinfection.

hypochlorite   Calcium, sodium, or lithium hypochlorite.

Imhoff cone   A cone-shaped graduated vessel used to measure the volume of settleable solids in various liquids of wastewater origin during various settling times.

impedance, resistance, and reactance   The relationship between impedance, resistance, and reactance is given by the following equation:

$$Z = R + jX$$

Where
  Z = impedance (Ω),
  R = resistance (Ω),
  X = reactance (Ω), and
  j is the imaginary unit $\sqrt{-1}$.

impeller   A rotating set of vanes designed to impel rotation of a mass of fluid.

incineration   Combustion or controlled burning of volatile organic matter in sludge and solid waste reducing the volume of the material while producing heat, dry inorganic ash, and gaseous emissions.

incinerator   A furnace or apparatus for incineration.

index   (1) An indicator, usually numerically expressed, of the relation of one phenomenon to another. (2) An indicating part of an instrument.

indicator   (1) A device that shows by an index, pointer, or dial the instantaneous value of such quantities as depth, pressure, velocity, stage, or the movements or positions of water-controlling devices; a gauge. See also *recorder*. (2) A substance giving a visible change, usually of color, at a desired point in a chemical reaction, generally at a prescribed end point.

indicator gauge   A gauge that shows, by means of an index, pointer, or dial the instantaneous value of such characteristics as depth, pressure, velocity, stage, discharge, or the movements or positions of waste-controlling devices. See also *indicator*, *recorder*.

industrial wastewater   Wastewater derived from industrial sources or processes.

infectious hepatitis   An acute viral inflammation of the liver characterized by jaundice, fever, nausea, vomiting, and abdominal discomfort; may be waterborne.

infiltration   (1) The flow or movement of water through the interstices or pores of a soil or other porous medium. (2) The quantity of groundwater that leaks into a pipe through joints, porous walls, or breaks. (3) The entrance of water from the ground into a gallery. (4) The absorption of liquid by the soil, either as it falls as precipitation or from a stream flowing over the surface.

inflow   In relation to sanitary sewers, the extraneous flow that enters a sanitary sewer from sources other than infiltration, such as roof leaders, basement drains, land drains, and manhole covers. See also *infiltration*.

influent   Water, wastewater, or other liquid flowing into a reservoir, basin, treatment plant, or treatment process. See also *effluent*.

inlet   (1) A surface connection to a drain pipe. (2) A structure at the diversion end of a conduit. (3) The upstream end of any structure through which water may flow. (4) A form of connection between the surface of the ground and a drain or sewer for the admission of surface or stormwater. (5) An intake.

inlet control   Control of the relationship between headwater elevation and discharge by the inlet or upstream end of any structure through which water may flow.

inorganic   All those combinations of elements that do not include organic carbon.

inorganic matter   Mineral-type compounds that are generally nonvolatile, not combustible, and not biodegradable. Most inorganic-type compounds or reactions are ionic in nature; therefore, rapid reactions are characteristic.

instrumentation   Use of technology to control, monitor, or analyze physical, chemical, or biological parameters.

interlock   A means of tying one or more circuits to another. It may be mechanical in nature but more likely involves the use of relays or solid state components.

intermittent chlorination   A technique of noncontinuous chlorination used to control biological fouling of surfaces in freshwater circuits, particularly those used for heat transfer.

intermittent filter   A natural or artificial bed of sand or other fine-grained material to the surface of which wastewater is applied intermittently in flooding doses and through which it passes; filtration is accomplished under aerobic conditions.

inventory   A detailed list showing quantities, descriptions, and values of property. It may also include units of measure and unit prices. The term is often confined to consumable materials but may also cover fixed assets. When an inventory covers all property of the enterprise and is priced as of a certain date, it is known as an appraisal.

ion   A charged atom, molecule, or radical that affects the transport of electricity through an electrolyte or, to a certain extent, through a gas. An atom or molecule that has lost or gained one or more electrons.

ion exchange   (1) A chemical process involving reversible interchange of ions between a liquid and a solid, but no radical change in structure of the solid. (2) A chemical process in which ions from two different molecules are exchanged. (3) The reversible transfer or sorption of ions from a liquid to a solid phase by replacement with other ions from the solid to the liquid. See also *regeneration*.

irrigation   The artificial application of water to lands to meet the water needs of growing plants not met by rainfall.

irrigation requirement   The quantity of water, exclusive of precipitation, that is required for crop production. It includes surface evaporation and other economically unavoidable water waste.

irrigation return water   Drainage water from irrigated farmlands, generally containing high concentrations of dissolved salts and other materials that have been leached out of the upper layers of the soil.

jacketed pump   A pump equipped with jackets around the cylinders, heads, and stuffing boxes through which steam or other heat may be forced to permit the handling of such materials as pitch, resin, and asphalt that are solid when cold but melt on heating; when the pump handles materials at high temperatures, cold water may be substituted for steam or heat.

Jackson turbidity unit (JTU)   A standard unit of turbidity based on the visual extinction of a candle flame when viewed through a column of turbid water containing suspended solids. It varies with the solids composition (barium sulfate, diatomaceous earth, and so on). The JTU has largely been replaced by the more reproducible nephelometric turbidity unit.

jar test   A laboratory procedure for evaluating coagulation, flocculation, and sedimentation processes in a series of parallel comparisons.

jet   The stream of water under pressure issuing from an orifice, nozzle, or tube.

joint   (1) A surface of contact between two bodies or masses of material of like or different character or composition. (2) A connection between two lengths of pipe, made either with or without the use of a third part. (3) A length or piece of pipe.

kinematic viscosity   Ratio of absolute viscosity, expressed in poises (grams per centimeter per second [g/cm·s]), and the density, in grams per cubic centimeter (g/cm³), at room temperature.

kinetics   The study of the rates at which changes occur in chemical, physical, and biological treatment processes.

Kjeldahl nitrogen (TKN)   The combined amount of organic and ammonia nitrogen.

Kraus process   A modification of the activated-sludge process in which aerobically conditioned supernatant liquor from anaerobic digesters is added to activated sludge aeration tanks to improve the settling characteristics of the sludge and to add an oxygen resource in the form of nitrates.

kilovar   One thousand reactive volt amps (see *power*).

kilovolt   One thousand volts.

kilowatt   One thousand watts.

kilowatt-hour   The units of electrical energy equal to one kilowatt of electrical power in an electrical circuit for 1 hour.

laboratory procedures   Modes of conducting laboratory processes and analytical tests consistent with validated standard testing techniques.

lag growth phase   The initial period following bacterial introduction during which the population grows slowly as the bacteria acclimates to the new environment.

lagoon   Any large holding or detention pond, usually with earthen dikes, used to contain wastewater while sedimentation and biological oxidation occur. See also *anaerobic lagoon*.

laminar flow   The flow of a viscous fluid in which particles of the fluid move in parallel layers, each of which has a constant velocity but is in motion relative to its neighboring layers. Also called *streamline flow, viscous flow*.

land application   The recycling, treatment, or disposal of wastewater or wastewater solids to the land under controlled conditions.

landfill   The disposal of solid wastes or sludges by placing on land, compacting, and covering with a thin layer of soil.

leachate   Liquid that has percolated through solid waste or other permeable material and extracted soluble dissolved or suspended materials from it.

leakage   Uncontrolled loss of water from artificial structures as a result of hydrostatic pressure.

leakage detector   A device or appliance, the principle of which is the audibility of water flowing through a leak. Most of these devices are marketed under descriptive trade names.

lethal concentration   The concentration of a test material that causes death of a specified percentage of a population, usually expressed as the median or 50% level (L_{50}).

lift station   A structure that contains pumps and appurtenant piping, valves, and other mechanical and electrical equipment for pumping water, wastewater, or other liquid. Also called a *pumping station*.

lighting panel   An enclosure carrying numerous low-voltage breakers, switches, and fuses servings lights or receptacles in an area.

lime   Any of a family of chemicals consisting essentially of calcium hydroxide made from limestone (calcite) composed almost wholly of calcium carbonate or a mixture of calcium and magnesium carbonate; used to increase pH to promote precipitation reactions or for lime stabilization to kill parthenogenic organisms.

lining   A protective covering over all or a portion of the perimeter of a conduit or reservoir intended to prevent seepage losses, withstand pressure, or resist erosion. In the case of conduits, lining is also sometimes installed to reduce friction losses.

lipids   A group of organic compounds that make up the fats and other esters with analogous properties.

liquefaction   (1) Act or process of liquefying or of rendering or becoming liquid; reduction to a liquid state. (2) Act or process of converting a solid or a gas to a liquid by changes in temperature or pressure, or the changing of the organic matter in wastewater from a solid to a soluble state.

liquid   A substance that flows freely; characterized by free movement of the constituent molecules among themselves, but without the tendency to separate from one another, which is characteristic of gases. Liquid and fluid are often used synonymously, but fluid has the broader significance of including both liquids and gases.

liquid chlorine   Elemental chlorine converted to a liquid state by compression and refrigeration of the dry, purified gas. Liquid chlorine is shipped under pressure in steel containers.

load   The amount of electrical power required at any specified point or points on an electrical system. Load originates at the power-consuming equipment (see *demand*).

load center   A point at which the load of a given area is assumed to be concentrated.

load diversity   The difference between the sum of the individual maximum demands of two or more individual loads and the coincident maximum demand of those loads.

load factor   The ratio of the average load in kilowatts supplied during a designated period to the peak or maximum load in kilowatts occurring in that period. Load factor, in percent, may be derived by multiplying the kilowatt-hours used in the period by

100 and then dividing by the product of the maximum demand in kilowatts and the number of hours in the period.

log growth phase   Initial stage of bacterial growth, during which there is an ample food supply, causing bacteria to grow at their maximum rate.

loss of head   (1) The decrease in energy between two points resulting from friction, bend, obstruction, expansion, or any other cause. It does not include changes in the elevation of the hydraulic grade unless the hydraulic and energy grades parallel each other. (2) The difference between the total heads at two points in a hydraulic system.

low-rate filter   A trickling filter designed to receive a small load of BOD per unit volume of filtering material and to have a low dosage rate per unit of surface area, usually 2 to 5 mgd/ac (2.2×10^{-5} to 5.4×10^{-5} m³/m²·s) generally without recirculation. The organic loading (BOD) rate is usually in the range of 5 to 25 lb/1 000 cu ft (80 to 400 g/m³). Also called a *standard rate filter*.

Manning formula   A formula for open-channel flow published by Manning in 1890. It gives the value of *c* in the Chezy formula. See also *Manning roughness coefficient*.

Manning roughness coefficient   The roughness coefficient in the Manning formula for determination of the discharge coefficient in the Chezy formula.

manometer   An instrument for measuring pressure. It usually consists of a U-shaped tube containing a liquid, the surface of which moves proportionally in one end of the tube with changes in pressure in the liquid in the other end; also, a tube-type of differential pressure gauge.

mass spectrometer   A device that permits observation of the masses of molecular fragments produced by destructible bombardment of the molecule with electrons in a vacuum; coupled with gas chromatography (GC-MS), mass spectrometry can yield very specific compound identification.

mass spectrometry   A means of sorting ions by separating them according to their masses.

mean   (1) The arithmetic average of a group of data. (2) The statistical average (50% point) determined by probability analysis.

mean cell residence time (MCRT)   The average time that a given unit of cell mass stays in the activated-sludge aeration tank. It is usually calculated as the total mixed liquor suspended solids in the aeration tank divided by the combination of solids in the effluent and solids wasted.

mechanical aeration   (1) The mixing, by mechanical means, of wastewater and activated sludge in the aeration tank of the activated-sludge process to bring fresh surfaces of liquid into contact with the atmosphere. (2) The introduction of atmospheric oxygen into a liquid by the mechanical action of paddle, paddle wheel, spray, or turbine mechanisms.

mechanical aerator   A mechanical device for the introduction of atmospheric oxygen into a liquid. See also *mechanical aeration*.

mechanically cleaned screen   A screen equipped with a mechanical cleaning apparatus for removal of retained solids.

mechanical rake   A machine-operated mechanism used for cleaning debris from racks located at the intakes of conduits supplying water to hydroelectric power plants, water supply systems, or for other uses, and conveying wastewater to pumps or treatment processes.

median   In a statistical array, the value having as many cases larger in value as cases smaller in value.

membrane filter test   A sample of water is passed through a sterile filter membrane. The filter is removed and placed on a culture medium and then incubated for a preset period of time. Coliform colonies, which have a pink to dark-red color with a metallic sheen, are then counted using the aid of a low-power binocular wide-field dissecting microscope. The membrane filter test is used to test for the presence and relative number of coliform organisms.

mercaptans   Aliphatic organic compounds that contain sulfur. They are noted for their disagreeable odor and are found in certain industrial wastes.

mercury gauge   A gauge in which the pressure of a fluid is measured by the height the fluid pressure will sustain a column of mercury.

mesh   One of the openings or spaces in a screen. The value of the mesh is usually given as the number of openings per linear inch. This gives no recognition to the diameter of the wire; thus, the mesh number does not always have a definite relationship to the size of the hole.

mesophilic   That group of bacteria that grow best within the temperature range of 20 to 40 °C (68 to 104 °F).

mesophilic digestion   Digestion by biological action at 27 to 38 °C (80 to 100 °F).

mesophilic range   Operationally, that temperature range most conducive to the maintenance of optimum digestion by mesophilic bacteria, generally accepted as between 27 and 38 °C (80 and 100 °F).

metabolism   (1) The biochemical processes in which food is utilized and wastes formed by living organisms. (2) All biochemical reactions involved in cell synthesis and growth.

metazoan   A group of animals having bodies composed of cells differentiated into tissues and organs and usually having a digestive cavity lined with specialized cells.

meter   An instrument for measuring some quantity such as the rate of flow of liquids, gases, or electric currents.

methane (CH_4)   A colorless, odorless, flammable, gaseous hydrocarbon present in natural gas and formed by the anaerobic decomposition of organic matter, or produced artificially by heating carbon monoxide and hydrogen over a nickel catalyst. See also *anaerobic digestion*.

methane bacteria   A specialized group of obligate anaerobic bacteria that decompose organic matter to form methane.

methane fermentation   A reaction sequence that produces methane during the anaerobic decomposition or organic waste. In the first phase, acid-forming bacteria produce acetic acid; in the second, the methane bacteria use this acid and carbon dioxide to produce methane. Fermentation results in the conversion of organic matter into methane gas.

mgd   Million gallons per day; a measure of flow equal to 1.547 cu ft/sec, 681 gpm, or 3 785 m^3/d.

mg/L   Milligrams per liter; a measure of concentration equal to and replacing ppm in the case of dilute concentrations.

microbial activity   The activities of microorganisms resulting in chemical or physical changes.

microbial film   A gelatinous film of microbial growth attached to or spanning the interstices of a support medium. Also called *biological slime*.

microorganisms   Very small organisms, either plant or animal, invisible or barely visible to the naked eye. Examples are algae, bacteria, fungi, protozoa, and viruses.

microscopic   Very small, generally between 0.5 and 100 mm, and visible only by magnification with an optical microscope.

microscopic examination   (1) The examination of water to determine the presence and amounts of plant and animal life, such as bacteria, algae, diatoms, protozoa, and crustacea. (2) The examination of water to determine the presence of microscopic solids. (3) The examination of microbiota in process water, such as the mixed liquor in an activated-sludge plant.

mist   Fine liquid droplets of such small size that gravity separation is hindered. Fog is a water mist.

mixed-flow pump   A centrifugal pump in which the head is developed partly by centrifugal force and partly by the lift of the vanes on the liquid. This type of pump has a single inlet impeller; the flow enters axially and leaves axially and radially.

mixed liquor   A mixture of raw or settled wastewater and activated sludge contained in an aeration tank in the activated-sludge process. See also *mixed liquor suspended solids*.

mixed liquor suspended solids (MLSS)   The concentration of suspended solids in activated-sludge mixed liquor, expressed in milligrams per liter (mg/L). Commonly used in connection with activated-sludge aeration units.

mixed liquor volatile suspended solids (MLVSS)   That fraction of the suspended solids in activated-sludge mixed liquor that can be driven off by combustion at 550 °C (1022 °F); it indicates the concentration of microorganisms available for biological oxidation.

mixed-media filter   A filter containing filtering media of different particle size or density.

mixing basin   A basin or tank in which agitation is applied to water, wastewater, or sludge to increase the dispersion rate of applied chemicals; also, tanks used for general mixing purposes.

mixing chamber   A chamber used to facilitate the mixing of chemicals with liquid or the mixing of two or more liquids of different characteristics. It may be equipped with a mechanical device that accomplishes the mixing.

mixing channel   A channel provided in a water or wastewater treatment plant; the hydraulic characteristics of the waterway or its construction features are such that chemicals or liquids are thoroughly mixed.

modified aeration   A modification of the activated-sludge process in which a shortened period of aeration (1.5 to 3 hours) is used with a reduced quantity of suspended solids (200 to 500 mg/L MLSS) in the mixed liquor. Sludge settling is usually poor; high suspended solids concentration may be expected in effluent.

moisture   Condensed or diffused water collected on or excluded to a surface.

moisture content   The quantity of water present in soil, wastewater sludge, industrial waste sludge, and screenings, usually expressed in percentage of wet weight.

mole   (1) Molecular weight of a substance, normally expressed in grams. (2) A device to clear sewers and pipelines. (3) A massive harbor work, with a core of earth or stone, extending from shore into deep water. It serves as a breakwater, a berthing facility, or a combination of the two.

monitoring   (1) Routine observation, sampling, and testing of designated locations or parameters to determine the efficiency of treatment or compliance with standards or requirements. (2) The procedure or operation of locating and measuring radioactive contamination by means of survey instruments that can detect and measure, as dose rate, ionizing radiations.

Monod equation   A mathematical expression first used by Monod in describing the relationship between the microbial growth rate and concentration of growth-limiting substrate.

most probable number (MPN)   That number of organisms per unit volume which, in accordance with statistical theory, would be more likely than any other number to yield the observed test result or would yield the observed test result with the greatest frequency. Expressed as density of organisms/100 mL. Results are computed from the number of positive findings of coliform group organisms resulting from multiple portion decimal dilution plantings. Used commonly for coliform bacteria.

motor controller   A specialized type of controller whose typical functions performed by a motor controller include starting, accelerating, stopping, reversing, and protecting motors.

moving average   Trend analysis tool for determining patterns or changes in treatment process. For example, a 7-day moving average would be the sum of the datum points for 7 days divided by 7.

mudballs   (1) Accretions of siliceous incrustations on the exterior surface of sand grains. From these incrustations grow numerous filamentous organisms over which there is a gelatinous coating. Mudballs are approximately spherical in shape and vary in size from that of a pea up to 1 or 2 in. (2.5 to 5.1 cm) or more in diameter. They are formed principally by the retention and gradual building up of growths that are not completely removed by the washing process. (2) Balls of sediment sometimes found in debris-laden flow and channel deposits.

mud blanket   A layer of flocculant material that forms on the surface of a sand filter.

multimedia filter beds   A filtration apparatus consisting of two or more media, such as anthracite and sand, through which wastewater flows and by which it is cleansed. Media may be intermixed or segregated.

multiple-hearth incinerator   A countercurrent-type of incinerator frequently used to dry and burn partially dried sludges. Heated air and products of combustion pass by finely pulverized sludge that is continuously raked to expose fresh surfaces.

multiple-stage sludge digestion   The progressive digestion of waste sludge in two or more tanks arranged in series.

multistage pump   A centrifugal pump with two or more sets of vanes or impellers connected in series in the same casing. Such a pump may be designated as two-stage, three-stage, or more, according to the number of sets of vanes used. The purpose is to increase the head of the discharging fluid.

municipal wastewater treatment   Generally includes the treatment of domestic, commercial, and industrial wastes.

nappe   The sheet or curtain of water overflowing a weir or dam. When freely overflowing any given structure, it has a well-defined upper and lower surface.

National Pollutant Discharge Elimination System (NPDES)   A permit that is the basis for the monthly monitoring reports required by most states in the United States.

negative head   (1) The loss of head in excess of the static head (a partial vacuum). (2) A condition of negative pressure produced by clogging of rapid sand filters near the end of a filter run.

negative pressure   A pressure less than the local atmospheric pressure at a given point.

nematode   Member of the phylum (Nematoda) of elongated cylindrical worms parasitic in animals or plants or free-living in soil or water.

nephelometer   An instrument for comparing turbidities of solutions by passing a beam of light through a transparent tube and measuring the ratio of the intensity of the shattered light to that of the incident light.

nephelometric turbidity unit (NTU)   Units of a turbidity measurement using a nephelometer.

net available head   The difference in pressure between the water in a power conduit before it enters the water wheel and the first free water surface in the conduit below the water wheel.

n factor   Values of the roughness coefficient used in Manning formula or Kutter formula. See also *roughness coefficient, Manning formula*.

nitrate (NO_3)   An oxygenated form of nitrogen.

nitrification   The oxidation of ammonia nitrogen to nitrate nitrogen in wastewater by biological or chemical reactions. See also *denitrification*.

nitrifying bacteria   Bacteria capable of oxidizing nitrogenous material.

nitrite (NO_2)   An intermediate oxygenated form of nitrogen.

nitrogen (N)   An essential nutrient that is often present in wastewater as ammonia, nitrate, nitrite, and organic nitrogen. The concentrations of each form and the sum (total nitrogen) are expressed as milligrams per liter (mg/L) elemental nitrogen. Also present in some groundwater as nitrate and in some polluted groundwater in other forms. See also *nutrient*.

nitrogen cycle   A graphical presentation of the conservation of matter in nature showing the chemical transformation of nitrogen through various stages of decomposition and assimilation. The various chemical forms of nitrogen as it moves among living and nonliving matter are used to illustrate general biological principles that are applicable to wastewater and sludge treatment.

nitrogenous oxygen demand (NOD)   A quantitative measure of the amount of oxygen required for the biological oxidation of nitrogenous material, such as ammonia nitrogen and organic nitrogen, in wastewater; usually measured after the carbonaceous oxygen demand has been satisfied. See also *biochemical oxygen demand, nitrification, second-stage BOD*.

nitrogen removal   The removal of nitrogen from wastewater through physical, chemical, or biological processes, or by some combination of these.

Nitrosomonas   A genus of bacteria that oxidize ammonia to nitrate.

Nocardia   Irregularly bent, short filamentous organisms that are characterized in an activated-sludge system when a dark chocolate mousse foam is present.

nonclogging impeller   An impeller of the open, closed, or semiclosed type designed with large passages for passing large solids.

nonsettleable solids   Suspended matter that will stay in suspension for an extended period of time. Such a period may be arbitrarily taken for testing purposes as 1 hour. See also *suspended solids*.

nonuniform flow   A flow in which the slope, cross-sectional area, and velocity change from section to section in the channel.

nozzle   (1) A short, cone-shaped tube used as an outlet for a hose or pipe. The velocity of the emerging stream of water is increased by the reduction in cross-sectional area of the nozzle. (2) A short piece of pipe with a flange on one end and a saddle flange on the other end. (3) A side outlet attached to a pipe by riveting, brazing, or welding.

nozzle aerator   An aerator consisting of a pressure nozzle through which water is propelled into the air in a fine spray. Also called *spray aerator*.

nutrient   Any substance that is assimilated by organisms and promotes growth; generally applied to nitrogen and phosphorus in wastewater, but also to other essential and trace elements.

odor control   Prevention or reduction of objectionable odors by chlorination, aeration, or other processes, or by masking with chemical aerosols.

odor threshold   The point at which, after successive dilutions with odorless water or air, the odor of a sample can barely be detected. The threshold odor is expressed quantitatively by the number of times the sample is diluted with odorless water or air.

off-peak power   That part of the available load or energy that can be produced at off-peak hours outside the load curve when the combined primary and secondary load has fallen below plant capacity.

ohm   The unit of measurement of electrical resistance. It is that resistance through which an electromotive force of one volt will produce a current of one ampere.

oil separation   (1) Removal of insoluble oils and floating grease from municipal wastewater. (2) Removal of soluble or emulsified oils from industrial wastewater.

open centrifugal pump   A centrifugal pump in which the impeller is built with a set of independent vanes.

open channel   Any natural or artificial water conduit in which water flows with a free surface.

open-channel flow   Flow of a fluid with its surface exposed to the atmosphere. The conduit may be an open channel or a closed conduit flowing partly full.

open impeller   An impeller without attached side walls.

operators   (1) Persons employed to operate a treatment facility. (2) Mechanism used to manipulate valve positions.

organic   Refers to volatile, combustible, and sometimes biodegradable chemical compounds containing carbon atoms (carbonaceous) bonded together with other elements. The principal groups of organic substances found in wastewater are proteins, carbohydrates, and fats and oils. See also *inorganic*.

organic loading   The amount of organic material, usually measured as BOD_5, applied to a given treatment process; expressed as weight per unit time per unit surface area or per unit weight.

organic nitrogen   Nitrogen chemically bound in organic molecules such as proteins, amines, and amino acids.

orifice   (1) An opening with a closed perimeter, usually of regular form, in a plate, wall, or partition through which water may flow; generally used for the purpose of measurement or control of such water. The edge may be sharp or of another configuration. (2) The end of a small tube such as a pitot tube or piezometer.

orifice plate   A plate containing an orifice. In pipes, the plate is usually inserted between a pair of flanges and the orifice is smaller in area than the cross section of the pipe.

orthophosphate   (1) A salt that contains phosphorus as $(PO_4)^{-3}$. (2) A product of hydrolysis of condensed (polymeric) phosphates. (3) A nutrient required for plant and animal growth. See also *nutrient, phosphorus removal.*

osmosis   The process of diffusion of a solvent through a semipermeable membrane from a solution of lower concentration to one of higher concentration.

outfall   (1) The point, location, or structure where wastewater or drainage discharges from a sewer, drain, or other conduit. (2) The conduit leading to the ultimate disposal area.

outlet   A point on the wiring system at which the current is taken to supply utilization equipment.

overflow rate   One of the criteria in the design of settling tanks for treatment plants; expressed as the settling velocity of particles that are removed in an ideal basin if they enter at the surface. It is expressed as a volume of flow per unit water surface area.

overflow weir   Any device or structure over which any excess water or wastewater beyond the capacity of the conduit or container is allowed to flow or waste.

overland flow   (1) The flow of water over the ground before it enters some defined channel. (2) A type of wastewater irrigation.

overturn   The phenomenon of vertical circulation that occurs in large bodies of water because of the increase in density of water above and below 39.2 °F (4 °C). In the spring, as the surface of the water warms above the freezing point, the water increases in density and tends to sink, producing vertical currents; in the fall, as the surface water becomes colder, it also tends to sink. Wind may also create such vertical currents.

oxidant   A chemical substance capable of promoting oxidation, for example, O_2, O_3, and Cl_2. See also *oxidation, reduction.*

oxidation   (1) A chemical reaction in which the oxidation number (valence) of an element increases because of the loss of one or more electrons by that element. Oxidation of an element is accompanied by simultaneous reduction of the other reactant. See also *reduction.* (2) The conversion or organic materials to simpler, more stable forms with the release of energy. This may be accomplished by chemical or biological means. (3) The addition of oxygen to a compound.

oxidation ditch   A secondary wastewater treatment facility that uses an oval channel with a rotor placed across it to provide aeration and circulation. The screened wastewater in the ditch is aerated by the rotor and circulated at approximately 1 to 2 ft/sec (0.3 m/s). See also *secondary treatment.*

oxidation pond   A relatively shallow body of wastewater contained in an earthen basin of controlled shape in which biological oxidation of organic matter is effected by natural or artificially accelerated transfer of oxygen.

oxidation process   Any method of wastewater treatment for the oxidation of the putrescible organic matter.

oxidation–reduction potential (ORP)   The potential required to transfer electrons from the oxidant to the reductant and used as a qualitative measure of the state of oxidation in wastewater treatment systems.

oxidized sludge   Sludge in which the organic matter has been stabilized by chemical or biological oxidation.

oxidized wastewater   Wastewater in which the organic matter has been stabilized.

oxygen (O)   A necessary chemical element. Typically found as O_2 and used in biological oxidation. It constitutes approximately 20% of the atmosphere.

oxygenation capacity   In treatment processes, a measure of the ability of an aerator to supply oxygen to a liquid.

oxygen consumed   A measure of the oxygen-consuming capability of inorganic and organic matter present in water or wastewater. See also *chemical oxygen demand*.

oxygen deficiency   (1) The additional quantity of oxygen required to satisfy the oxygen requirement in a given liquid; usually expressed in milligrams per liter (mg/L). (2) Lack of oxygen.

oxygen transfer   (1) Exchange of oxygen between a gaseous and a liquid phase. (2) The amount of oxygen absorbed by a liquid compared to the amount fed into the liquid through an aeration or oxygenation device; usually expressed as percent.

oxygen uptake rate   The oxygen used during biochemical oxidation, typically expressed as mg O_2/L/h in the activated sludge process.

oxygen utilization   (1) The portion of oxygen effectively used to support aerobic treatment processes. (2) The oxygen used to support combustion in the degradation of sludge by incineration or wet-air oxidation.

ozonation   The process of contacting water, wastewater, or air with ozone for purposes of disinfection, oxidation, or odor control.

ozone (O₃)   Oxygen in a molecular form with three atoms of oxygen forming each molecule.

paddle aerator   A device, similar in form to a paddle wheel, that is used in the aeration of water.

panel board   One or more panel units designed for assembly into a single panel, including buses, and with or without switched and/or automatic overcurrent devices. Panel boards are used to control light, heat, or power circuits of small individual or grouped loads. They are designed to be set in a cabinet box or in or against a wall or partition and are accessible from the front only (see *switchboard*).

Parshall flume   A calibrated device developed by Parshall for measuring the flow of liquid in an open conduit consisting essentially of a contracting length, a throat, and an expanding length. At the throat is a sill over which the flow passes at Belanger's critical depth. The upper and lower heads are each measured at a definite distance from the sill. The lower head need not be measured unless the sill is submerged more than about 67%.

partial pressure   The pressure exerted by each gas independently of the others in a mixture of gases. The partial pressure of each gas is proportional to the amount (percent by volume) of that gas in the mixture.

particles   Generally, discrete solids suspended in water or wastewater that can vary widely in size, shape, density, and charge.

parts per million (ppm)   The number of weight or volume units of a minor constituent present with each 1 million units of a solution or mixture. The more specific term, milligrams per liter (mg/L), is preferred.

pathogenic bacteria   Bacteria that cause disease in the host organism by their parasitic growth.

pathogens   Pathogenic or disease-producing organisms.

peak   (1) The maximum quantity that occurs over a relatively short period of time. Also called *peak demand*, *peak load*. (2) The highest load carried by an electric generating system during any specific period. It is usually expressed in kilowatts (kW).

peak load   (1) The maximum average load carried by an electric generating plant or system for a short time period such as 1 hour or less. (2) The maximum demand for water placed on a pumping station, treatment plant, or distribution system; expressed as a rate. (3) The maximum rate of flow of wastewater to a pumping station or treatment plant. Also called *peak demand*.

period   (1) The interval required for the completion of a recurring event. (2) Any specified duration of time.

peripheral weir   The outlet weir extending around the inside of the circumference of a circular settling tank over which the effluent discharges.

permeability   (1) The property of a material that permits appreciable movement of water through it when it is saturated; the movement is actuated by hydrostatic pressure of the magnitude normally encountered in natural subsurface water. Perviousness is sometimes used in the same sense as permeability. (2) The capacity of a rock or rock material to transmit a fluid. See also *permeability coefficient*.

permeability coefficient   A coefficient expressing the rate of flow of a fluid through a cross section of permeable material under a hydraulic or pressure gradient. The standard coefficient of permeability used in the hydrologic work of the U.S. Geological Survey (USGS), known also as the *Meinzer unit*, is defined as the rate of flow of water in gallons per day (gpd) at 60 °F through a cross section of 1 ft (0.3 m) under a hydraulic gradient of 100%. See also *field permeability coefficient*.

pervious   Possessing a texture that permits water to move through perceptibly under the head differences ordinarily found in subsurface water. See also *permeability*.

pH   A measure of the hydrogen-ion concentration in a solution, expressed as the logarithm (base 10) of the reciprocal of the hydrogen-ion concentration in gram moles per liter (g/mole/L). On the pH scale (0 to 14), a value of 7 at 25 °C (77 °F) represents a neutral condition. Decreasing values indicate increasing hydrogen-ion concentration (acidity); increasing values indicate decreasing hydrogen-ion concentration (alkalinity).

phase   Any portion of a physical system separated by a definite physical boundary from the rest of the system. The three physical phases are solid, liquid, and gas; colloids are the dispersed phase and liquids are the continuous phase.

phenolic compounds   Hydroxyl derivatives of benzene. The simplest phenolic compound is hydroxyl benzene (C_6H_5OH).

phosphate   A salt or ester of phosphoric acid. See also *orthophosphate, phosphorus*.

phosphorus   An essential chemical element and nutrient for all life forms. Occurs in orthophosphate, pyrophosphate, tripolyphosphate, and organic phosphate forms. Each of these forms and their sum (total phosphorus) is expressed as milligrams per liter (mg/L) elemental phosphorus. See also *nutrient*.

phosphorus removal   The precipitation of soluble phosphorus by coagulation and subsequent flocculation and sedimentation.

photosynthesis   The synthesis of complex organic materials, especially carbohydrates, from carbon dioxide, water, and inorganic salts with sunlight as the source of energy and with the aid of a catalyst, such as chlorophyll.

photosynthetic bacteria   Bacteria that obtain their energy for growth from light by photosynthesis.

physical analysis   The examination of water and wastewater to determine physical characteristics such as temperature, turbidity, color, odors, and taste.

physical-chemical treatment   Treatment of wastewater by unit processes other than those based on microbiological activity. Unit processes commonly included are precipitation with coagulants, flocculation with or without chemical flocculents, filtration, adsorption, chemical oxidation, air stripping, ion exchange, reverse osmosis, and several others.

physical treatment   Any treatment process involving only physical means of solid–liquid separation, for example, screens, racks, clarification, and comminutors. Chemical and biological reactions do not play an important role in treatment.

phytoplankton   Plankton consisting of plants, such as algae.

pin floc   Small floc particles that settle poorly.

pipe   A closed conduit that diverts or conducts water or wastewater from one location to another.

pipe diameter   The nominal or commercially designated inside diameter of a pipe, unless otherwise stated.

pipe fittings   Connections, appliances, and adjuncts designed to be used in connection with pipes; examples are elbows and bends to alter the direction of a pipe; tees and crosses to connect a branch with a main; plugs and caps to close an end; and bushings, diminishers, or reducing sockets to couple two pipes of different dimensions.

pipe gallery   (1) Any conduit for pipe, usually of a size to allow a person to walk through. (2) A gallery provided in a treatment plant for the installation of conduits and valves and used as a passageway to provide access to them.

piping system   A system of pipes, fittings, and appurtenances within which a fluid flows.

piston pump   A reciprocating pump in which the cylinder is tightly fitted with a reciprocating piston.

plant hydraulic capacity   The level of flow into a plant above which the system is hydraulically overloaded.

plastic media   Honeycomb-like products, manufactured from plastics of various compositions, with high surface area:volume ratios that are used in trickling filters in place of crushed stone. The product is available in large modules fabricated from sheets that may be cut to size on-site, and small discrete pieces to be loosely packed in the filter bed. See also *trickling filter*.

plate press   A filter press consisting of a number of parallel plate units lined with filter cloth that rests on drainage channels in the plates. Pressure is exerted by the pumping of solids into chambers created between the cloths. The operation is carried out in batches.

plug flow   Flow in which fluid particles are discharged from a tank or pipe in the same order in which they entered it. The particles retain their discrete identities and remain in the tank for a time equal to the theoretical detention time.

plumbing   (1) The pipes, fixtures, and other apparatus inside a building for bringing in the water supply and removing the liquid and waterborne wastes. (2) The installation of the foregoing pipes, fixtures, and other apparatus.

plumbing fixtures   Receptacles that receive liquid, water, or wastewater and discharge them into a drainage system.

plunger pump   A reciprocating pump with a plunger that does not come in contact with the cylinder walls, but enters and withdraws from it through packing glands. Such packing may be inside or outside the center, according to the design of the pump.

pneumatic ejector   A device for raising wastewater, sludge, or other liquid by alternately admitting it through an inward swinging check valve into the bottom of an airtight pot and then discharging it through an outward swinging check value by admitting compressed air to the pot above the liquid.

point gauge   A sharp-pointed rod attached to a graduated staff or vernier scale used for measuring the elevation of the surface of water. The point is lowered until the tip barely touches the water and forms a streak in flowing water and a meniscus jump in still water. It can also be used in a still well and can operate on an electric current with a buzzer or light that will operate when contact with the water is made.

polishing   A general term for those treatment processes that are applied after conventional ones. See also *advanced waste treatment, tertiary treatment.*

pollution   (1) Specific impairment of water quality by agricultural, domestic, or industrial wastes (including thermal and atomic wastes) to a degree that has an adverse effect on any beneficial use of the water. (2) The addition to a natural body of water any material that diminishes the optimal economic use of a water body by the population it serves and has an adverse effect on the surrounding environment.

polychlorinated biphenyls (PCBs)   A class of aromatic organic compounds with two six-carbon unsaturated rings, with chlorine atoms substituted on each ring and more than two such chlorine atoms per molecule of PCB. They are typically stable, resist both chemical and biological degradation, and are toxic to many biological species.

polyelectrolyte flocculants   Polymeric organic compounds used to induce or enhance the flocculation of suspended and colloidal solids and thereby facilitate sedimentation or the dewatering of sludges.

polyelectrolytes   Complex polymeric compounds, usually composed of synthetic macromolecules that form charged species (ions) in solution; water-soluble polyelectrolytes are used as flocculants; insoluble polyelectrolytes are used as ion exchange resins. See also *polymers*.

polymers   Synthetic organic compounds with high molecular weights and composed of repeating chemical units (monomers); they may be polyelectrolytes, such as water-soluble flocculents or water-insoluble ion exchange resins, or insoluble uncharged materials, such as those used for plastic or plastic-lined pipe and plastic trickling filter media.

polyvinyl chloride (PVC)   An artificial polymer made from vinyl chloride monomer ($CH_2:CHCl$); frequently used in pipes, sheets, and vessels for transport, containment, and treatment in water and wastewater facilities. See also *polymers*.

population dynamics   The ever-changing numbers of microscopic organisms within the activated sludge process.

population equivalent   The estimated population that would contribute a given amount of a specific waste parameter (BOD_5, suspended solids, or flow); usually applied to industrial waste. Domestic wastewater contains material that consumes, on the average, 0.17 lb of oxygen/cap/d (0.08 kg/cap·d), as measured by the standard BOD test. For example, if an industry discharges 1 000 lb of BOD/d (454 kg/d), its waste is equivalent to the domestic wastewater from 6 000 persons (1 000/0.17 = approximately 6 000).

pore   As applied to stone, soil, or other material, any small interstice or open space, generally one that allows the passage or adsorption of liquid or gas.

pore space   Open space in rock or granular material.

porosity   (1) The quality of being porous or containing interstices. (2) The ratio of the aggregate volume of interstices in a rock or soil to its total volume; usually stated as a percentage.

positive-displacement pump   Pump type in which liquid is induced to flow from the supply source through an inlet pipe and inlet valve. Water is brought into the pump chamber by a vacuum created by the withdrawal of a piston or pistonlike device, which, on its return, displaces a certain volume of water contained in the chamber and forces it to flow through the discharge valve and pipe.

postaeration   The addition of air to plant effluent to increase the oxygen concentration of treated wastewater.

postchlorination   The application of chlorine to wastewater following treatment.

power (apparent)   The mathematical product of the volts and amps of a circuit. The product generally is divided by 1000 and designated as kilovolt amperes (kVA).

power (electric)   The time rate of transferring or using electrical energy, usually expressed in kilowatts (kW).

power (reactive)   That portion of *apparent power* that does no work. It is usually measured in kilovars. Reactive power must be supplied to most types of magnetic equipment, such as motors, ballasts, transformers, and relays. Typically, it is supplied by generators or by electrostatic equipment such as capacitors.

power requirements   The rate of energy input needed to operate a piece of equipment, a treatment plant, or other facility or system. The form of energy may be electrical, fossil fuel, other types, or a combination.

preaeration   A preparatory treatment of wastewater consisting of aeration to remove gases, add oxygen, promote flotation of grease, and aid coagulation.

prechlorination   The application of chlorine to wastewater at or near the treatment plant entrance. Often used after bar screens and grit chambers to control odors in primary settling tanks.

precipitate   (1) To condense and cause to fall as precipitation, as water vapor condenses and falls as rain. (2) The separation from solution as a precipitate. (3) The substance that is precipitated.

preliminary treatment   Unit operations, such as screening, comminution, and grit removal, that prepare the wastewater for subsequent major treatment.

press filter   A press operated mechanically for partially dewatering sludge. See also *filter press, plate press.*

pressure   (1) The total load or force acting on a surface. (2) In hydraulics, unless otherwise stated, the pressure per unit area or intensity of pressure above local atmospheric pressure expressed in pounds per square inch (psi) or kilograms per square centimeter (kg/cm^2).

pressure filter   (1) An enclosed vessel having a vertical or horizontal cylinder of iron, steel, wood, or other material containing granular media through which liquid is forced under pressure. (2) A mechanical filter for partially dewatering sludge. See also *filter press, plate press.*

pressure gauge   A device for registering the pressure of solids, liquids, or gases. It may be graduated to register pressure in any units desired.

pressure-relief valve   Valve that opens automatically when the pressure reaches a preset limit to relieve stress on a pipeline.

pressure tank   A tank used in connection with a water distribution system for a single household, for several houses, or for a portion of a larger water system that is airtight and holds both air and water and in which air is compressed and the pressure so created is transmitted to the water.

pretreatment   Treatment of industrial wastewater at its source before discharge to municipal collection systems.

primary effluent   The liquid portion of wastewater leaving the primary treatment process.

primary sedimentation tank   The first settling tank for the removal of settleable solids through which wastewater is passed in a treatment works. Sometimes called a *primary clarifier*.

primary sludge   Sludge obtained from a primary sedimentation tank.

primary treatment   (1) The first major treatment in a wastewater treatment facility, used for the purpose of sedimentation. (2) The removal of a substantial amount of suspended matter, but little or no colloidal and dissolved matter. (3) Wastewater treatment processes usually consisting of clarification with or without chemical treatment to accomplish solid–liquid separation.

primary voltage   The voltage of the circuit supplying power to a transformer, as opposed to the output voltage or load-supplied voltage, which is called *secondary voltage*. In power supply practice, the primary is almost always the high-voltage side and the secondary the low-voltage side of the transformer.

propeller pump   A centrifugal pump that develops most of its head by the propelling or lifting action of the vanes on the liquid. Also called an *axial-flow pump*.

propeller-type impeller   An impeller of the straight axial-flow type.

proportional weir   A special type of weir in which the discharge through the weir is directly proportional to the head.

protozoa   Small one-celled animals including amoebae, ciliates, and flagellates.

publicly owned treatment works   Wastewater treatment plant.

pump   A mechanical device for causing flow, for raising or lifting water or other fluid, or for applying pressure to fluids.

pump curve   A curve or curves showing the interrelation of speed, dynamic head, capacity, brake horsepower, and efficiency of a pump.

pump efficiency   The ratio of energy converted into useful work to the energy applied to the pump shaft, or the energy difference in the water at the discharge and suction nozzles divided by the power input at the pump shaft.

pumping head   The sum of the static head and friction head on a pump discharging a given quantity of water.

pumping station   (1) A facility housing relatively large pumps and their appurtenances. *Pump house* is the usual term for shelters for small water pumps. (2) A facility containing lift pumps to facilitate wastewater collection or reclaimed water distribution.

pump pit   A dry well or chamber, below ground level, in which a pump is located.

pump stage   The number of impellers in a centrifugal pump; for example, a single-stage pump has one impeller; a two-stage pump has two impellers.

putrefaction   Biological decomposition, usually of organic matter, with the production of foul-smelling products associated with anaerobic conditions.

putrescibility   (1) The relative tendency of organic matter to undergo decomposition in the absence of oxygen. (2) The susceptibility of wastewaters, effluent, or sludge to putrefaction. (3) The stability of a polluted, raw, or partially treated wastewater.

quicklime   A calcined material, the major part of which is calcium oxide, or calcium oxide in natural association with a lesser amount of magnesium oxide. It is capable of combining with water, that is, being slaked.

raceway   Any channel for holding wires, cables, or bus bars that is designed expressly and solely for that purpose. Raceways may be of metal or insulating materials and the term includes rigid metal conduit, nonmetallic conduit, flexible metal conduit, and electrical metallic tubing. Raceways may be located beneath the floor or on or above the surface (refer to the National Electric Code for approved raceways).

rack   A device fixed in place and used to remove suspended or floating solids from wastewater. It is composed of parallel bars that are evenly spaced. See also *screen*.

radial flow   The direction of flow across a tank from center to periphery or vice versa.

radiation   (1) The emission and propagation of energy through space or through a material medium; also the energy so propagated. (2) The dispersion of energy by electromagnetic waves rather than by conduction and convection.

range   A measure of the variability of a quantity; the difference between the largest and smallest values in the sequence of values of the quantity.

rate   (1) The speed at which a chemical reaction occurs. (2) Flow volume per unit time. See also *kinetics*.

rate-of-flow controller   An automatic device that controls the rate of flow of a fluid.

rate-of-flow recorder   A recorder for registering the rate of flow of water; generally, used with a rapid sand filter.

raw sludge   Settled sludge promptly removed from sedimentation tanks before decomposition has much advanced.

raw wastewater   Wastewater before it receives any treatment.

reaction rate   The rate at which a chemical reaction progresses. See also *kinetics, rate*.

reactor   The container, vessel, or tank in which a chemical or biological reaction is carried out.

recalcining   Recovery of lime from water and wastewater treatment sludge.

recarbonation   (1) The process of introducing carbon dioxide as a final stage in the lime-soda ash softening process to convert carbonates to bicarbonates and thereby stabilize the solution against precipitation of carbonates. (2) The addition of carbon dioxide to the effluent of an advanced wastewater treatment ammonia air stripping process to lower the pH. (3) The diffusion of carbon dioxide gas through a liquid to replace the carbon dioxide removed by the addition of lime. (4) The diffusion of carbon dioxide gas through a liquid to render the liquid stable with respect to precipitation or dissolution of alkaline constituents.

receiving water   A river, lake, ocean, or other watercourse into which wastewater or treated effluent is discharged.

receptacle   A contact device installed at the outlet for connection of a single attachment plug. A single receptacle is a single contact device with no other contact device on the same yoke. A multiple receptacle is a single device containing two or more receptacles.

reciprocating pump   A type of displacement pump consisting essentially of a closed cylinder containing a piston or plunger as the displacing mechanism. Liquid is drawn into the cylinder through an inlet valve and forced out through an outlet valve. When the piston acts on the liquid in one end of the cylinder, the pump is termed *single-action*; when it acts in both ends, it is termed *double-action*.

recirculation   (1) In the wastewater field, the return of all or a portion of the effluent in a trickling filter to maintain a uniform high rate through the filter. Return of a portion of the effluent to maintain minimum flow is sometimes called *recycling*. (2) The return of effluent to the incoming flow. (3) The return of the effluent from a process, factory, or operation to the incoming flow to reduce the water intake. The incoming flow is called *makeup water*.

reclaimed wastewater   Wastewater used for some beneficial purpose usually after some degree of treatment.

recorder   (1) A device that makes a graph or other record of the stage, pressure, depth, velocity, or the movement or position of water-controlling devices, usually as a function of time. See also *indicator*. (2) The person who records the observational data.

recording gauge   An automatic instrument for measuring and recording graphically and continuously. Also called a *register*.

rectangular weir   A weir having a notch that is rectangular in shape.

recycle   (1) To return water after some type of treatment for further use; generally implies a closed system. (2) To recover useful values from segregated solid waste.

recycling   (1) An operation in which a substance is passed through the same series of processes, pipes, or vessels more than once. (2) The conversion of solid waste into usable materials or energy.

reduce   The opposite of oxidize. The action of a substance to decrease the positive valence of an ion.

reduction   The addition of electrons to a chemical entity decreasing its valence. See also *oxidation*.

refractory   Brick or similar material that lines a furnace or incinerator.

regeneration   (1) In ion exchange, the process of restoring an ion exchange material to the state used for adsorption. (2) The periodic restoration of exchange capacity of ion exchange media used in water treatment.

relative humidity   (1) The amount of water vapor in the air; expressed as a percentage of the maximum amount that the air could hold at the given temperature. (2) The ratio of the actual water vapor pressure to the saturation vapor pressure.

relay   An electrical device that is designed to interpret input conditions in a prescribed manner and, after specified conditions are met, to respond to cause electrical operation or similar abrupt change in associated control circuits. The most common form of relay uses a coil and set of contacts. When current flows in the coil, contacts are opened or closed, depending on their arrangement. Relays are said to be normally open or normally closed.

relief valve   A valve that releases air from a pipeline automatically without loss of water, or introduces air into a line automatically if the internal pressure becomes less than that of the atmosphere.

removal efficiency   A measure of the effectiveness of a process in removing a constituent, such as BOD or TSS. Removal efficiency is calculated by subtracting the effluent value from the influent value and dividing it by the influent value. Multiply the answer by 100 to convert to a percentage.

repair   An element of maintenance, as distinguished from replacement or retirement.

replacement   Installation of new or alternate equipment in place of existing equipment for a variety of reasons, such as obsolescence, total disrepair, improvement, or modification.

replacement cost   (1) The actual or estimated cost of duplication with a property of equal utility and desirability. (2) The cost of replacing property.

residue   (1) The equilibrium quantity of a compound or element remaining in an organism after uptake and clearance. (2) The dry solid remaining after evaporation.

resistance   The property of an electrical circuit or device that opposes current flow, thereby causing conversion of electrical energy to heat or radiant energy.

respiration   Intake of oxygen and discharge of carbon dioxide as a result of biological oxidation.

retention time   The theoretical time required to displace the contents of a tank or unit at a given rate of discharge (volume divided by the rate of discharge). Also called *detention time*.

return sludge   Settled activated sludge returned to mix with incoming raw or primary settled wastewater. More commonly called *return activated sludge*.

reverse osmosis   An advanced method used in water and wastewater treatment that relies on a semipermeable membrane to separate the water from its impurities. An external force is used to reverse the normal osmotic flow resulting in movement of the water from a solution of higher solute concentration to one of lower concentration. Also called *hyperfiltration*.

revolving screen   A screen or rack in the form of a cylinder or a continuous belt that is revolved mechanically. The screenings are removed by water jets, automatic scrapers, or manually.

Reynolds' number   A dimensionless quantity used to characterize the type of flow in a hydraulic structure where resistance to motion depends on the viscosity of the liquid in conjunction with inertia. It is equal to the ratio of inertial forces to viscous forces. The number is chiefly applicable to closed systems of flow, such as pipes or conduits where there is no free water surface, or to bodies fully immersed in the fluid so the free surface need not be considered.

riprap   Broken stone or boulders placed compactly or irregularly on dams, levees, dikes, or similar embankments for protection of earth surfaces against the action of waves or currents.

rising time   The time necessary for removal, by flotation, of suspended or aggregated colloidal substances.

rotary distributor   A movable distributor made up of horizontal arms that extend to the edge of the circular trickling filter bed, revolve about a central post, and distribute liquid over the bed through orifices in the arms. The jet action of the discharging liquid normally supplies the motive power. See also *distributor*.

rotary dryer A long, slowly revolving, steel cylinder with its long axis slightly inclined, through which passes the material to be dried in hot air. The material passes through from inlet to outlet, tumbling about.

rotary pump A type of displacement pump consisting essentially of elements rotating in a pump case that is closely fit. The rotation of these elements alternately draws in and discharges the water being pumped. Such pumps act with neither suction nor discharge valves, operate at almost any speed, and do not depend on centrifugal forces to lift the water.

rotary valve A valve consisting of a casing more or less spherical in shape and a gate that turns on trunnions through 90 deg when opening or closing and having a cylindrical opening of the same diameter as that of the pipe it serves.

rotating biological contactor (RBC) A device for wastewater treatment composed of large, closely spaced plastic discs that are rotated about a horizontal shaft. The discs alternately move through the wastewater and the air and develop a biological growth on their surfaces.

rotating distributor A distributor consisting of rotating or reciprocating perforated pipes or troughs from which liquid is discharged in the form of a spray or in a thin sheet at uniform rates over the surface area to be wetted.

rotifer Minute, multicellular aquatic animals with rotating cilia on the head and forked tails. Rotifers help stimulate microfloral activity and decomposition, enhance oxygen penetration, and recycle mineral nutrients.

roughing filter A trickling filter used to remove an initial portion of the soluble BOD, usually about 50%, but not to provide complete removal.

roughness coefficient A factor in many engineering equations for computing the average velocity of flow of water in a conduit or channel. It represents the effect of the roughness of the confining material on the energy losses in the flowing water.

safety valve A valve that automatically opens when prescribed conditions, usually pressure, are exceeded in a pipeline or other closed receptacle containing liquids or gases. It prevents such conditions from being exceeded and causing damage.

Salmonella A genus of aerobic, rod-shaped, usually motile bacteria that are pathogenic for man and other warm-blooded animals.

sampler A device used with or without flow measurement to obtain a portion of liquid for analytical purposes. May be designed for taking single samples (grab), composite samples, continuous samples, or periodic samples.

sand filter A bed of sand through which water is passed to remove fine suspended particles. Commonly used in tertiary wastewater treatment plants and sludge drying beds.

sanitary sewer   A sewer that carries liquid and waterborne wastes from residences, commercial buildings, industrial plants, and institutions together with minor quantities of ground, storm, and surface water that are not admitted intentionally. See also *wastewater*.

Sarcadina   Species of amoeba found in wastewater. Does not play a significant role in the activated-sludge process other than as an indication of start up or the passing of a toxic influence.

saturated air   Air containing all the water vapor that it is capable of holding at a given temperature and pressure.

saturated liquid   Liquid that contains at a given temperature as much of a solute as it can retain in the presence of an excess of that solute.

scraper   (1) Device used to remove solids from a clarifier to a sump. (2) Mechanism to remove dewatered solids from a belt filter press or conveyor.

screen   A device with openings, generally of uniform size, used to retain or remove suspended or floating solids in flow stream preventing them from passing a given point in a conduit. The screening element may consist of parallel bars, rods, wires, grating, wire mesh, or perforated plate.

screening   A preliminary treatment process that removes large suspended or floating solids from raw wastewater to prevent subsequent plugging of pipes or damage to pumps.

screenings   (1) Material removed from liquids by screens. (2) Broken rock, including the dust, of a size that will pass through a given screen depending on the character of the stone.

screenings grinder   A device for grinding, shredding, or macerating material removed from wastewater by screens.

screw-feed pump   A pump with either a horizontal or vertical cylindrical casing in which operates a runner with radial blades like those of a ship's propeller. See also *vertical screw pump*.

scrubbing   Removal of suspended solids and undesirable gases from gaseous emissions.

scum   (1) The extraneous or foreign matter that rises to the surface of a liquid and forms a layer or film there. (2) A residue deposited on a container or channel at the water surface. (3) A mass of solid matter that floats on the surface.

scum baffle   A vertical baffle dipping below the surface of wastewater in a tank to prevent the passage of floating matter.

scum breaker   A device installed in a sludge digestion tank to break up scum.

scum chamber   A space provided in a sludge digestion tank for accumulated scum rising from the digestion unit.

scum collector   A mechanical device for skimming and removing scum from the surface of settling tanks.

scum removal   Separation of floating grease and oil from wastewater usually during preliminary or primary treatment.

scum trough   A trough placed in a primary sedimentation tank to intercept scum and convey it out of the tank.

Secchi disk   Tool to measure the clarity of the water.

secondary effluent   (1) The liquid portion of wastewater leaving secondary treatment. (2) An effluent that, with some exceptions, contains not more than 30 mg/L each (on a 30-day average basis) of BOD_5 and suspended solids.

secondary sedimentation tank   A settling tank following secondary treatment designed to remove by gravity part of the suspended matter. Also called a *secondary clarifier*.

secondary treatment   (1) Generally, a level of treatment that produces secondary effluent. (2) Sometimes used interchangeably with the concept of biological wastewater treatment, particularly the activated-sludge process. Commonly applied to treatment that consists chiefly of clarification followed by a biological process with separate sludge collection and handling.

secondary voltage   The output or load-supplied voltage of a transformer or substation (see *primary voltage*).

second-stage BOD   That part of the oxygen demand associated with the biochemical oxidation of nitrogenous material. As the term implies, the oxidation of the nitrogenous materials usually does not start until a portion of the carbonaceous material has been oxidized during the first stage.

sedimentation   (1) The process of subsidence and decomposition of suspended matter or other liquids by gravity. It is usually accomplished by reducing the velocity of the liquid below the point at which it can transport the suspended material. Also called settling. It may be enhanced by coagulation and flocculation. (2) Solid–liquid separation resulting from the application of an external force, usually settling in a clarifier under the force of gravity. It can be variously classed as discrete, flocculent, hindered, and zone sedimentation.

sedimentation tank   A basin or tank in which wastewater containing settleable solids is retained for removal of the suspended matter by gravity. Also called a *sedimentation basin, settling basin, settling tank,* or *clarifier*.

seed sludge   In biological treatment, the inoculation of the unit process with biologically active sludge resulting in acceleration of the initial stage of the process.

self-cleansing velocity   The minimum velocity necessary to keep solids in suspension in sewers, thus preventing their deposition and subsequent nuisance from stoppages and odors of decomposition.

separate sewer system   A sewer system carrying sanitary wastewater and other water-carried wastes from residences, commercial buildings, industrial plants, and institutions, as well as minor quantities of ground, storm, and surface water that are not intentionally admitted. See also *combined sewer, wastewater.*

septage   The sludge produced in individual on-site wastewater disposal systems such as septic tanks and cesspools.

septic   (1) Anaerobic. (2) Putrid, rotten, foul smelling; anaerobic.

septicity   A condition produced by growth of anaerobic organisms.

septic wastewater   Wastewater undergoing anaerobic decomposition.

service   Conductors and equipment for delivering energy from the electrical supply system to the wiring system of the premises served.

service charge   The rate charged by the utility for rendering service, usually used as a ready-to-serve charge.

service conductors   Supply conductors that extend from the street main or from transformers to the service equipment of the premises served.

service equipment   Necessary equipment usually consisting of a circuit breaker or switch and fuses and their accessories that is located near the point of entrance of the supply conductors to a building or other structure and intended to constitute the main control and means of cut-off of the electrical supply.

settleability   The tendency of suspended solids to settle.

settleability test   A determination of the settleability of solids in a suspension by measuring the volume of solids settled out of a measured volume of sample in a specified interval of time, usually reported in milliliters per liter (mL/L). Also called the *Imhoff cone test.*

settleable solids   (1) That matter in wastewater that will not stay in suspension during a preselected settling period, such as 1 hour, but settles to the bottom. (2) In the Imhoff cone test, the volume of matter that settles to the bottom of the cone in 1 hour. (3) Suspended solids that can be removed by conventional sedimentation.

settleometer   A 2-L or larger beaker used to conduct the settleability test.

settling   The process of subsidence and deposition of suspended matter carried by a liquid. It is usually accomplished by reducing the velocity of the liquid below the point at which it can transport the suspended material. Also called *sedimentation*.

settling tank   A tank or basin in which water, wastewater, or other liquid containing settleable solids is retained for a sufficient time, and in which the velocity of flow is sufficiently low to remove by gravity a part of the suspended matter. See also *sedimentation tank*.

settling time   Time necessary for the removal of suspended or colloidal substances by gravitational settling, aggregation, or precipitation.

settling velocity   Velocity at which subsidence and deposition of settleable suspended solids in wastewater will occur.

sheet flow   Flow in a relatively thin sheet of generally uniform thickness.

short-circuiting   A hydraulic condition occurring in parts of a tank where the time of travel is less than the flow-through time.

side contraction   The contraction of the nappe or reduction in width of water overflowing a weir brought about by the detachment of the sides of the nappe or jet of water passing over the sides of the weir.

side-water depth   The depth of water measured along a vertical exterior wall.

single-action pump   A reciprocating pump in which the suction inlet admits water to only one side of the plunger or piston and the discharge is intermittent.

single-stage digestion   Digestion limited to a single tank for the entire digestion period.

single-suction impeller   An impeller with one suction inlet.

skimming   (1) The process of diverting water from the surface of a stream or conduit by means of a shallow overflow. (2) The process of diverting water from any elevation in a reservoir by means of an outlet at a different elevation or by any other skimming device in order to obtain the most palatable drinking water. (3) The process of removing grease or scum from the surface of wastewater in a tank.

skimmings   Grease, solids, liquids, and scum skimmed from wastewater settling tanks.

slake   To become mixed with water so that a true chemical combination takes place, as in the slaking of lime.

slimes   (1) Substances of a viscous organic nature, usually formed from microbiological growth, that attach themselves to other objects forming a coating. (2) The coating of biomass (humus, schmutzdecke, sluff) that accumulates in trickling filters or sand filters and periodically sloughs away to be collected in clarifiers. See also *biofilm*.

slope   (1) The inclination of gradient from the horizontal of a line or surface. The degree of inclination is usually expressed as a ratio such as 1:25, indicating unit rise in 25 units of horizontal distance; or in a decimal fraction (0.04); degrees (2 deg 18 min); or percent (4%). (2) Inclination of the invert of a conduit expressed as a decimal or as feet (meters) per stated length measured horizontally in feet. (3) In plumbing, the inclination of a conduit, usually expressed in inches per foot (meter) length of pipe.

sloughing   The disattachment of slime and solids accumulated on the media of trickling filters and RBCs in contact areas. Sloughed material usually is removed subsequently in clarifiers. See also *slimes*.

slow sand filter   A filter for the purification of water in which water without previous treatment is passed downward through a filtering medium consisting of a layer of sand from 24- to 40-in. (0.6- to 1-m) thick. The filtrate is removed by an underdrainage system and the filter is cleaned by scraping off the clogged sand and eventually replacing it. It is characterized by a slow rate of filtration, commonly 3 to 6 mgd/ac (28 to 56 ML/ha·d) of filter area. Its effectiveness depends on the biological mat (or schmutzdecke) that forms in the top few millimeters.

sludge   (1) The accumulated solids separated from liquids during the treatment process that have not undegone a stabilization process. (2) The removed material resulting from chemical treatment, coagulation, flocculation, sedimentation, flotation, or biological oxidation of water or wastewater. (3) Any solid material containing large amounts of entrained water collected during water or wastewater treatment. See also *activated sludge, settleable solids*.

sludge age   Average residence time of suspended solids in a biological treatment system equal to the total weight of suspended solids in the system divided by the total weight of suspended solids leaving the systems.

sludge blanket   Accumulation of sludge hydrodynamically suspended within an enclosed body of water or wastewater.

sludge boil   An upwelling of water and sludge deposits caused by release of decomposition gases in the sludge deposits.

sludge circulation   The overturning of sludge in sludge digestion tanks by mechanical or hydraulic means or by the use of gas recirculation to disperse scum layers and promote digestion.

sludge collector   A mechanical device for scraping the sludge on the bottom of a settling tank to a sump from which it can be drawn.

sludge concentration   Any process of reducing the water content of sludge leaving the sludge in a fluid condition.

sludge density index (SDI)   A measure of the degree of compaction of a sludge after settling in a graduated container, expressed as mL/g. The sludge volume index (SVI) is the reciprocal of the sludge density index.

sludge dryer   A device for removing a large percentage of moisture from sludge or screenings by heat.

sludge drying   The process of removing a large percentage of moisture from sludge by drainage or evaporation by any method.

sludge-gas holder   A tank used to store gas collected from sludge digestion tanks for the purpose of stabilizing the flow of gas to the burners, maintaining a nearly constant pressure, and supplying gas during periods when the digestion tanks are temporarily out-of-service or when gas production is low.

sludge-gas utilization   Using the gas produced by the anaerobic digestion of sludge for beneficial purposes such as heating sludge, mixing sludge, drying sludge, heating buildings, incineration, or fueling engines.

sludge pressing   The process of dewatering sludge by subjecting it to pressure, usually within a cloth fabric through which the water passes and in which the solids are retained.

sludge reaeration   The continuous aeration of sludge after its initial aeration for the purpose of improving or maintaining its condition.

sludge reduction   The reduction in quantity and change in character of sludge as the result of digestion.

sludge solids   Dissolved and suspended solids in sludge.

sludge thickener   A tank or other equipment designed to concentrate wastewater sludges.

sludge thickening   The increase in solids concentration of sludge in a sedimentation tank, DAF, gravity thickener, centrifuge or gravity belt thickener.

sludge utilization   The use of sludges resulting from industrial wastewater treatment as soil builders and fertilizer admixtures.

sludge volume index (SVI)   The ratio of the volume (in milliliters) of sludge settled from a 1000-mL sample in 30 minutes to the concentration of mixed liquor (in milligrams per liter [mg/L]) multiplied by 1000.

slurry   A thick, watery mud or any substance resembling it, such as lime slurry.

soda ash   A common name for commercial sodium carbonate (Na_2CO_3).

sodium bisulfite (NaHSO$_3$)   A salt used for reducing chlorine residuals; a strong reducing agent; typically found in white powder or granular form in strengths up to 44%. At a strength of 38%, 1.46 parts will consume 1 part of chlorine residual.

sodium carbonate (Na₂CO₃)   A salt used in water treatment to increase the alkalinity or pH of water or to neutralize acidity. Also called *soda ash*.

sodium hydroxide (NaOH)   A strong caustic chemical used in treatment processes to neutralize acidity, increase alkalinity, or raise the pH value. Also known as *caustic soda, sodium hydrate, lye,* and *white caustic.*

sodium hypochlorite (NaOCl)   A water solution of sodium hydroxide and chlorine in which sodium hypochlorite is the essential ingredient.

sodium metabisulfite (Na₂S₂O₅)   A cream-colored powder used to conserve chlorine residual; 1.34 parts of $Na_2S_2O_5$ will consume 1 part of chlorine residual.

soil absorption capacity   In subsurface effluent disposal, the ability of the soil to absorb water. See *soil absorption test.*

soil absorption test   A test for determining the suitability of an area for subsoil effluent disposal by measuring the rate at which the undisturbed soil will absorb water per unit of surface.

soil horizon   A layer or section of the soil profile, more or less well defined, occupying a position approximately parallel to the soil surface, and having characteristics that have been produced through the operation of soil-building processes.

soil infiltration rate   The maximum rate at which a soil, in a given condition at a given time, can absorb water.

soil porosity   The percentage of the soil (or rock) volume that is not occupied by solid particles, including all pore space filled with air and water. The total porosity may be calculated from the following formula: percent pore space = (1 − volume weight/ specific gravity) × 100.

solids disposal   Any process for the ultimate disposal of solid wastes or sludges by incineration, landfilling, soil conditioning, or other means.

solids inventory   Amount of sludge in the treatment system typically expressed as kilogram (tons). Inventory of plant solids should be tracked through the use of a mass balance set of calculations.

solids loading   Amount of solids applied to a treatment process per unit time per unit volume.

solids retention time (SRT)   The average time of retention of suspended solids in a biological waste treatment system, equal to the total weight of suspended solids leaving the system, per unit time.

sparger   An air diffuser designed to give large bubbles, used singly or in combination with mechanical aeration devices.

species   A subdivision of a genus having members differing from other members of the same genus in minor details.

specific gravity   The ratio of the mass of a body to the mass of an equal volume of water at a specific temperature, typically 20 °C (68 °F).

specific oxygen uptake rate   Measures the microbial activity in a biological system expressed in mg O_2/g·h of VSS. Also called *respiration rate*.

specific resistance   The relative resistence a sludge offers to the draining of its liquid component.

specific speed   A speed or velocity of revolution, expressed in revolutions per minute (rpm), at which the runner of a given type or turbine would operate if it were so reduced in size and proportion that it would develop 1 hp under a 1-ft head. The quantity is used in determining the proper type and character of turbine to install at a hydroelectric power plant under given conditions.

spiral-air flow diffusion   A method of diffusing air in the grit chamber or aeration tank of the activated-sludge process where, by means of properly designed baffles and the proper location of diffusers, a spiral helical movement is given to the air and the tank liquor.

splitter box   (1) A division box that splits the incoming flow into two or more streams. (2) A device for splitting and directing discharge from the head box to two separate points of application.

spray aerator   An aerator consisting of a pressure nozzle through which water is propelled into the air in a fine spray.

spray irrigation   A method of land treatment for disposing of some organic wastewaters by spraying them, usually from pipes equipped with fixed or moving spray nozzles. See also *land application*.

stabilization pond   A type of oxidation pond in which biological oxidation of organic matter is effected by natural or artificially accelerated transfer of oxygen to the water from air.

staged digestion   The progressive digestion of waste in two or more tanks arranged in series, usually divided into primary digestion with mixed contents and secondary digestion where quiescent conditions prevail and supernatant liquor is collected.

staged treatment   (1) Any treatment in which similar processes are used in series or stages. (2) In the activated-sludge process, two or more stages consisting of a clarifying stage and a biological stage, or two biological stages. (3) In anaerobic digestion, an operation in which sludge is completely mixed in the first tank and pumped to a second tank for separation of the supernatant liquor from the solids.

staged trickling filter   A series of trickling filters through which wastewater passes successively with or without intermediate sedimentation.

stale wastewater   Wastewater containing little or no oxygen, but as yet free from putrefaction. See also *septic wastewater*.

stalked ciliates   Small, one-celled organisms possessing cilia (hair-like projections used for feeding) that are not motile. They develop at lower prey densities, long SRTs, and low F:M ratios.

Standard Methods   (1) An assembly of analytical techniques and descriptions commonly accepted in water and wastewater treatment (*Standard Methods for the Examination of Water and Wastewater*) published jointly by the American Public Health Association, the American Water Works Association, and the Water Environment Federation. (2) Validated methods published by professional organizations and agencies covering specific fields or procedures. These include, among others, the American Public Health Association, American Public Works Association, American Society of Civil Engineers, American Society of Mechanical Engineers, American Society for Testing and Materials, American Water Works Association, U.S. Bureau of Standards, U.S. Standards Institute (formerly American Standards Association), U.S. Public Health Service, Water Environment Federation, and U.S. Environmental Protection Agency.

standard pressure   Atmospheric pressure at sea level under standard conditions.

static head   Vertical distance between the free level of the source of supply and the point of free discharge or the level of the free surface.

static level   (1) The elevation of the water table or pressure surface when it is not influenced by pumping or other forms of extraction from the groundwater. (2) The level of elevation to which the top of a column of water would rise, if afforded the opportunity to do so, from an artesian aquifer, basin, or conduit under pressure. Also called *hydrostatic level*.

static suction head   The vertical distance from the source of supply, when its level is above the pump, to the center line of the pump.

static suction lift   The vertical distance between the center of the suction end of a pump and the free surface of the liquid being pumped. Static lift does not include friction losses in the suction pipes. Static suction head includes lift and friction losses.

steady flow   (1) A flow in which the rate, or quantity of water passing a given point per unit of time, remains constant. (2) Flow in which the velocity vector does not change in either magnitude or direction with respect to time at any point or section.

steady nonuniform flow   A flow in which the quantity of water flowing per unit of time remains constant at every point along the conduit, but the velocity varies along the conduit because of a change in the hydraulic characteristics.

step aeration   A procedure for adding increments of settled wastewater along the line of flow in the aeration tanks of an activated-sludge plant. Also called *step feed*.

stoichiometric   Pertaining to or involving substances that are in the exact proportions required for a given reaction.

straggler floc   Large (6-mm or larger) floc particles that have poor settling characteristics.

submerged weir   A weir that, when in use, has a water level on the downstream side at an elevation equal to, or higher than, the weir crest. The rate of discharge is affected by the tailwater. Also called a *drowned weir*.

submergence   (1) The condition of a weir when the elevation of the water surface on the downstream side is equal to or higher than that of the weir crest. (2) The ratio, expressed as a percentage, of the height of the water surface downstream from a weir above the weir crest to the height of the water surface upstream above the weir crest. The distances upstream or downstream from the crest at which such elevations are measured are important, but have not been standardized. (3) In water power engineering, the ratio of tailwater elevation to the headwater elevation when both are higher than the crest. The overflow crest of the structure is the datum of reference. The distances upstream or downstream from the crest at which headwater and tailwater elevations are measured are important, but have not been standardized. (4) The depth of flooding over a pump suction inlet.

substation   An assembly of equipment for switching and/or changing or regulating the voltage electrical supply.

substrate   (1) Substances used by organisms in liquid suspension. (2) The liquor in which activated sludge or other matter is kept in suspension.

suction head   (1) The head at the inlet to a pump. (2) The head below atmospheric pressure in a piping system.

suction lift   The vertical distance from the liquid surface in an open tank or reservoir to the center line of a pump drawing from the tank or reservoir and set higher than the liquid surface.

suctoreans   Ciliates that are stalked in the adult stage and have rigid tentacles to catch prey.

sulfate-reducing bacteria   Bacteria capable of assimilating oxygen from sulfate compounds, reducing them to sulfides. See also *sulfur bacteria*.

sulfur bacteria   Bacteria capable of using dissolved sulfur compounds in their growth; bacteria deriving energy from sulfur or sulfur compounds.

sulfur cycle   A graphical presentation of the conservation of matter in nature showing the chemical transformation of sulfur through various stages of decomposition and assimilation. The various chemical forms of sulfur as it moves among living and non-living matter is used to illustrate general biological principles that are applicable to wastewater and sludge treatment.

sump   A tank or pit that receives drainage and stores it temporarily, and from which the discharge is pumped or ejected.

sump pump   A mechanism used for removing water or wastewater from a sump or wet well; it may be energized by air, water, steam, or electric motor. Ejectors and sub-merged centrifugal pumps, either float or manually controlled, are often used for the purpose.

supernatant   (1) The liquid remaining above a sediment or precipitate after sedimentation. (2) The most liquid stratum in a sludge digester.

supersaturation   (1) An unstable condition of a vapor in which its density is greater than that normally in equilibrium under the given conditions. (2) The condition existing in a given space when it contains more water vapor than is needed to cause saturation; that is, when its temperature is below that required for condensation to take place. This condition probably can occur only when water or ice is immediately present, and when the space contains no dust or condensation nuclei. (3) An unstable condition of a solution in which it contains a solute at a concentration exceeding saturation.

suppressed weir   A weir with one or both sides flush with the channel of approach. This prevents contraction of the nappe adjacent to the flush side. The suppression may occur on one end or both ends.

surface aeration   The absorption of air through the surface of a liquid.

surface overflow rate   A design criterion used for sizing clarifiers; typically expressed as the flow volume per unit amount of clarifier space ($m^3/m^2 \cdot s$ [gpd/sq ft]).

surfactant   A surface-active agent, such as ABS or LAS, that concentrates at interfaces, forms micelles, increases solution, lowers surface tension, increases adsorption, and may decrease flocculation.

surge   (1) A momentary increase in flow (in an open conduit) or pressure (in a closed conduit) that passes longitudinally along the conduit, usually because of sudden changes in velocity or quantity. (2) Any periodic, usually abrupt, change in flow, temperature, pH, concentration, or similar factor.

surge suppressor   A device used in connection with automatic control of pumps to minimize surges in a pipeline.

surge tank   A tank or chamber located at or near a hydroelectric powerhouse and connected with the penstock above the turbine. When the flow of water delivered to the turbine is suddenly decreased, the tank absorbs the water that is held back and cushions the increased pressure on the penstock caused by the rapid deceleration of the water flowing in it; also, when the flow delivered to the turbine is suddenly increased, the tank supplies the increased quantity of water required until the flow in the penstock has been accelerated sufficiently. Also used in connection with pumping systems.

suspended matter   (1) Solids in suspension in water, wastewater, or effluent. (2) Solids in suspension that can be readily removed by standard filtering procedures in a laboratory.

suspended solids   (1) Insoluble solids that either float on the surface of, or are in suspension in, water, wastewater, or other liquids. (2) Solid organic or inorganic particles (colloidal, dispersed, coagulated, or flocculated) physically held in suspension by agitation or flow. (3) The quantity of material removed from wastewater in a laboratory test, as prescribed in *Standard Methods* and referred to as *nonfilterable residue*.

switchboard   A large panel or assembly of panels on which switches, overcurrent, and/or other protective devices such as buses and instruments are mounted. Switchboards are generally accessible from the rear as well as from the front and are not intended to be installed in cabinets.

synergism   Interaction between two entities producing an effect greater than a simple additive one. See also *antagonism*.

tapered aeration   The method of supplying varying quantities of air into the different parts of an aeration tank in the activated-sludge process, more at the inlet, less near the outlet, in approximate proportion to the oxygen demand of the mixed liquor under aeration.

tee   A pipe fitting, either cast or wrought, that has one side outlet at right angles to the run. A single-outlet branch pipe.

temperature   (1) The thermal state of a substance with respect to its ability to transmit heat to its environment. (2) The measure of the thermal state on some arbitrarily chosen numerical scale. See also *Celsius, centigrade, Fahrenheit*.

temporary hardness   Hardness that can be removed by boiling; more properly called *carbonate hardness*. See also *carbonate hardness, hardness*.

tertiary effluent   The liquid portion of wastewater leaving tertiary treatment.

tertiary treatment   The treatment of wastewater beyond the secondary or biological stage; term normally implies the removal of nutrients, such as phosphorus and nitrogen, and a high percentage of suspended solids; term now being replaced by *advanced waste treatment*. See also *advanced waste treatment*.

thermal stratification   The formation of layers of different temperatures in bodies of water.

thermophilic digestion   Digestion occurring at a temperature approaching or within the thermophilic range, generally between 43 and 60 °C (110 and 140 °F).

thermophilic range   That temperature range most conducive to maintenance of optimum digestion by thermophilic bacteria, generally accepted as between 49 and 57 °C (120 and 135 °F). See also *thermophilic digestion*.

thickeners   Any equipment or process, after gravity sedimentation, that increases the concentration of solids in sludges with or without the use of chemical flocculents.

threshold odor   The minimum odor of the water sample that can barely be detected after successive dilutions with odorless water. Also called *odor threshold*.

threshold odor number   The greatest dilution of a sample with odor-free water that yields a definitely perceptible odor.

titration   The determination of a constituent in a known volume of solution by the measured addition of a solution of known strength to completion of the reaction as signalled by observation of an end point.

tolerance   The ability of an organism to withstand exposure to a specific compound; a tolerance level may be defined as a period of exposure or a level of exposure (concentration) that is withstood.

total carbon (TC)   A quantitative measure of both total inorganic and total organic carbon as determined instrumentally by chemical oxidation to carbon dioxide and subsequent infrared detection in a carbon analyzer. See also *total organic carbon*.

total dissolved solids (TDS)   The sum of all dissolved solids (volatile and nonvolatile).

total dynamic discharge head   Total dynamic head plus the dynamic suction head or minus the dynamic suction lift.

total dynamic head (TDH)   The difference between the elevation corresponding to the pressure at the discharge flange of a pump and the elevation corresponding to the vacuum or pressure at the suction flange of the pump, corrected to the same datum plane, plus the velocity head at the discharge flange of the pump minus the velocity head at the suction flange of the pump.

total head   (1) The sum of the pressure, velocity, and position heads above a datum. The height of the energy line above a datum. (2) The difference in elevation between the surface of the water at the source of supply and the elevation of the water at the outlet, plus velocity head and lost head. (3) The high distance of the energy line above the datum; energy head. (4) In open channel flow, the depth plus the velocity head.

total organic carbon (TOC)   The amount of carbon bound in organic compounds in a sample. Because all organic compounds have carbon as the common element, total organic carbon measurements provide a fundamental means of assessing the degree of organic pollution.

total oxygen demand (TOD)   A quantitative measure of all oxidizable material in a sample water or wastewater as determined instrumentally by measuring the depletion of oxygen after high-temperature combustion. See also *chemical oxygen demand, total organic carbon.*

total pumping head   The measure of the energy increase imparted to each pound of liquid as it is pumped, and therefore, the algebraic difference between the total discharge head and the total suction head.

total solids (TS)   The sum of dissolved and suspended solid constituents in water or wastewater.

total suspended solids (TSS)   The amount of insoluble solids floating and in suspension in the wastewater. Also referred to as *total nonfilterable residue.*

toxicant   A substance that kills or injures an organism through chemical, physical, or biological action; examples include cyanides, pesticides, and heavy metals.

toxicity   The adverse effect that a biologically active substance has, at some concentration, on a living entity.

toxic wastes   Wastes that can cause an adverse response when they come in contact with a biological entity.

trace nutrients   Substances vital to bacterial growth. Trace nutrients are defined in this text as nitrogen, phosphorus, and iron.

transformer   An electromagnetic device for changing the voltage of alternating current electricity.

trap   (1) A device used to prevent a material flowing or carried through a conduit from reversing its direction of flow or movement, or from passing a given point. (2) A device to prevent the escape of air from sewers through a plumbing fixture or catch basin.

trash   Debris that may be removed from reservoirs, combined sewers, and storm sewers by coarse racks.

trash rack   A grid or screen placed across a waterway to catch floating debris.

trickling filter   Secondary treatment process where wastewater trickles over rock or honeycombed-shaped plastic media. Biomass and slimes containing microorganisms form on the media and utilize the organic matter for growth and energy.

tri-halomethanes (THM)   Derivatives of methane (CH_4) in which three halogen atoms (chlorine, bromine, or iodine) are substituted for three of the hydrogen atoms.

trough   A structure, usually with a length several times its transverse dimensions, used to hold or transport water or other liquids.

tube settler   A series of tubes, about 2 in. in diameter or 2-in. square, placed in a sedimentation tank to improve the solids removal efficiency.

tubing   (1) Flexible pipe of small diameter, usually less than 2 in. (2) A special grade of high-test pipe fitted with couplings and fittings of special design.

turbidimeter   An instrument for measurement of turbidity in which a standard suspension is used for reference.

turbidity   (1) A condition in water or wastewater caused by the presence of suspended matter and resulting in the scattering and absorption of light. (2) Any suspended solids imparting a visible haze or cloudiness to water that can be removed by filtration. (3) An analytical quantity usually reported in turbidity units determined by measurements of light scattering. See also *formazine turbidity unit, nephelometric turbidity unit.*

turbine pump   A centrifugal pump in which fixed guide vanes partially convert the velocity energy of the water to pressure head as the water leaves the impeller.

turbulence   (1) The fluid property that is characterized by irregular variation in the speed and direction of movement of individual particles or elements of the flow. (2) A state of flow of water in which the water is agitated by cross currents and eddies, as opposed to laminar, streamline, or viscous flow. See also *turbulent flow.*

turbulent flow   (1) The flow of a liquid past an object such that the velocity at any fixed point in the fluid varies irregularly. (2) A type of fluid flow in which there is an unsteady motion of the particles and the motion at a fixed point varies in no definite manner. Also called *eddy flow* or *sinuous flow.*

turnover   The phenomenon of vertical circulation that occurs in large bodies of water. It results from the increase in density of water above and below 39.2 °F (4 °C), the temperature of minimum density. In the spring, as the surface of the water warms above the freezing point, the water increases in density and tends to sink, producing vertical currents; in the fall, as the surface water becomes colder, it also tends to sink. Wind may also create such vertical currents. Also called *overturn*.

two-staged digestion   The biological decomposition of organic matter in sludge followed by solids–liquid separation of the digested sludge. Two-stage digestion uses two compartments or two tanks to separate the violent initial digestion period from the slower final period to enhance both the digestion and the solids–liquid separation after digestion.

ultimate biochemical oxygen demand (BOD_u)   (1) Commonly, the total quantity of oxygen required to completely satisfy the first-stage BOD. (2) More strictly, the quantity of oxygen required to completely satisfy both the first-stage and second-stage BOD.

ultimate disposal   The final release of a biologically and chemically stable wastewater or sludge into the environment.

ultraviolet radiation (UV)   Light waves shorter than the visible blue-violet waves of the spectrum.

ultraviolet ray   Light rays beyond the violet of the spectrum; these are invisible to humans.

underdrain   A drain that carries away groundwater or the drainage from prepared beds to which water or wastewater has been applied.

unsteady nonuniform flow   Flow in which the velocity and the quantity of water flowing per unit time at every point along the conduit varies with respect to time and position.

upflow   Term used to describe treatment units in which flow enters at the bottom and exits at the top.

upflow clarifier   A treatment unit in which liquid containing suspended solids is passed upward through a blanket of settling sludge; mixing, flocculation, and solids removal are all accomplished in the same unit.

upflow coagulation   Coagulation achieved by passing liquid, to which coagulating chemicals may have been added, upward through a blanket of settling sludge.

upflow filter   A gravity or pressure filtration system in which the wastewater flows upward, generally first through a coarse medium and then through a fine medium, before discharging.

upflow tank   A sedimentation tank in which water or wastewater enters near the bottom and rises vertically, usually through a blanket of previously settled solids. The clarified liquid flows out at the top and settled sludge flows out the bottom; a vertical-flow tank.

user   The party who is billed, usually for sewer service from a single connection; has no reference to the number of persons served. Also called a *customer*.

user charge   Charge made to users of wastewater services supplied.

utilization equipment   Equipment that uses electrical energy for mechanical, chemical, heating, lighting, or similar useful purposes.

vacuum breaker   A device for relieving a vacuum or partial vacuum formed in a pipeline, thereby preventing backsiphoning.

vacuum filter   (1) A filter used to accomplish sludge dewatering and consisting of a cylindrical drum mounted on a horizontal axis, covered with filter media, and revolving partially submerged in a dilute sludge mixture. A vacuum is maintained under the media for the larger part of a revolution to extract moisture. The dewatered cake that is formed is scraped off mechanically for disposal. See also *vacuum filtration*. (2) A diatomaceous earth filter open to the atmosphere and on the inlet side of a pump.

vacuum filtration   A usually continuous filtration operation that is generally accomplished on a rotating cylindrical drum. As the drum rotates, part of its circumference is subject to an internal vacuum that draws sludge to the filter medium and removes water for subsequent treatment. The dewatered sludge cake is released by a scraper.

vacuum pump   (1) A pump for creating a partial vacuum in a closed space. (2) A pump in which water is forced up a pipe by the difference in pressure between the atmosphere and a partial vacuum. (3) An air compressor used in connection with steam condensers and for improving the suction head on other pumps. The compressor takes its suction at low absolute pressure, performs a large number of compressions, and generally discharges at atmospheric pressure.

valence   An integer representing the number of hydrogen atoms with which one atom of an element (or one radical) can combine (negative valence), or the number of hydrogen atoms the atom or radical can displace (positive valence).

valve   (1) A device installed in a pipeline to control the magnitude and direction of the flow. It consists essentially of a shell and a disk or plug fitted to the shell. (2) In a pump, a waterway, passage through which is controlled by a mechanism.

vapor   (1) The gaseous form of any substance. (2) A visible condensation such as fog, mist, or steam that is suspended in air.

vaporization   The process by which a substance such as water changes from the liquid or solid state to the gaseous state.

vapor pressure   (1) Pressure exerted by a vapor in a confined space. It is a function of the temperature. (2) The partial pressure of water vapor in the atmosphere. See also *relative humidity*. (3) The partial pressure of any liquid.

velocity head   (1) The vertical distance or height through which a body would have to fall freely, under the force of gravity, to acquire the velocity it possesses. It is equal to the square of the velocity divided by twice the acceleration of gravity. (2) The theoretical vertical height through which a liquid body may be raised by its kinetic energy. It is equal to the share of the velocity divided by twice the acceleration caused by gravity.

velocity meter   A vaned water meter that operates on the principle that the vanes of the wheel move at approximately the same velocity as the flowing water.

Venturi meter   A differential meter for measuring the flow of water or other fluid through closed conduits or pipes. It consists of a Venturi tube and one of several proprietary forms of flow-registering devices. The difference in velocity heads between the entrance and the contracted throat is an indication of the rate of flow.

vertical pump   (1) A reciprocating pump in which the piston or plunger moves in a vertical direction. (2) A centrifugal pump in which the pump shaft is in a vertical position.

vertical screw pump   A pump, similar in shape, characteristics, and use to a horizontal screw pump, but which has the axis of its runner in a vertical position.

virus   The smallest (10 to 300 mm in diameter) life form capable of producing infection and diseases in man and animals.

viscosity   The molecular attractions within a fluid that make it resist a tendency to deform under applied forces.

V-notch weir   A triangular weir.

void   A pore or open space in rock or granular material not occupied by solid matter. It may be occupied by air, water, or other gaseous or liquid material. Also called *interstice* or *void space*.

volatile   Capable of being evaporated at relatively low temperatures.

volatile acids   Fatty acids containing six or fewer carbon atoms. They are soluble in water and can be steam-distilled at atmospheric pressure. They have pungent odors and are often produced during anaerobic decomposition.

volatile solids (VS)   Materials, generally organic, that can be driven off from a sample by heating, usually to 550 °C (1022 °F); nonvolatile inorganic solids (ash) remain.

volatile suspended solids (VSS)   That fraction of suspended solids, including organic matter and volatile inorganic salts, that will ignite and burn when placed in an electric muffle furnace at 550 °C (1022 °F) for 60 minutes.

volt   The unit of electromotive force or electrical pressure (analogous to water pressure). It is the electromotive force that, if steadily applied to a circuit having a resistance of one ohm, will produce a current of one ampere.

voltage (of a circuit)   The root mean square (effective) difference in potential between any two points of the circuit concerned. On various systems such as a three-phase, 4-wire and a single-phase, 3-wire, there may be various circuits of varying voltages.

volumetric   Pertaining to measurement by volume.

volute pump   A centrifugal pump with a casing made in the form of a spiral or volute as an aid to the partial conversion of the velocity energy into pressure head as the water leaves the impellers.

vortex   A revolving mass of water in which the stream lines are concentric circles and in which the total head for each stream line is the same.

washout   Condition whereby excessive influent flows (typically at peak flow conditions) cause the solids in the aeration basins and/or clarifiers to be carried over into downstream processes or discharged to the receiving stream.

waste activated sludge (WAS)   Solids removed from the activated-sludge process to prevent an excessive buildup in the system.

wastewater   The spent or used water of a community or industry containing dissolved and suspended matter.

water column   (1) The water above the valve in a set of pumps. (2) A measure of head or pressure in a closed pipe or conduit.

water vapor   The gaseous form of water; molecules of water present as a gas in an atmosphere of other gases. Movement takes place from higher to lower vapor pressure regions to maintain vapor pressure equilibrium. Also called *aqueous vapor*.

watt   The electrical unit of power. Power is the measure of the rate of doing work. A watt is the rate of energy transfer from one ampere flowing under a pressure of one volt at a unity power factor. It is analogous to horsepower or foot-pounds per minute of mechanical power. One horsepower is equivalent to approximately 746 W.

weir   A device that has a crest and some side containment of known geometric shape, such as a V, trapezoid, or rectangle, and is used to measure flow of liquid. The liquid surface is exposed to the atmosphere. Flow is related to the upstream height of water above the crest, position of crest with respect to downstream water surface, and geometry of the weir opening.

weir overflow rate   The amount of flow applied to a treatment process (typically a clarifier) per linear measure of weir (gpd/lin ft).

wet-air oxidation   A method of sludge disposal that involves the oxidation of sludge solids in water suspension under high pressure and temperature. Also called the *wet oxidation process*.

wire-to-water efficiency   The ratio of mechanical output of a pump to the electrical input at the meter.

Conversion Factors

Table 1.   Conversion from customary to metric units (in alphabetical order).

Multiply Customary Unit	Abbreviation	By	To Obtain Metric Unit
acre	ac	4.047×10^3	m^2
		0.404 7	ha (hectare)
acre-foot	ac-ft	1233	m^3
atmosphere	atm	101.3	kPa (kilopascal)
bar	bar	100.0	kPa
British thermal unit	Btu	1.055	kJ (kilojoule)
British thermal units per cubic foot	Btu/cu ft	37.26	kJ/m^3
British thermal units per gallon	Btu/gal	278.7	kJ/m^3
		0.278 7	kJ/L
British thermal units per hour	Btu/hr	0.2931	W (watt)
British thermal units per hour per square foot	Btu/hr/sq ft	3.155	$J/m^2{\cdot}s$
British thermal units per pound	Btu/lb	2.326	kJ/kg
British thermal units per ton	Btu/ton	1.163	J/kg
calorie international	cal	4.187	J (joule)
15°C		4.186	J
20°C		4.182	J
calories per second	cal/sec	4.187	W
centipoise	cP	1.000×10^{-3}	Pa·s
cubic feet per gallon	cu ft/gal	7.482	m^3/m^3
		7.482×10^{-3}	m^3/L
cubic feet per hour	cfh or cu ft/hr	7.867×10^{-6}	m^3/s
		7.867×10^{-3}	L/s
cubic feet per hour per square foot	cfh/sq ft	8.467×10^{-5}	$m^3/m^2{\cdot}s$
		304.8	$L/m^2{\cdot}h$
cubic feet per million gallons	cu ft/mil. gal	7.482	mL/m^3
cubic feet per minute	cfm or cu ft/min	$4.719\ 10^{-4}$	m^3/s
		0.471 9	L/s
cubic feet per minute per foot	cfm/ft	1.549	L/m ·s
cubic feet per minute per thousand cubic feet	cfm/1000 cu ft	1.667×10^{-2}	$L/m^3{\cdot}s$
cubic feet per minute per thousand gallons	cfm/1000 gal	0.124 7	$L/m^3{\cdot}s$

cubic feet per pound	cu ft/lb	6.243×10^{-2}	m^3/kg
cubic feet per second	cfs	2.832×10^{-2}	m^3/s
		28.32	L/s
cubic feet per second per acre	cfs/ac	6.997×10^{-6}	$m^3/m^2 \cdot s$
		69.97	L/ha·s
cubic feet per second per square mile	cfs/sq mile	1.093×10^{-8}	$m^3/m^2 \cdot s$
		0.109 3	L/ha·s
cubic foot	cu ft	2.832×10^{-2}	m^3
		28.32	L
cubic inch	cu in.	16.39×10^{-6}	m^3
		16.39	mL
cubic yard	cu yd	0.764 6	m^3
degrees Fahrenheit	°F	0.555 6(°F − 32)	°C (degrees Celsius)
Rankine	°R	0.555 6	K (Kelvin)
feet of head	ft	0.304 8	m
feet per hour	ft/hr	8.467×10^{-5}	m/s
feet per minute	ft/min or fpm	5.080	mm/s
feet per second	ft/sec or fps	0.304 8	m/s
feet per second squared	ft/sec²	0.304 8	m/s^2
feet per year	ft/yr	0.304 8	m/a (meters per annum)
foot	ft	0.304 8	m
foot-head	ft-head	2989	Pa
foot-pound	ft-lb	1.356	N·m
foot-pounds per inch	ft-lb/in.	53.38	J/m
foot-pounds per minute	ft-lb/min	2.259×10^{-2}	W
foot-pounds per second	ft-lb/sec	1.355	W
gallon	gal	3.785×10^{-3}	m^3
		3.785	L
gallons per day	gpd	4.381×10^{-5}	L/s
		3.785×10^{-3}	m^3/d
gallons per day per acre	gpd/ac	1.083×10^{-11}	$m^3/m^2 \cdot s$
		1.083×10^{-8}	$L/m^2 \cdot s$
		9.353	L/ha·d
gallons per day per capita	gpd/cap	3.785	L/cap·d

Table 1.   Conversion from customary to metric units (in alphabetical order) (*continued*).

Multiply Customary Unit	Abbreviation	By	To Obtain Metric Unit
gallons per day per linear foot	gpd/ft	1.437×10^{-7}	$m^3/m \cdot s$
		1.437×10^{-4}	$L/m \cdot s$
		1.242×10^{-2}	$m^3/m \cdot d$
gallons per day per foot of manhole diameter per foot-head	gpd/ft/ft-head	4.074×10^{4}	$(mL/m \cdot d)/m$
gallons per day per inch of diameter per mile	gpd/in. dia/mile	92.60	$(mL/m \cdot d)/m$
gallons per day per mile	gpd/mile	2.720×10^{-11}	$m^3/m \cdot s$
		2.352	$mL/m \cdot d$
gallons per day per square foot	gpd/sq ft	4.715×10^{-7}	$m^3/m^2 \cdot s$ or m/s
		4.074×10^{-2}	$m^3/m^2 \cdot d$ or m/d
		40.74	$L/m^2 \cdot d$
gallons per day per thousand square feet	gpd/1000 sq ft	4.074×10^{-2}	$L/m^2 \cdot d$
gallons per hour	gph	1.051×10^{-6}	m^3/s
		1.051	mL/s
gallons per hour per foot of diameter per foot-head	gph/ft dia/ft-head	11.32	$(mL/m \cdot s)/m$
gallons per hour per inch diameter per thousand feet	gph/in. dia/1000 ft	488.9	$(mL/m \cdot h)/m$
gallons per million gallons	gal/mil. gal	1.000	mL/m^3
gallons per minute	gpm	6.308×10^{-5}	m^3/s
		6.308×10^{-2}	L/s
		5.451	m^3/d
gallons per minute per acre	gpm/ac	0.155 8	$L/ha \cdot s$
gallons per minute per cubic foot	gpm/cu ft	2.228	$L/m^3 \cdot s$
gallons per minute per foot	gpm/ft	2.070×10^{-4}	$m^3/m \cdot s$
		0.207 0	$L/m \cdot s$
gallons per minute per square foot	gpm/sq ft	6.791×10^{-4}	$m/m \cdot s$
		0.679 1	$L/m^2 \cdot s$
		58.67	$m^3/m^2 \cdot d$
gallons per pound	gal/lb	8344	mL/kg
gallons per ton	gal/ton	4.173	mL/kg
grain	gr	64.80	mg

grains per gallon	gr/gal	17.12 g/m³
		17.12 mg/L
hectare	ha	1.000×10^4 m²
horsepower	hp	0.745 7 kW
horsepower-hour	hp-hr	2.685 MJ
horsepower-hours per pound	hp-hr/lb	5.919 MJ/kg
horsepower-hours per thousand cubic feet	hp-hr/1000 cu ft	26.34 W/m³
horsepower-hours per thousand gallons	hp-hr/1000 gal	0.710 3 kJ/L
horsepower-hours per thousand pounds	hp-hr/1000 lb	5.919 kJ/kg
horsepower per gallon per minute	hp/gpm	11.82 MJ/m³
		11.82 kJ/L
horsepower per million gallons	hp/mil. gal	0.197 W/m³
		1.970×10^{-4} W/L
horsepower per thousand gallons	hp/1000 gal	0.197 kW/m³
inch	in.	25.40 mm
inches of water	in. H₂O	0.248 8 kPa
inches of mercury	in. Hg	3.377 kPa
inches per foot	in./ft	8.333 %
inches per hour	in./hr	25.40 mm/h
inches per year	in./yr	25.40 mm/a (millimeters per annum)
kilowatt-hour	kWh	3.600 MJ
kilowatt-hours per day	kWh/d	41.67 W
kilowatt-hours per gallon	kWh/gal	951.1 MJ/m³
		951.1 MJ/L
kilowatt-hours per million gallons	kWh/mil. gal	0.951 1 J/m³
		0.951 1 J/L
kilowatt-hours per pound	kWh/lb	7.936 MJ/kg
kilowatt-hours per thousand pounds	kWh/1000 lb	7.936 kJ/kg
kilowatt-hours per ton	kWh/ton	3.969 kJ/kg
kip	kip	4.448 kN (kilonewton)
kip-feet	kip-ft	1.356 kN·m
kip-feet per second	kip-ft/sec	1.356 kN·m/s
kips per foot	kip/ft	14.59 kN/m
kips per square foot	kip/sq ft	47.88 kPa

Table 1.   Conversion from customary to metric units (in alphabetical order) *(continued)*.

Multiply Customary Unit	Abbreviation	By	To Obtain Metric Unit
micron	μ	1.000	μm
mil	mil	25.40	μm
mile	mile	1.609	km
miles per hour	mph	0.446 9	m/s
		1.609	km/h
million gallons	mil. gal	3.785×10^3	m^3
		3.785	ML
million gallons per day	mgd	4.383×10^{-2}	m^3/s
		43.83	L/s
		3.785×10^3	m^3/d
		3.785	ML/d
million gallons per day per acre	mgd/ac	1.083×10^{-5}	$m^3/m^2 \cdot s$
		9.353×10^3	$m^3/ha \cdot d$
		9.353	ML/ha·d
most probable number per hundred milliliters	MPN/100 mL	10.00	MPN/L
parts per billion	ppb	1.00 (app.)	μg/L
		1.000	μg/kg
parts per million	ppm	1.00 (app.)	mg/L
		1.000	mg/kg
pound (force)	lbf	4.448	N
pound (mass)	lb	0.453 6	kg
pound-foot	lb-ft	1.356	N·m
pound-feet per second	lb-ft/sec	1.356	N·m/s
pounds per day per capita	lb/d/cap	0.453 6	kg/cap·d
pounds per cubic foot	lb/cu ft	16.02	kg/m^3
pounds per thousand cubic feet	lb/1000 cu ft	16.02	g/m^3
pounds per cubic yard	lb/cu yd	0.593 3	kg/m^3
pounds per day	lb/d	5.250	mg/s
		0.453 6	kg/d

pounds per day per acre	lb/d/ac	0.112 1	g/m²·d
pounds per day per acre-foot	lb/d/ac-ft	0.367 7	g/m³·d
pounds per day per cubic foot	lb/d/cu ft	16.02	kg/m³·d
pounds per day per pound	lb/d/lb	11.57	mg/kg·s
pounds per day per square foot	lb/d/sq ft	56.51	mg/m²·s
		4.882	kg/m²·d
pounds per day per thousand cubic feet	lb/d/1000 cu ft	1.602×10^{-2}	kg/m³·d
pounds per day per thousand square feet	lb/d/1000 sq ft	4.882	g/m²·d
pounds per foot	lb/ft	1.488	kg/m
pounds per foot per foot of diameter	lb/ft/ft dia	47.88	(N/m)/m
pounds per gallon	lb/gal	0.119 8	kg/L
pounds per hour	lb/hr	1.260×10^{-4}	kg/s
		0.453 6	kg/h
pounds per hour per square foot	lb/hr/sq ft	4.882	kg/m²·h
		1.356	g/m²·s
pounds per hour per cubic foot	lb/hr/cu ft	4.450×10^{-6}	kg/L·s
		16.02	kg/m³·h
pounds per million gallons	lb/mil. gal	0.119 8	g/m³
		0.119 8	mg/L
pounds per pound	lb/lb	1000	g/kg
pounds per square foot (mass dose)	lb/sq ft	4.883	kg/m²
pounds per square foot (force)	lb/sq ft	47.88	Pa
pounds per square inch (force)	psi	6895	Pa
pounds per thousand cubic feet	lb/1000 cu ft	16.02	g/m³
pounds per thousand gallons	lb/1000 gal	0.119 8	g/m³
pounds per ton	lb/ton	0.500	g/kg
pounds per year per acre	lb/yr/ac	1.121	kg/ha·a
pounds per year per cubic foot	lb/yr/cu ft	16.02	kg/m³·a
pounds per year per square foot	lb/yr/sq ft	4.882	kg/m²·a
pound-seconds per square foot	lb-sec/sq ft	47.88	Pa·s
pound-square feet per second	lb-sq ft/sec	4.883	kg·m²/s
revolutions per minute	rpm	1.000	r/min
revolutions per second	rps	1.000	r/s
slug	slug	14.59	kg

Table 1.   Conversion from customary to metric units (in alphabetical order) (*continued*).

Multiply Customary Unit	Abbreviation	By	To Obtain Metric Unit
square foot	sq ft	9.290×10^{-2}	m^2
square feet per capita	sq ft/cap	9.290×10^{-2}	m^2/cap
square feet per cubic foot	sq ft/cu ft	3.281	m^2/m^3
square feet per foot	sq ft/ft	0.304 8	m^2/m
square feet per second	sq ft/sec	9.290×10^{-2}	m^2/s
square inch	sq in.	645.2	mm^2
square mile	sq mile	2.590	km^2
square yard	sq yd	0.836 1	m^2
ton (long)	ton	1.016	Mg
ton (short)	ton	0.907 2	Mg
		907.2	kg
tons per acre	ton/ac	0.224 2	kg/m^2
		2.242	Mg/ha
tons per cubic yard	ton/cu yd	1.187	Mg/m^3
watt-hour	W-hr	3.600	J
watts per square foot	W/sq ft	10.76	W/m^2
yard	yd	0.914 4	m

Table 2. Conversion from metric to customary units (in alphabetical order).

Multiply Metric Unit	Abbreviation	By	To Obtain Customary Unit
capita per hectare	cap/ha	0.404 7	cap/ac
capita per square kilometer	cap/km^2	2.590	cap/sq mile
cubic meter	m^3	8.107×10^{-4}	ac-ft
		6.290	barrel (oil)
		8.386	barrel (water)
		28.38	bushel (U.S. dry)
		35.31	cu ft
		1.308	cu yd
		264.2	gal
		2.642×10^{-4}	mil. gal
cubic meters per cubic meter	m^3/m^3	0.133 7	cu ft/gal
		1.337×10^5	cu ft/mil. gal
cubic meters per cubic meter per second	m^3/m^3·s	5.999×10^4	cfm/1000 cu ft
		8.019×10^3	cfm/1000 gal
cubic meters per day	m^3/d	0.183 5	gpm
		2.642×10^{-4}	mgd
		264.2	gpd
cubic meters per hectare per day	m^3/ha·d	106.9	gpd/ac
cubic meters per hectare per second	m^3/ha·s	1.069×10^{-4}	mgd/ac
		14.29	cfs/ac
cubic meters per kilogram	m^3/kg	9.147×10^3	cfs/sq mile
		16.02	cu ft/lb
cubic meters per meter per day	m^3/m·d	80.53	gpd/ft
cubic meters per meter per second	m^3/m·s	646.0	cfm/ft
		6.959×10^6	gpd/ft
		3.676×10^{10}	gpd/mile

Table 2.   Conversion from metric to customary units (in alphabetical order) (continued).

Multiply Metric Unit	Abbreviation	By	To Obtain Customary Unit
cubic meters per second	m^3/s	1.271×10^5	cfh or cu ft/hr
		2.119×10^3	cfm
		35.32	cfs
		9.515×10^5	gph
		1.585×10^4	gpm
		22.83	mgd
cubic meters per square meter per day	$m^3/m^2 \cdot d$	24.55	gpd/sq ft
		1.704×10^{-2}	gpm/sq ft
cubic meters per square meter per hour	$m^3/m^2 \cdot h$	3.281	cfh/sq ft
cubic meters per square meter per second	$m^3/m^2 \cdot s$	1.429×10^3	cfs/ac
		9.149×10^7	cfs/sq mile
		9.238×10^{10}	gpd/ac
		2.121×10^6	gpd/sq ft
		1.473×10^3	gpm/sq ft
		9.238×10^4	mgd/ac
		1.181×10^4	cfh/sq ft
cubic millimeter	mm^3	6.102×10^{-5}	cu in.
degrees Celsius	°C	1.800 (°C)+ 32	°F
Kelvin	K	1.800	°R
grams per cubic meter	g/m^3	8.344	lb/mil.gal
		6.243×10^{-2}	lb/1000 cu ft
		8.347	lb/1000 gal
grams per cubic meter per day	$g/m^3 \cdot d$	2.720	lb/d/ac-ft
grams per cubic meter per second	$g/m^3 \cdot s$	0.224 7	lb/hr/cu ft
grams per kilogram	g/kg	1.000×10^{-3}	lb/lb
		2.000	lb/ton
grams per liter	g/L	0.0624	lb/cu ft
grams per meter	g/m	0.672 0	lb/1000 ft
grams per second	g/s	7.937	lb/hr
		8.922	lb/d/ac
grams per square meter per day	$g/m^2 \cdot d$	0.204 8	lb/d/1000 sq ft

grams per square meter per second	g/m²·s	17.70	lb/d/sq ft
		0.737 3	lb/hr/sq ft
hectare	ha	2.471	ac (acre)
joule	J	0.7376	ft-lb
joules per cubic meter	J/m³	1.051×10^{-3}	kWh/mil. gal
joules per kilogram	J/kg	4.299×10^{-4}	Btu/lb
		0.859 8	Btu/ton
		1.689×10^{-4}	hp-hr/1000 lb
joules per liter	J/L	8.461×10^{-5}	hp/gpm
		1.051	kWh/mil. gal
joules per meter	J/m	1.873×10^{-2}	ft-lb/in.
joules per square meter per second	J/m²·s	0.317 1	Btu/hr/sq ft
kilogram	kg	2.205	lb (mass)
		1.102×10^{-3}	ton
		6.854×10^{-2}	slug
kilogram meter	kg·m	0.672 0	lb-ft
kilogram meter per second	kg·m/s	0.672 0	lb-ft/sec
kilograms per capita per day	kg/cap·d	2.205	lb/cap/d
kilograms per cubic meter	kg/m³	6.243×10^{-2}	lb/cu ft
		1.686	lb/cu yd
		834 7	lb/mil. gal
		62.42	lb/1000 cu ft
kilograms per cubic meter per day	kg/m³·d	6.242×10^{-2}	lb/cu ft
		6.242×10^{-2}	lb/d/1000 cu ft
kilograms per cubic meter per hour	kg/m³·h	6.242×10^{-2}	lb/hr/cu ft
kilograms per cubic meter per second	kg/m³·s	224.8	lb/hr/cu ft
kilograms per day	kg/d	2.205	lb/d
kilograms per hectare per day	kg/ha·d	0.892 2	lb/d/ac
kilograms per hectare per year (annum)	kg/ha·a	0.892 2	lb/yr/ac
kilograms per hour	kg/h	2.205	lb/hr
kilograms per kilogram per second	kg/kg·s	8.64×10^{4}	lb/d/lb
		3.613×10^{-2}	lb/cu in.
kilograms per liter	kg/L	8.347	lb/gal
		8.347×10^{6}	lb/mil. gal

Table 2. Conversion from metric to customary units (in alphabetical order) (*continued*).

Multiply Metric Unit	Abbreviation	By	To Obtain Customary Unit
kilograms per meter	kg/m	0.672 0	lb/ft
kilograms per second	kg/s	1.905×10^5	lb/d
		7.936×10^3	lb/hr
kilograms per square meter	kg/m²	0.204 8	lb/sq ft (mass dose)
kilograms per square meter per day	kg/m²·d	0.204 8	lb/d/sq ft
kilograms per square meter per hour	kg/m²·h	0.204 8	lb/hr/sq ft
kilograms per square meter per second	kg/m²·s	1.770×10^4	lb/d/sq ft
kilograms per square meter per year (annum)	kg/m²·a	0.204 8	lb/yr/sq ft
kilojoule	kJ	0.947 8	Btu
kilojoules per cubic meter	kJ/m³	2.684×10^{-2}	Btu/cu ft
		1.051	kWh/mil. gal
		1.408×10^{-3}	hp-hr/1000 gal
kilojoules per kilogram	kJ/kg	0.430 0	Btu/lb
		859.8	Btu/ton
		0.252 0	kWh/ton
		0.126 0	kWh/1000 lb
		1.689×10^{-4}	hp-hr/lb
kilojoules per liter	kJ/L	1.410	Btu/gal
kilometer	km	0.621 4	mile
kilometers per hour	km/h	0.621 4	mph
kilonewton	kN	0.224 8	kip
kilonewton per meter	kN/m	0.737 5	kip-ft
kilonewtons per meter per second	kN/m·s	0.737 5	kip-ft/sec
kilopascal	kPa	9.872×10^{-3}	atm
		20.89	psf (force)
		0.145 0	psi (force)
		4.019	in. H₂O
		0.296 1	in. Hg
kilowatt	kW	1.341	hp
kilowatt per cubic meter	kW/m³	5.076	hp/1000 gal
liter	L	3.532×10^{-2}	cu ft
		0.264 2	gal

Unit name	Symbol	Factor	To
liters per capita	L/cap	3.532×10^{-2}	cu ft/cap
liters per capita per day	L/cap·d	0.264 2	gpd/cap
liters per cubic meter	L/m³	133.7	cu ft/mil. gal
liters per cubic meter per second	L/m³·s	0.488 8	cfm/cu ft
		59.99	cfm/1000 cu ft
		8.019	cfm/1000 gal
liters per hectare per second	L/ha·s	1.429×10^{-2}	cfs/ac
		9.149	cfs/sq mile
		6.418	gpm/ac
liters per kilogram	L/kg	239.6	gal/ton
liters per meter	L/m	425.2	gal/mile
liters per meter per second	L/m·s	0.645 9	cfm/ft
		4.832	gpm/ft
		6.958×10^{3}	gpd/ft
liters per second	L/s	127.1	cfh
		2.119	cfm
		3.532×10^{-2}	cfs
		2.282×10^{4}	gpd
		15.85	gpm
		2.283×10^{-2}	mgd
liters per square meter per day	L/m²·d	2.455×10^{-2}	gpd/sq ft
liters per square meter per second	L/m²·s	11.82	cfh/sq ft
		2.121×10^{3}	gpd/sq ft
		9.237×10^{-3}	gpd/ac
		1.473	gpm/sq ft
megagram or metric ton	Mg or t	1.102	ton (short)
		0.984 3	ton (long)
megagram per cubic meter	Mg/m³	0.842 5	ton/cu yd
megagram per hectare	Mg/ha	0.446 0	ton/ac
megajoule	MJ	0.372 5	hp-hr
megajoules per cubic meter	MJ/m³	8.460×10^{-2}	hp/gpm
		1.051×10^{-3}	kWh/gal
megajoules per kilogram	MJ/kg	168.9	hp-hr/1000 lb
		0.126 0	kWh/lb

Table 2.   Conversion from metric to customary units (in alphabetical order) (*continued*).

Multiply Metric Unit	Abbreviation	By	To Obtain Customary Unit
megajoules per liter	MJ/L	1.051	kWh/gal
megaliter	ML	0.264 2	mil. gal
megaliters per day	ML/d	0.264 2	mgd
megaliters per hectare per day	ML/ha·d	0.106 9	mgd/ac
meter	m	3.281	ft
meters per hour	m/h	3.281	ft/hr
meters per second	m/s	3.281	fps
		2.238	mph
meters per second squared	m/s^2	3.281	ft/sec^2
micrograms per liter	µg/L	1.00 (app.)	ppb
micrometer	µm	1.000	micron
		3.937×10^{-2}	mil
miligram	mg	1.543×10^{-2}	grain
milligrams per liter	mg/L	5.841×10^{-2}	grains per gallon
		1.00 (app.)	ppm
milliliter	mL	6.102×10^{-2}	cu in.
milliliters per cubic meter	mL/m^3	0.133 7	cu ft/mil. gal
milliliters per kilogram	mL/kg	1.198×10^{-4}	gal/lb
		0.239 6	gal/ton
milliliters per meter per day per meter	(mL/m·d)/m	2.450×10^{-3}	gpd/ft/ft-head
milliliters per second	mL/s	0.951 5	gph
millimeter	mm	3.937×10^{-2}	in.
millimeters per hour	mm/h	3.937×10^{-2}	in./hr
millimeters per second	mm/s	0.196 9	fpm
		11.81	ft/hr
most probable number per cubic meter	MPN/m^3	1.000×10^{-4}	MPN/100 mL
most probable number per liter	MPN/L	0.100 0	MPN/100 mL

Unit	Symbol	Factor	To obtain
newton	N	0.224 8	lbf (force)
newton meter	N·m	0.737 6	ft-lb
newton meters per meter	N·m/m	0.224 8	lb-ft/ft dia
pascal	Pa	9.872×10^{-6}	atm
		3.345×10^{-4}	ft (head)
		2.089×10^{-2}	lb/sq ft
		1.450×10^{-4}	psi
pascal second	Pa·s	2.089×10^{-2}	lb-sec/sq ft
square kilometer	km²	0.386 1	sq mile
square meter	m²	2.471×10^{-4}	ac
		10.76	sq ft
square meters per cubic meter	m²/m³	3.048	sq ft/cu ft
square meters per meter	m²/m	3.281	sq ft/ft
square meter per second	m²/s	10.76	sq ft/sec
square millimeter	mm²	1.550×10^{-3}	sq in.
ton (metric)	t	1.102×10^{-3}	lb
		0.984 3	ton (long)
		1.102	ton (short)
watt	W	3.412	Btu/hr
		947.8	Btu/sec
		44.27	ft-lb/min
		0.738 0	ft-lb/sec
		1.341×10^{-3}	hp
watts per cubic meter	W/m³	2.400×10^{-2}	kWh/d
		5.076	hp/mil. gal
watts per liter	W/L	3.797×10^{-2}	hp/1000 cu ft
		5.076×10^{3}	hp/mil. gal
watts per square meter	W/m²	9.290×10^{-2}	W/sq ft

Index

H